WCDMA FOR UMTS

Radio Access for Third Generation
Mobile Communications

Third Edition

Edited by

Harri Holma and Antti Toskala
Both of
Nokia, Finland

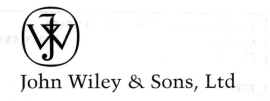

John Wiley & Sons, Ltd

Other Wiley Editorial Offices

John Wiley & Sons Inc., 111 River Street, Hoboken, NJ 07030, USA

Jossey-Bass, 989 Market Street, San Francisco, CA 94103-1741, USA

Wiley-VCH Verlag GmbH, Boschstr. 12, D-69469 Weinheim, Germany

John Wiley & Sons Australia Ltd, 33 Park Road, Milton, Queensland 4064, Australia

John Wiley & Sons (Asia) Pte Ltd, 2 Clementi Loop #02-01, Jin Xing Distripark, Singapore 129809

John Wiley & Sons Canada Ltd, 22 Worcester Road, Etobicoke, Ontario, Canada M9W 1L1

Wiley also publishes its books in a variety of electronic formats. Some content that appears in
print may not be available in electronic books.

British Library Cataloguing in Publication Data

A catalogue record for this book is available from the British Library

ISBN 0-470-87096-6 (CB)

Typeset in 10/12pt Times by Thomson Press (India) Limited, New Delhi.
Printed and bound in Great Britain by TJ International, Padstow, Cornwall.
This book is printed on acid-free paper responsibly manufactured from sustainable forestry
in which at least two trees are planted for each one used for paper production.

Contents

Preface   xv

Acknowledgements   xix

Abbreviations   xxi

1 Introduction   **1**

Harri Holma, Antti Toskala and Ukko Lappalainen
1.1   WCDMA in Third Generation Systems   1
1.2   Air Interfaces and Spectrum Allocations for Third Generation Systems   2
1.3   Schedule for Third Generation Systems   5
1.4   Differences between WCDMA and Second Generation Air Interfaces   6
1.5   Core Networks and Services   9
References   10

2 UMTS Services and Applications   **11**

Harri Holma, Martin Kristensson, Jouni Salonen and Antti Toskala
2.1   Introduction   11
2.2   Person-to-Person Circuit Switched Services   12
    2.2.1   AMR Speech Service   12
    2.2.2   Video Telephony   15
2.3   Person-to-Person Packet Switched Services   17
    2.3.1   Images and Multimedia   17
    2.3.2   Push-to-Talk over Cellular (PoC)   19
    2.3.3   Voice over IP (VoIP)   21
    2.3.4   Multiplayer Games   22
2.4   Content-to-person Services   23
    2.4.1   Browsing   23
    2.4.2   Audio and Video Streaming   25
    2.4.3   Content Download   25
    2.4.4   Multimedia Broadcast Multicast Service, MBMS   26
2.5   Business Connectivity   28
2.6   IP Multimedia Sub-system, IMS   30

2.7   Quality of Service Differentiation   31
2.8   Capacity and Cost of Service Delivery   37
         2.8.1   Capacity per Subscriber   37
         2.8.2   Cost of Capacity Delivery   38
2.9   Service Capabilities with Different Terminal Classes   40
2.10   Location Services in WCDMA   40
         2.10.1   Location Services   40
         2.10.2   Cell Coverage Based Location Calculation   41
         2.10.3   Observed Time Difference Of Arrival, OTDOA   42
         2.10.4   Assisted GPS   44
References   45

3 Introduction to WCDMA   47

Peter Muszynski and Harri Holma
3.1   Introduction   47
3.2   Summary of the Main Parameters in WCDMA   47
3.3   Spreading and Despreading   49
3.4   Multipath Radio Channels and Rake Reception   52
3.5   Power Control   55
3.6   Softer and Soft Handovers   58
References   60

4 Background and Standardisation of WCDMA   61

Antti Toskala
4.1   Introduction   61
4.2   Background in Europe   61
         4.2.1   Wideband CDMA   62
         4.2.2   Wideband TDMA   63
         4.2.3   Wideband TDMA/CDMA   63
         4.2.4   OFDMA   64
         4.2.5   ODMA   64
         4.2.6   ETSI Selection   64
4.3   Background in Japan   65
4.4   Background in Korea   65
4.5   Background in the United States   66
         4.5.1   W-CDMA N/A   66
         4.5.2   UWC-136   66
         4.5.3   cdma2000   66
         4.5.4   TR46.1   67
         4.5.5   WP-CDMA   67
4.6   Creation of 3GPP   67
4.7   How does 3GPP Operate?   69
4.8   Creation of 3GPP2   70
4.9   Harmonisation Phase   70

4.10   IMT-2000 Process in ITU   70
4.11   Beyond 3GPP Release '99   72
References   73

5 Radio Access Network Architecture   **75**

Fabio Longoni, Atte Länsisalmi and Antti Toskala
5.1   System Architecture   75
5.2   UTRAN Architecture   78
       5.2.1   The Radio Network Controller   79
       5.2.2   The Node B (Base Station)   80
5.3   General Protocol Model for UTRAN Terrestrial Interfaces   80
       5.3.1   General   80
       5.3.2   Horizontal Layers   80
       5.3.3   Vertical Planes   81
5.4   Iu, the UTRAN–CN Interface   82
       5.4.1   Protocol Structure for Iu CS   82
       5.4.2   Protocol Structure for Iu PS   84
       5.4.3   RANAP Protocol   85
       5.4.4   Iu User Plane Protocol   86
       5.4.5   Protocol Structure of Iu BC, and the SABP Protocol   87
5.5   UTRAN Internal Interfaces   88
       5.5.1   RNC–RNC Interface (Iur Interface) and the RNSAP Signalling   88
       5.5.2   RNC–Node B Interface and the NBAP Signalling   91
5.6   UTRAN Enhancements and Evolution   93
       5.6.1   IP Transport in UTRAN   93
       5.6.2   Iu Flex   93
       5.6.3   Stand Alone SMLC and Iupc Interface   94
       5.6.4   Interworking between GERAN and UTRAN, and the Iur-g Interface   94
       5.6.5   All IP RAN Concept   94
5.7   UMTS Core Network Architecture and Evolution   95
       5.7.1   Release '99 Core Network Elements   95
       5.7.2   Release 5 Core Network and IP Multimedia Sub-system   96
References   98

6 Physical Layer   **99**

Antti Toskala
6.1   Introduction   99
6.2   Transport Channels and their Mapping to the Physical Channels   100
       6.2.1   Dedicated Transport Channel   101
       6.2.2   Common Transport Channels   101
       6.2.3   Mapping of Transport Channels onto the Physical Channels   103
       6.2.4   Frame Structure of Transport Channels   104
6.3   Spreading and Modulation   104
       6.3.1   Scrambling   104

	6.3.2	Channelisation Codes	105
	6.3.3	Uplink Spreading and Modulation	105
	6.3.4	Downlink Spreading and Modulation	110
	6.3.5	Transmitter Characteristics	113
6.4		User Data Transmission	114
	6.4.1	Uplink Dedicated Channel	114
	6.4.2	Uplink Multiplexing	117
	6.4.3	User Data Transmission with the Random Access Channel	119
	6.4.4	Uplink Common Packet Channel	120
	6.4.5	Downlink Dedicated Channel	120
	6.4.6	Downlink Multiplexing	122
	6.4.7	Downlink Shared Channel	124
	6.4.8	Forward Access Channel for User Data Transmission	125
	6.4.9	Channel Coding for User Data	126
	6.4.10	Coding for TFCI Information	127
6.5		Signalling	127
	6.5.1	Common Pilot Channel (CPICH)	127
	6.5.2	Synchronisation Channel (SCH)	128
	6.5.3	Primary Common Control Physical Channel (Primary CCPCH)	128
	6.5.4	Secondary Common Control Physical Channel (Secondary CCPCH)	130
	6.5.5	Random Access Channel (RACH) for Signalling Transmission	131
	6.5.6	Acquisition Indicator Channel (AICH)	131
	6.5.7	Paging Indicator Channel (PICH)	131
	6.5.8	Physical Channels for the CPCH Access Procedure	132
6.6		Physical Layer Procedures	133
	6.6.1	Fast Closed Loop Power Control Procedure	133
	6.6.2	Open Loop Power Control	134
	6.6.3	Paging Procedure	134
	6.6.4	RACH Procedure	135
	6.6.5	CPCH Operation	136
	6.6.6	Cell Search Procedure	137
	6.6.7	Transmit Diversity Procedure	138
	6.6.8	Handover Measurements Procedure	139
	6.6.9	Compressed Mode Measurement Procedure	140
	6.6.10	Other Measurements	142
	6.6.11	Operation with Adaptive Antennas	143
	6.6.12	Site Selection Diversity Transmission	144
6.7		Terminal Radio Access Capabilities	145
References			148

7 Radio Interface Protocols   149

Jukka Vialén and Antti Toskala

7.1	Introduction	149
7.2	Protocol Architecture	149
7.3	The Medium Access Control Protocol	151

		7.3.1	MAC Layer Architecture	151
		7.3.2	MAC Functions	152
		7.3.3	Logical Channels	153
		7.3.4	Mapping Between Logical Channels and Transport Channels	154
		7.3.5	Example Data Flow Through the MAC Layer	154
	7.4	The Radio Link Control Protocol		155
		7.4.1	RLC Layer Architecture	156
		7.4.2	RLC Functions	157
		7.4.3	Example Data Flow Through the RLC Layer	158
	7.5	The Packet Data Convergence Protocol		160
		7.5.1	PDCP Layer Architecture	160
		7.5.2	PDCP Functions	161
	7.6	The Broadcast/Multicast Control Protocol		161
		7.6.1	BMC Layer Architecture	161
		7.6.2	BMC Functions	161
	7.7	Multimedia Broadcast Multicast Service		162
	7.8	The Radio Resource Control Protocol		164
		7.8.1	RRC Layer Logical Architecture	164
		7.8.2	RRC Service States	165
		7.8.3	RRC Functions and Signalling Procedures	168
	7.9	Early UE Handling Principles		183
	References			183

8 Radio Network Planning   185

Harri Holma, Zhi-Chun Honkasalo, Seppo Hämäläinen, Jaana Laiho,
Kari Sipilä and Achim Wacker

8.1	Introduction		185
8.2	Dimensioning		186
	8.2.1	Radio Link Budgets	187
	8.2.2	Load Factors	190
	8.2.3	Capacity Upgrade Paths	202
	8.2.4	Capacity per km^2	203
	8.2.5	Soft Capacity	204
	8.2.6	Network Sharing	207
8.3	Capacity and Coverage Planning and Optimisation		208
	8.3.1	Iterative Capacity and Coverage Prediction	208
	8.3.2	Planning Tool	209
	8.3.3	Case Study	210
	8.3.4	Network Optimisation	214
8.4	GSM Co-planning		217
8.5	Inter-operator Interference		219
	8.5.1	Introduction	219
	8.5.2	Uplink vs. Downlink Effects	220
	8.5.3	Local Downlink Interference	221
	8.5.4	Average Downlink Interference	223

8.5.5   Path Loss Measurements   223
8.5.6   Solutions to Avoid Adjacent Channel Interference   225
8.6   WCDMA Frequency Variants   226
8.6.1   Introduction   226
8.6.2   Differences Between Frequency Variants   226
8.6.3   WCDMA1900 in an Isolated 5 MHz Block   228
References   229

9 Radio Resource Management   231

Harri Holma, Klaus Pedersen, Jussi Reunanen, Janne Laakso and Oscar Salonaho

9.1   Interference-Based Radio Resource Management   231
9.2   Power Control   232
9.2.1   Fast Power Control   232
9.2.2   Outer Loop Power Control   239
9.3   Handovers   245
9.3.1   Intra-frequency Handovers   245
9.3.2   Inter-system Handovers Between WCDMA and GSM   254
9.3.3   Inter-frequency Handovers within WCDMA   258
9.3.4   Summary of Handovers   259
9.4   Measurement of Air Interface Load   261
9.4.1   Uplink Load   261
9.4.2   Downlink Load   263
9.5   Admission Control   264
9.5.1   Admission Control Principle   264
9.5.2   Wideband Power-Based Admission Control Strategy   265
9.5.3   Throughput-Based Admission Control Strategy   267
9.6   Load Control (Congestion Control)   267
References   268

10 Packet Scheduling   269

Jeroen Wigard, Harri Holma, Renaud Cuny, Nina Madsen, Frank Frederiksen
and Martin Kristensson

10.1   Transmission Control Protocol (TCP)   269
10.2   Round Trip Time   276
10.3   User-specific Packet Scheduling   278
10.3.1   Common Channels (RACH/FACH)   279
10.3.2   Dedicated Channel (DCH)   280
10.3.3   Downlink Shared Channel (DSCH)   282
10.3.4   Uplink Common Packet Channel (CPCH)   282
10.3.5   Selection of Transport Channel   282
10.3.6   Paging Channel States   286
10.4   Cell-specific Packet Scheduling   286
10.4.1   Priorities   288
10.4.2   Scheduling Algorithms   289

        10.4.3   Packet Scheduler in Soft Handover   289
10.5   Packet Data System Performance   291
        10.5.1   Link Level Performance   291
        10.5.2   System Level Performance   292
10.6   Packet Data Application Performance   294
        10.6.1   Introduction to Application Performance   295
        10.6.2   Person-to-person Applications   296
        10.6.3   Content-to-person Applications   300
        10.6.4   Business Connectivity   302
        10.6.5   Conclusions on Application Performance   305
References   306

11 High-speed Downlink Packet Access   307

Antti Toskala, Harri Holma, Troels Kolding, Preben Mogensen,
Klaus Pedersen and Karri Ranta-aho

11.1   Release '99 WCDMA Downlink Packet Data Capabilities   307
11.2   HSDPA Concept   308
11.3   HSDPA Impact on Radio Access Network Architecture   310
11.4   Release 4 HSDPA Feasibility Study Phase   311
11.5   HSDPA Physical Layer Structure   311
        11.5.1   High-speed Downlink Shared Channel (HS-DSCH)   312
        11.5.2   High-speed Shared Control Channel (HS-SCCH)   315
        11.5.3   Uplink High-speed Dedicated Physical Control Channel
                  (HS-DPCCH)   317
        11.5.4   HSDPA Physical Layer Operation Procedure   318
11.6   HSDPA Terminal Capability and Achievable Data Rates   320
11.7   Mobility with HSDPA   321
        11.7.1   Measurement Event for Best Serving HS-DSCH Cell   322
        11.7.2   Intra-Node B HS-DSCH to HS-DSCH Handover   322
        11.7.3   Inter-Node B HS-DSCH to HS-DSCH Handover   323
        11.7.4   HS-DSCH to DCH Handover   324
11.8   HSDPA Performance   326
        11.8.1   Factors Governing Performance   326
        11.8.2   Spectral Efficiency, Code Efficiency and Dynamic Range   326
        11.8.3   User Scheduling, Cell Throughput and Coverage   330
        11.8.4   HSDPA Network Performance with Mixed Non-HSDPA
                  and HSDPA Terminals   334
11.9   Terminal Receiver Aspects   337
11.10   Evolution Beyond Release 5   338
        11.10.1   Multiple Receiver and Transmit Antenna Techniques   338
        11.10.2   High Speed Uplink Packet Access (HSUPA)   339
11.11   Conclusion   344
References   345

12 Physical Layer Performance   **347**

Harri Holma, Jussi Reunanen, Leo Chan, Preben Mogensen,
Klaus Pedersen, Kari Horneman, Jaakko Vihriälä and Markku Juntti
12.1   Introduction   347
12.2   Cell Coverage   347
        12.2.1   Uplink Coverage   350
        12.2.2   Downlink Coverage   354
12.3   Downlink Cell Capacity   360
        12.3.1   Downlink Orthogonal Codes   360
        12.3.2   Downlink Transmit Diversity   365
        12.3.3   Downlink Voice Capacity   367
12.4   Capacity Trials   369
        12.4.1   Single Cell Capacity Trials   369
        12.4.2   Multicell Capacity Trials   383
        12.4.3   Summary   385
12.5   3GPP Performance Requirements   387
        12.5.1 E_b/N_0 Performance   387
        12.5.2   RF Noise Figure   390
12.6   Performance Enhancements   391
        12.6.1   Smart Antenna Solutions   391
        12.6.2   Multiuser Detection   398
References   407

13 UTRA TDD Modes   **411**

Antti Toskala, Harri Holma, Otto Lehtinen and Heli Väätäjä
13.1   Introduction   411
        13.1.1   Time Division Duplex (TDD)   411
        13.1.2   Differences in the Network Level Architecture   413
13.2   UTRA TDD Physical Layer   413
        13.2.1   Transport and Physical Channels   414
        13.2.2   Modulation and Spreading   415
        13.2.3   Physical Channel Structures, Slot and Frame Format   415
        13.2.4   UTRA TDD Physical Layer Procedures   421
13.3   UTRA TDD Interference Evaluation   425
        13.3.1   TDD–TDD Interference   425
        13.3.2   TDD and FDD Co-existence   426
        13.3.3   Unlicensed TDD Operation   429
        13.3.4   Conclusions on UTRA TDD Interference   429
13.4   HSDPA Operation with TDD   430
13.5   Concluding Remarks and Future Outlook on UTRA TDD   431
References   431

14 cdma2000433

Antti Toskala
14.1Introduction433
14.2Logical Channels435
14.2.1*Physical Channels*435
14.3Multicarrier Mode Spreading and Modulation436
14.3.1*Uplink Spreading and Modulation*436
14.3.2*Downlink Spreading and Modulation*436
14.4User Data Transmission438
14.4.1*Uplink Data Transmission*438
14.4.2*Downlink Data Transmission*439
14.4.3*Channel Coding for User Data*440
14.5Signalling441
14.5.1*Pilot Channel*441
14.5.2*Synch Channel*441
14.5.3*Broadcast Channel*441
14.5.4*Quick Paging Channel*442
14.5.5*Common Power Control Channel*442
14.5.6*Common and Dedicated Control Channels*442
14.5.7*Random Access Channel (RACH) for Signalling Transmission*442
14.6Physical Layer Procedures442
14.6.1*Power Control Procedure*442
14.6.2*Cell Search Procedure*443
14.6.3*Random Access Procedure*443
14.6.4*Handover Measurements Procedure*444
References445

Index447

Preface

Second generation telecommunication systems, such as GSM, enabled voice traffic to go wireless: the number of mobile phones exceeds the number of landline phones and the mobile phone penetration exceeds 80 % in countries with the most advanced wireless markets. The data handling capabilities of second generation systems are limited, however, and third generation systems are needed to provide the high bit rate services that enable high quality images and video to be transmitted and received, and to provide access to the web with higher data rates. These third generation mobile communication systems are referred to in this book as UMTS (Universal Mobile Telecommunication System). WCDMA (Wideband Code Division Multiple Access) is the main third generation air interface in the world and deployment has been started in Europe and Asia, including Japan and Korea, in the same frequency band, around 2 GHz. WCDMA will be deployed also in the USA in the US frequency bands. During the writing of this third edition, the largest WCDMA operators have reached the 6 million subscribers milestone and GSM/WCDMA multimode terminals are being sold in more than 50 countries. Though less than 10 million subscribers is still small compared to the GSM subscriber base, the growth rate is expected to follow a similar track to GSM in the early days, and eventually the subscribers currently using PDC or GSM will emigrate to WCDMA as the terminals on offer and service coverage continue to improve. The large market for WCDMA and its flexible multimedia capabilities will create new business opportunities for manufacturers, operators, and the providers of content and applications. This book gives a detailed description of the WCDMA air interface and its utilisation. The contents are summarised in Figure 1. Chapter 1 introduces the third generation air interfaces, the spectrum allocation, the time schedule, and the main differences from second generation air interfaces. Chapter 2 presents example UMTS applications, concept phones and the quality of service classes. Chapter 3 introduces the principles of the WCDMA air interface, including spreading, Rake receiver, power control and handovers. Chapter 4 presents the background to WCDMA, the global harmonisation process and the standardisation. Chapters 5–7 give a detailed presentation of the WCDMA standard, while Chapters 8–12 cover the utilisation of the standard and its performance. Chapter 5 describes the architecture of the radio access network, interfaces within the radio access network between base stations and radio network controllers (RNC), and the interface between the radio access network and the core network. Chapter 6 covers the physical layer (Layer 1), including spreading, modulation, user data and signalling transmission, and the main physical layer procedures of power control, paging, transmission diversity and handover measurements. Chapter 7 introduces the radio interface protocols, consisting of the data link layer (Layer 2) and the network layer (Layer 3). Chapter 8 presents the

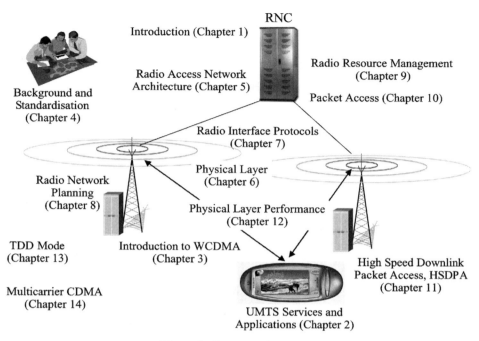

Figure 1. Contents of this book

guidelines for radio network dimensioning, gives an example of detailed capacity and coverage planning, and covers GSM co-planning. Chapter 9 covers the radio resource management algorithms that guarantee the efficient utilisation of the air interface resources and the quality of service. These algorithms are power control, handovers, admission and load control. Chapter 10 depicts packet access and presents the performance of packet protocols of WCDMA. Chapter 11 presents the significant Release 5 feature, High-Speed Downlink Packet Access, HSDPA, and its performance. Chapter 12 analyses the coverage and capacity of the WCDMA air interface with bit rates up to 2 Mbps. Chapter 13 introduces the time division duplex (TDD) mode of the WCDMA air interface and its differences from the frequency division duplex (FDD) mode. In addition to WCDMA, third generation services can also be provided with EDGE or with multicarrier CDMA. EDGE is the evolution of GSM for high data rates within the GSM carrier spacing. Multicarrier CDMA is the evolution of IS-95 for high data rates using three IS-95 carriers, and is introduced in Chapter 14.

The second edition contained coverage of the recently introduced key features of 3GPP Release 5 specifications, such as High-Speed Downlink Packet Access, HSDPA and IP Multimedia Sub-system (IMS).

The third edition of the book continues to deepen the coverage of several existing topics, both based on field experiences and more detailed simulation studies. The third edition covers the main updates in 3GPP standard Release 6. Chapter 2 introduces example packet-based person-to-person services, including Push-to-talk over Cellular (PoC), Real time videosharing and multiplayer games. In Chapter 4, standardisation related milestones have been updated and the 3GPP way of working has been described to improve understanding of how things get done in standardisation. In Chapter 6, the beamforming measurements have been added, as well as a discussion of the terminal capabilities available commercially for WCDMA as of today. The new Layer 2/3 related 3GPP items finalised or about to be finalised, early UE handling and Multimedia Broadcast Multicast Service (MBMS), have been added to Chapter 7, along with additional signalling examples. Chapter 9 covers handover measurements from the field. Chapter 10 has been completely rewritten to reflect the latest understanding of the application end-to-end performance over WCDMA, including measurement results from the commercial networks. HSDPA performance has been studied in more depth in Chapter 11. The next step in the WCDMA evolution, High Speed Uplink Packet Access (HSUPA), is covered in Chapter 11. For the TDD description in Chapter 13, the 1.28 Mcps TDD (known also as Chinese TD-SCDMA) has been covered in more detail. In general also the feedback received from readers has been taken into account to sharpen the details where necessary, which the authors are happy to acknowledge. In Chapter 14, minor additions have been made to reflect the development on the 3GPP2 side.

This book is aimed at operators, network and terminal manufacturers, service providers, university students and frequency regulators. A deep understanding of the WCDMA air interface, its capabilities and its optimal usage is the key to success in the UMTS business.

This book represents the views and opinions of the authors, and does not necessarily represent the views of their employers.

Acknowledgements

The editors would like to acknowledge the time and effort put in by their colleagues in contributing to this book. Besides the editors, the contributors were Leo Chan, Renaud Cuny, Frank Frederiksen, Zhi-Chun Honkasalo, Seppo Hämäläinen, Markku Juntti, Troels Kolding, Martin Kristensson, Janne Laakso, Jaana Laiho, Ukko Lappalainen, Otto Lehtinen, Fabio Longoni, Atte Länsisalmi, Nina Madsen, Preben Mogensen, Peter Muszynski, Jari Mäkinen, Klaus Pedersen, Kari Horneman, Karri Ranta-aho, Jussi Reunanen, Oscar Salonaho, Jouni Salonen, Kari Sipilä, Jukka Vialen, Heli Väätäjä, Jaakko Vihriälä, Achim Wacker and Jeroen Wigard.

While we were developing this book, many of our colleagues from different Nokia sites in three continents offered their help in suggesting improvements and finding errors. Also, a number of colleagues from other companies have helped us in improving the quality of the book. The editors are grateful for the comments received from Heikki Ahava, Erkka Ala-Tauriala, David Astely, Erkki Autio, Kai Heikkinen, Kari Heiska, Kimmo Hiltunen, Klaus Hugl, Alberg Höglund, Kaisu Iisakkila, Ann-Louise Johansson, Susanna Kallio, Ilkka Keskitalo, Pasi Kinnunen, Tero Kola, Petri Komulainen, Lauri Laitinen, Anne Leino, Arto Leppisaari, Pertti Lukander, Esko Luttinen, Jonathan Moss, Olli Nurminen, Tero Ojanperä, Lauri Oksanen, Kari Pehkonen, Mika Rinne, David Soldani, Rauno Ruismäki, Kimmo Terävä, Mitch Tseng, Antti Tölli and Veli Voipio.

The team at John Wiley & Sons, Ltd participating in the production of this book provided excellent support and worked hard to keep the demanding schedule. The editors especially would like to thank Sarah Hinton and Mark Hammond for assistance with practical issues in the production process, and especially the copy-editor, for her efforts in smoothing out the engineering approach to the English language expressions.

We are extremely grateful to our families, as well as the families of all the authors, for their patience and support, especially during the late night and weekend editing sessions near different production milestones.

Special thanks are due to our employer, Nokia Networks, for supporting and encouraging such an effort and for providing some of the illustrations in this book.

Finally, we would like to acknowledge the efforts of our colleagues in the wireless industry for the great work done within the 3rd Generation Partnership Project (3GPP) to produce the global WCDMA standard in merely a year and thus to create the framework for this book. Without such an initiative this book would never have been possible.

The editors and authors welcome any comments and suggestions for improvements or changes that could be implemented in forthcoming editions of this book. Feedback should be sent to the editors' email addresses: harri.holma@nokia.com and antti.toskala@nokia.com.

Abbreviations

3GPP	3rd Generation partnership project (produces WCDMA standard)
3GPP2	3rd Generation partnership project 2 (produced cdma2000 standard)
AAL2	ATM Adaptation Layer type 2
AAL5	ATM Adaptation Layer type 5
ACELP	Algebraic code excitation linear prediction
ACIR	Adjacent channel interference ratio, caused by the transmitter non-idealities and imperfect receiver filtering
ACK	Acknowledgement
ACIR	Adjacent channel interference ratio
ACLR	Adjacent channel leakage ratio, caused by the transmitter non-idealities, the effect of receiver filtering is not included
ACTS	Advanced communication technologies and systems, EU research projects framework
AICH	Acquisition indication channel
ALCAP	Access link control application part
AM	Acknowledged mode
AMD	Acknowledged mode data
AMR	Adaptive multirate (speech codec)
AMR-NB	Narrowband AMR
AMR-WB	Wideband AMR
ARIB	Association of radio industries and businesses (Japan)
ARP	Allocation and retention priority
ARQ	Automatic repeat request
ASC	Access service class
ASN.1	Abstract syntax notation one
ATM	Asynchronous transfer mode
AWGN	Additive white Gaussian noise
BB SS7	Broad band signalling system #7
BCCH	Broadcast channel (logical channel)
BCH	Broadcast channel (transport channel)
BCFE	Broadcast control functional entity
BCH	Broadcast channel (transport channel)
BER	Bit error rate
BLER	Block error rate
BMC	Broadcast/multicast control protocol

BoD	Bandwidth on demand
BPSK	Binary phase shift keying
BS	Base station
BSS	Base station subsystem
BSC	Base station controller
CA-ICH	Channel assignment indication channel
CB	Cell broadcast
CBC	Cell broadcast center
CBS	Cell broadcast service
CCCH	Common control channel (logical channel)
CCH	Common transport channel
CCH	Control channel
CD-ICH	Collision detection indication channel
CDF	Cumulative distribution function
CDMA	Code division multiple access
CFN	Connection frame number
CIR	Carrier to interference ratio
CM	Connection management
CN	Core network
C-NBAP	Common NBAP
CODIT	Code division test bed, EU research project
CPCH	Common packet channel
CPICH	Common pilot channel
CQI	Channel quality indicator
CRC	Cyclic redundancy check?
CRNC	Controlling RNC
C-RNTI	Cell-RNTI, radio network temporary identity
CS	Circuit Switched
CSCF	Call state control function
CSICH	CPCH status indication channel
CTCH	Common traffic channel
CWTS	China wireless telecommunications standard group
DCA	Dynamic channel allocation
DCCH	Dedicated control channel (logical channel)
DCFE	Dedicated control functional entity
DCH	Dedicated channel (transport channel)
DECT	Digital enhanced cordless telephone
DF	Decision feedback
DL	Downlink
D-NBAP	Dedicated NBAP
DNS	Domain name system
DPCCH	Dedicated physical control channel
DPDCH	Dedicated physical data channel
DRNC	Drift RNC
DRX	Discontinuous reception
DS-CDMA	Direct spread code division multiple access
DSCH	Downlink shared channel

DSL	Digital subscriber line
DTCH	Dedicated traffic channel
DTX	Discontinuous transmission
E-DCH	Enhanced uplink DCH
EDGE	Enhanced data rates for GSM evolution
EFR	Enhance full rate
EGSM	Extended GSM
EIRP	Equivalent isotropic radiated power
EP	Elementary Procedure
ETSI	European Telecommunications Standards Institute
FACH	Forward access channel
FBI	Feedback information
FCC	Federal communication commission
FCS	Fast cell selection
FDD	Frequency division duplex
FDMA	Frequency division multiple access
FER	Frame error ratio
FP	Frame protocol
FRAMES	Future radio wideband multiple access system, EU research project
FTP	File transfer protocol
GERAN	GSM/EDGE Radio Access Network
GGSN	Gateway GPRS support node
GMSC	Gateway MSC
GPRS	General packet radio system
GPS	Global positioning system
GSIC	Groupwise serial interference cancellation
GSM	Global system for mobile communications
GTP-U	User plane part of GPRS tunnelling protocol
HARQ	Hybrid automatic repeat request
HLR	Home location register
HSDPA	High speed downlink packet access
HS-DPCCH	Uplink high speed dedicated physical control channel
HS-DSCH	High speed downlink shared channel
HS-SCCH	High speed shared control channel
HSUPA	High speed uplink packet access
HSS	Home subscriber server
HTTP	Hypertext transfer protocol
IC	Interference cancellation
ID	Identity
IETF	Internet engineering task force
IMEISV	International Mobile Station Equipment Identity and Software Version
IMS	IP multimedia sub-system
IMSI	International mobile subscriber identity
IMT-2000	International mobile telephony, 3rd generation networks are referred as IMT-2000 within ITU
IN	Intelligent network
IP	Internet protocol

IPDL	Idle periods in downlink
IPI	Inter-path interference
IRC	Interference rejection combining
IS-2000	IS-95 evolution standard, (cdma2000)
IS-136	US-TDMA, one of the 2nd generation systems, mainly in Americas
IS-95	cdmaOne, one of the 2nd generation systems, mainly in Americas and in Korea
ISDN	Integrated services digital network
ISI	Inter-symbol interference
ITU	International telecommunications union
ITUN	SS7 ISUP Tunnelling
Iu BC	Iu broadcast
L2	Layer 2
LAI	Location area identity
LAN	Local area network
LCS	Location services
LP	Low pass
MAC	Medium access control
MAI	Multiple access interference
MAP	Maximum a posteriori
MBMS	Multimedia broadcast multicast service
MCCH	MBMS point-to-multipoint control channel
MCS	Modulation and coding scheme
MCU	Multipoint control unit
ME	Mobile equipment
MF	Matched filter
MGCF	Media gateway control function
MGW	Media gateway
MIMO	Multiple input multiple output
MLSD	Maximum likelihood sequence detection
MM	Mobility management
MMS	Multimedia message
MMSE	Minimum mean square error
MOS	Mean opinion score
MPEG	Motion picture experts group
MR-ACELP	Multirate ACELP
MRF	Media resource function
MS	Mobile station
MSC/VLR	Mobile services switching centre/visitor location register
MT	Mobile termination
MTCH	MBMS point-to-multipoint control channel
MTP3b	Message transfer part (broadband)
MUD	Multiuser detection
NAS	Non access stratum
NBAP	Node B application part
NRT	Non-real time
ODMA	Opportunity driven multiple access
OFDMA	Orthogonal frequency division multiple access

O&M	Operation and maintenance
OSS	Operations support system
OTDOA	Observed time difference of arrival
OVSF	Orthogonal variable spreading factor
PAD	Padding
PC	Power control
PCCC	Parallel concatenated convolutional coder
PCCCH	Physical common control channel
PCCH	Paging channel (logical channel)
PCCPCH	Primary common control physical channel
PCH	Paging channel (transport channel)
PCPCH	Physical common packet channel
PCS	Persona communication systems, 2nd generation cellular systems mainly in Americas, operating partly on IMT-2000 band
PDC	Personal digital cellular, 2nd generation system in Japan
PDCP	Packet data converge protocol
PDP	Packet data protocol
PDSCH	Physical downlink shared channel
PDU	Protocol data unit
PEP	Performance enhancement proxy
PER	Packed encoding rules
PHY	Physical layer
PI	Page indicator
PIC	Parallel interference cancellation
PICH	Paging indicator channel
PLMN	Public land mobile network
PNFE	Paging and notification control function entity
POC	Push-to-talk over cellular
PRACH	Physical random access channel
PS	Packet switched
PSCH	Physical shared channel
PSTN	Public switched telephone network
P-TMSI	Packet-TMSI
PU	Payload unit
PVC	Pre-defined Virtual Connection
QAM	Quadrature amplitude modulation
QoS	Quality of service
QPSK	Quadrature phase shift keying
RAB	Radio access bearer
RACH	Random access channel
RAI	Routing area identity
RAN	Radio access network
RANAP	RAN application part
RB	Radio bearer
RF	Radio frequency
RLC	Radio link control
RNC	Radio network controller

RNS	Radio network sub-system
RNSAP	RNS application part
RNTI	Radio network temporary identity
RRC	Radio resource control
RRM	Radio resource management
RSSI	Received signal strength indicator
RSVP	Resource reservation protocol
RT	Real time
RTCP	Real time transport control protocol
RTP	Real time protocol
RTSP	Real time streaming protocol
RU	Resource unit
SAAL-NNI	Signalling ATM adaptation layer for network to network interfaces
SAAL-UNI	Signalling ATM adaptation layer for user to network interfaces
SABP	Service Area Broadcast Protocol
SAP	Service access point
SAP	Session announcement protocol
SAS	Stand alone SMLC
SCCP	Signalling connection control part
SCCPCH	Secondary common control physical channel
SCH	Synchronisation channel
SCTP	Simple control transmission protocol
SDD	Space division duplex
SDP	Session description protocol
SDU	Service data unit
SEQ	Sequence
SF	Spreading Factor
SFN	System frame number
SGSN	Serving GPRS support node
SIP	Session initiation protocol
SHO	Soft handover
SIB	System information block
SIC	Successive interference cancellation
SID	Silence indicator
SINR	Signal-to-noise ratio where noise includes both thermal noise and interference
SIP	Session initiation protocol
SIR	Signal to interference ratio
SM	Session management
SMS	Short message service
SMLC	Serving mobile location centre
SN	Sequence number
SNR	Signal to noise ratio
SQ-PIC	Soft quantised parallel interference cancellation
SRB	Signalling radio bearer
SRNC	Serving RNC
SRNS	Serving RNS
SS7	Signalling System #7

SSCF	Service specific co-ordination function
SSCOP	Service specific connection oriented protocol
SSDT	Site selection diversity transmission
STD	Switched transmit diversity
STTD	Space time transmit diversity
TCH	Traffic channel
TCP	Transport control protocol
TCTF	Target channel type field
TD/CDMA	Time division CDMA, combined TDMA and CDMA
TDD	Time division duplex
TDMA	Time division multiple access
TD-SCDMA	Time division synchronous CDMA, 1.28 Mcps TDD
TE	Terminal equipment
TF	Transport format
TFCI	Transport format combination indicator
TFCS	Transport format combination set
TFI	Transport format indicator
TFRC	Transport format and resource combination
THP	Traffic handling priority
TMSI	Temporary mobile subscriber identity
TPC	Transmission power control
TR	Transparent mode
TS	Technical specification
TSTD	Time switched transmit diversity
TTA	Telecommunications Technology Association (Korea)
TTC	Telecommunication Technology Commission (Japan)
TTI	Transmission time interval
TxAA	Transmit adaptive antennas
UDP	User datagram protocol
UE	User equipment
UL	Uplink
UM	Unacknowledged mode
UMTS	Universal mobile telecommunication services
URA	UTRAN registration area
URL	Universal resource locator
U-RNTI	UTRAN RNTI
USCH	Uplink shared channel
USIM	UMTS subscriber identity module
US-TDMA	IS-136, one of the 2nd generation systems mainly in USA
UTRA	UMTS Terrestrial radio access (ETSI)
UTRA	Universal Terrestrial radio access (3GPP)
UTRAN	UMTS Terrestrial radio access network
VAD	Voice activation detection
VoIP	Voice over IP
VPN	Virtual private network
WAP	Wireless application protocol
WARC	World administrative radio conference

WCDMA	Wideband CDMA, Code division multiple access
WLL	Wireless local loop
WML	Wireless markup language
WWW	World wide web
XHTML	Extensible hypertext markup language
ZF	Zero forcing

1

Introduction

Harri Holma, Antti Toskala and Ukko Lappalainen

1.1 WCDMA in Third Generation Systems

Analog cellular systems are commonly referred to as first generation systems. The digital systems currently in use, such as GSM, PDC, cdmaOne (IS-95) and US-TDMA (IS-136), are second generation systems. These systems have enabled voice communications to go wireless in many of the leading markets, and customers are increasingly finding value also in other services, such as text messaging and access to data networks, which are starting to grow rapidly.

Third generation systems are designed for multimedia communication: with them person-to-person communication can be enhanced with high quality images and video, and access to information and services on public and private networks will be enhanced by the higher data rates and new flexible communication capabilities of third generation systems. This, together with the continuing evolution of the second generation systems, will create new business opportunities not only for manufacturers and operators, but also for the providers of content and applications using these networks.

In the standardisation forums, WCDMA technology has emerged as the most widely adopted third generation air interface. Its specification has been created in 3GPP (the 3rd Generation Partnership Project), which is the joint standardisation project of the standardisation bodies from Europe, Japan, Korea, the USA and China. Within 3GPP, WCDMA is called UTRA (Universal Terrestrial Radio Access) FDD (Frequency Division Duplex) and TDD (Time Division Duplex), the name WCDMA being used to cover both FDD and TDD operation.

Throughout this book, the chapters related to specifications use the 3GPP terms UTRA FDD and TDD, the others using the term WCDMA. This book focuses on the WCDMA FDD technology. The WCDMA TDD mode and its differences from the WCDMA FDD mode are presented in Chapter 13, which includes a description of TD-SCDMA.

WCDMA for UMTS, third edition.   Edited by Harri Holma and Antti Toskala
© 2004 John Wiley & Sons, Ltd   ISBN: 0-470-87096-6

1.2   Air Interfaces and Spectrum Allocations for Third Generation Systems

Work to develop third generation mobile systems started when the World Administrative Radio Conference (WARC) of the ITU (International Telecommunications Union), at its 1992 meeting, identified the frequencies around 2 GHz that were available for use by future third generation mobile systems, both terrestrial and satellite. Within the ITU these third generation systems are called International Mobile Telephony 2000 (IMT-2000). Within the IMT-2000 framework, several different air interfaces are defined for third generation systems, based on either CDMA or TDMA technology, as described in Chapter 4. The original target of the third generation process was a single common global IMT-2000 air interface. Third generation systems are closer to this target than were second generation systems: the same air interface – WCDMA – is to be used in Europe and Asia, including Japan and Korea, using the frequency bands that WARC-92 allocated for the third generation IMT-2000 system at around 2 GHz. In North America, however, that spectrum has already been auctioned for operators using second generation systems, and no new spectrum is available for IMT-2000. Thus, third generation services there must be implemented within the existing bands, and also WCDMA can be deployed in the existing band in North America. The global IMT-2000 spectrum is not available in countries that follow the US PCS spectrum allocation. Some of the Latin American countries, like Brazil, plan to follow the European spectrum allocation at 2 GHz.

In addition to WCDMA, the other air interfaces that can be used to provide third generation services are EDGE and cdma2000. EDGE (Enhanced Data Rates for GSM Evolution) can provide third generation services with bit rates up to 500 kbps within a GSM carrier spacing of 200 kHz [1]. EDGE includes advanced features that are not part of GSM to improve spectrum efficiency and to support the new services. cdma2000 can be used as an upgrade solution for the existing IS-95 operators and will be presented in more detail in Chapter 14.

The expected frequency bands and geographical areas where these different air interfaces are likely to be applied are shown in Figure 1.1. Within each region there are local exceptions in places where multiple technologies are already being deployed.

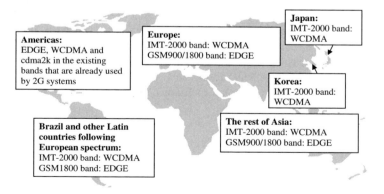

Figure 1.1. Expected air interfaces and spectrums for providing third generation services

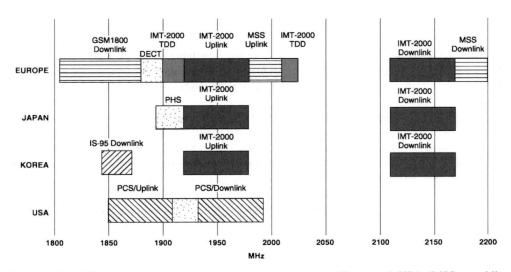

Figure 1.2. 2 GHz band spectrum allocation in Europe, Japan, Korea and USA (MSS = mobile satellite spectrum)

The spectrum allocation in Europe, Japan, Korea and the USA is shown in Figure 1.2 and in Table 1.1. In Europe and in most of Asia the IMT-2000 (or WARC-92) bands of 2×60 MHz (1920–1980 MHz plus 2110–2170 MHz) will be available for WCDMA FDD. The availability of the TDD spectrum varies: in Europe it is expected that 25 MHz will be available for licensed TDD use in the 1900–1920 MHz and 2020–2025 MHz bands. The rest of the unpaired spectrum is expected to be used for unlicensed TDD applications (SPA: Self Provided Applications) in the 2010–2020 MHz band. FDD systems use different frequency bands for uplink and for downlink, separated by the duplex distance, while TDD systems utilise the same frequency for both uplink and downlink.

Table 1.1. Existing frequency allocations around 2 GHz

	Uplink	Downlink	Total
GSM1800	1710–1785	1805–1880	2×75 MHz
UMTS-FDD	1920–1980	2110–2170	2×60 MHz
UMTS-TDD	1900–1920 and	2010–2025	$20 + 15$ MHz
Americas PCS	1850–1910	1930–1990	2×60 MHz

Also in Japan and Korea, as in the rest of Asia, the WARC-92 bands will be made available for IMT-2000. Japan has deployed PDC as a second generation system, while in Korea, IS-95 is used for both cellular and PCS operation. The PCS spectrum allocation in Korea is different from the US PCS spectrum allocation, leaving the IMT-2000 spectrum fully available in Korea. In Japan, part of the IMT-2000 TDD spectrum is used by PHS, the cordless telephone system.

In China, there are reservations for PCS or WLL (Wireless Local Loop) use on one part of the IMT-2000 spectrum, though these have not been assigned to any operators. Depending on the regulation decisions, up to 2×60 MHz of the IMT-2000 spectrum will be available for WCDMA FDD use in China. The TDD spectrum will also be made available in China.

In the USA no new spectrum has yet been made available for third generation systems. Third generation services can be implemented within the existing PCS spectrum. For the US PCS band, all third generation alternatives can be considered: EDGE, WCDMA and cdma2k.

EDGE can be deployed within the existing GSM900 and GSM1800 frequencies where those frequencies are in use. These GSM frequencies are not available in Korea and Japan. The total band available for GSM900 operation is 2×25 MHz plus EGSM 2×10 MHz, and for GSM1800 operation, 2×75 MHz. EGSM refers to the extension of the GSM900 band. The total GSM band is not available in all countries using the GSM system.

The first IMT-2000 licences were granted in Finland in March 1999, and followed by Spain in March 2000. No auction was conducted in Finland or in Spain. Also, Sweden granted the licenses without auction in December 2000. However, in other countries, such as the UK, Germany and Italy, an auction similar to the US PCS spectrum auctions was conducted.

A few example UMTS licenses are shown in Table 1.2 in Japan and in Europe. The number of UMTS operators per country is between three and six.

Table 1.2. Example UMTS licenses

Country	Number of operators	Number of FDD carriers (2×5 MHz) per operator	Number of TDD carriers (1×5 MHz) per operator
Finland	4	3	1
Japan	3	3	0
Spain	4	3	1
UK	5	2–3	0–1
Germany	6	2	0–1
Netherlands	5	2–3	0–1
Italy	5	2	1
Austria	5	2	0–2

More frequencies have been identified for IMT-2000 in addition to the WARC-92 frequency bands mentioned above. At the ITU-R WRC-2000 in May 2000 the following frequency bands were also identified for IMT-2000:

- 1710–1885 MHz;
- 2500–2690 MHz;
- 806–960 MHz.

It is worth noting that some of the bands listed, especially below 2 GHz, are partly used with systems like GSM. The main new spectrum in Europe for IMT-2000 will be 2500–2690 MHz. The duplex arrangement of that spectrum is under discussion.

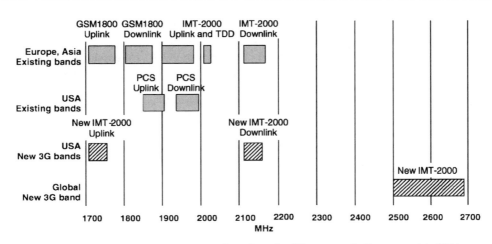

Figure 1.3. New expected spectrum allocations for 3G systems in Europe and in USA

In the USA, the 1.7/2.1 GHz spectrum is going to be available soon [2] and FCC has released the ruling for the operation on that band defining, e.g., spectrum masks and other necessary technical details for equipment development to start. That spectrum can be efficiently used for delivering third generation services with WCDMA. The 1.7 GHz band can be used for FDD uplink in line with GSM1800 arrangement and the 2.1 GHz used for WCDMA downlink would be in line with WARC-92 band arrangement. The main spectrum allocations for third generation services are shown in Figure 1.3 and in Table 1.3. Do note that 3GPP has also recently specified WCDMA performance requirements for the US and Japanese 800 MHz bands.

Table 1.3. New frequency allocations for third generation services

	Uplink	Downlink	Total
New band in US[1]	1710–1770	2110–2170	2×60 MHz
New IMT-2000	2500–2690		190 MHz
	(Frequency arrangement under discussion)		

[1]The Federal Communication Commission (FCC) ruling covers initial spectrum of 2×45 MHz in 1710–1755 and 2110–2155. The auction process is to be held later, but WCDMA (release independent) requirements have been completed as of March 2004 for this band.

1.3 Schedule for Third Generation Systems

European research work on WCDMA was initiated in the European Union research projects CODIT [3] and FRAMES [4], and also within large European wireless communications companies, at the start of the 1990s [5]. Those projects also produced WCDMA trial systems to evaluate link performance [6] and generated the basic understanding of WCDMA necessary for standardisation. In January 1998 the European standardisation body ETSI decided upon WCDMA as the third generation air interface [7]. Detailed standardisation work has been carried out as part of the 3GPP standardisation process. The first full set of specifications was completed at the end of 1999.

The first commercial network was opened in Japan during 2001 for commercial use in key areas, and in Europe at the beginning of 2002 for a pre-commercial testing phase. During 2003 we have seen a few more networks opening; however, the real large scale network opening is expected to take place later during 2H/2004 with a wider selection of WCDMA terminals available. The expected schedule is presented in Figure 1.4. This schedule relates to FDD mode operation. The TDD mode is expected to follow much later, and the first TDD networks will probably be based on the 3GPP Release 4 or 5 version of the specifications. In Japan, the schedule for TDD operation is also unclear due to the unavailability of the TDD spectrum.

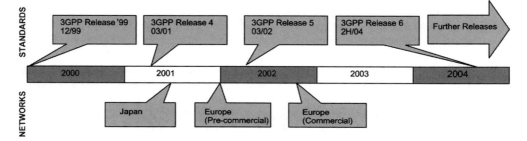

Figure 1.4. Standardisation and commercial operation schedule for WCDMA

Looking back at the history of GSM, we note that since the opening of the first GSM network in July 1991 (Radiolinja, Finland) several countries have reached more than 50 % cellular phone penetration. In some countries as much as 80 % penetration has been reached and the global GSM subscriber count has exceeded one billion. Early GSM experiences showed that once there were small sized attractive terminals available with low power consumption, the growth rates were very high. WCDMA is foreseen to follow the same trend.

Second generation systems could already enable voice traffic to go wireless; now third generation systems face the challenge of making a new set of data services go wireless as well.

1.4 Differences between WCDMA and Second Generation Air Interfaces

In this section the main differences between the third and second generation air interfaces are described. GSM and IS-95 (the standard for cdmaOne systems) are the second generation air interfaces considered here. Other second generation air interfaces are PDC in Japan and US-TDMA mainly in the Americas; these are based on TDMA (time division multiple access) and have more similarities with GSM than with IS-95. The second generation systems were built mainly to provide speech services in macro cells. To understand the background to the differences between second and third generation systems, we need to look at the new requirements of the third generation systems which are listed below:

- Bit rates up to 2 Mbps;

- Variable bit rate to offer bandwidth on demand;

- Multiplexing of services with different quality requirements on a single connection, e.g. speech, video and packet data;

- Delay requirements from delay-sensitive real time traffic to flexible best-effort packet data;

- Quality requirements from 10 % frame error rate to 10^{-6} bit error rate;

- Co-existence of second and third generation systems and inter-system handovers for coverage enhancements and load balancing;

- Support of asymmetric uplink and downlink traffic, e.g. web browsing causes more loading to downlink than to uplink;

- High spectrum efficiency;

- Co-existence of FDD and TDD modes.

Table 1.4 lists the main differences between WCDMA and GSM, and Table 1.5 those between WCDMA and IS-95. In this comparison only the air interface is considered. GSM also covers services and core network aspects, and this GSM platform will be used together with the WCDMA air interface: see the next section regarding core networks.

Table 1.4. Main differences between WCDMA and GSM air interfaces

	WCDMA	GSM
Carrier spacing	5 MHz	200 kHz
Frequency reuse factor	1	1–18
Power control frequency	1500 Hz	2 Hz or lower
Quality control	Radio resource management algorithms	Network planning (frequency planning)
Frequency diversity	5 MHz bandwidth gives multipath diversity with Rake receiver	Frequency hopping
Packet data	Load-based packet scheduling	Time slot based scheduling with GPRS
Downlink transmit diversity	Supported for improving downlink capacity	Not supported by the standard, but can be applied

The differences in the air interface reflect the new requirements of the third generation systems. For example, the larger bandwidth of 5 MHz is needed to support higher bit rates. Transmit diversity is included in WCDMA to improve the downlink capacity to support the asymmetric capacity requirements between downlink and uplink. Transmit diversity is not

Table 1.5. Main differences between WCDMA and IS-95 air interfaces

	WCDMA	IS-95
Carrier spacing	5 MHz	1.25 MHz
Chip rate	3.84 Mcps	1.2288 Mcps
Power control frequency	1500 Hz, both uplink and downlink	Uplink: 800 Hz, downlink: slow power control
Base station synchronisation	Not needed	Yes, typically obtained via GPS
Inter-frequency handovers	Yes, measurements with slotted mode	Possible, but measurement method not specified
Efficient radio resource management algorithms	Yes, provides required quality of service	Not needed for speech only networks
Packet data	Load-based packet scheduling	Packet data transmitted as short circuit switched calls
Downlink transmit diversity	Supported for improving downlink capacity	Not supported by the standard

supported by the second generation standards. The mixture of different bit rates, services and quality requirements in third generation systems requires advanced radio resource management algorithms to guarantee quality of service and to maximise system throughput. Also, efficient support of non-real time packet data is important for the new services.

The main differences between WCDMA and IS-95 are discussed below. Both WCDMA and IS-95 utilise direct sequence CDMA. The higher chip rate of 3.84 Mcps in WCDMA enables higher bit rates. The higher chip rate also provides more multipath diversity than the chip rate of 1.2288 Mcps, especially in small urban cells. The importance of diversity for system performance is discussed in Sections 9.2.1.2 and 12.2.1.3. Most importantly, increased multipath diversity improves the coverage. The higher chip rate also gives a higher trunking gain, especially for high bit rates, than do narrowband second generation systems.

WCDMA has fast closed loop power control in both uplink and downlink, while IS-95 uses fast power control only in uplink. The downlink fast power control improves link performance and enhances downlink capacity. It requires new functionalities in the mobile, such as SIR estimation and outer loop power control, that are not needed in IS-95 mobiles.

The IS-95 system was targeted mainly for macro cellular applications. The macro cell base stations are located on masts or rooftops where the GPS signal can be easily received. IS-95 base stations need to be synchronised and this synchronisation is typically obtained via GPS. The need for a GPS signal makes the deployment of the indoor and micro cells more problematic, since GPS reception is difficult without line-of-sight connection to the GPS satellites. Therefore, WCDMA is designed to operate with asynchronous base stations where no synchronisation from GPS is needed. The asynchronous base stations make the WCDMA handover slightly different from that of IS-95.

Inter-frequency handovers are considered important in WCDMA, to maximise the use of several carriers per base station. In IS-95 inter-frequency measurements are not specified, making inter-frequency handovers more difficult.

Experiences from second generation air interfaces have been important in the development of the third generation interface, but there are many differences, as listed above. In order to make the fullest use of the capabilities of WCDMA, a deep understanding of the

WCDMA air interface is needed, from the physical layer to network planning and performance optimisation.

1.5 Core Networks and Services

There are three basic solutions for the core network to which WCDMA radio access networks can be connected. The basis of the second generation has been either the GSM core network or one based on IS-41. Both will naturally be important options in third generation systems. An emerging alternative is GPRS with an all-IP-based core network. The most typical connections between the core networks and the air interfaces are illustrated in Figure 1.5. Other connections are also possible and are expected to appear in the standardisation forums in due course.

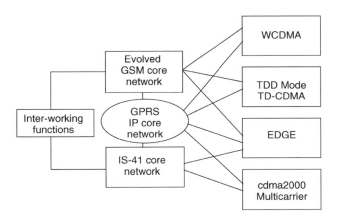

Figure 1.5. Core network relation to the third generation air interface alternatives

The market needs will determine which combinations will be used by the operators. It is expected that operators will remain with their second generation core network for voice services and will then add packet data functionalities on top of that.

Because of the different technologies and frequency allocations, global roaming will continue to require specific arrangements between operators, such as multimode and multiband handset and roaming gateways between the different core networks. To the end user the operator arrangements will not be visible, and global roaming terminals will probably emerge for those consumers willing to pay for global service.

In the long run the development proceeds towards all-IP networks where all the services are delivered via packet switched networks. GSM utilises mainly circuit switched services, like voice, short messages, WAP and email. 3GPP Release '99, together with packet core network, enables a large number of new packet switched services, while voice is still carried with the circuit switched network. With the introduction of IP Multimedia Sub-system (IMS) in 3GPP Releases 5 and 6 specifications, basically all services can be provided from packet switched network simplifying the network maintenance and service creation. The IMS is covered in Chapter 5. This development is shown in Figure 1.6.

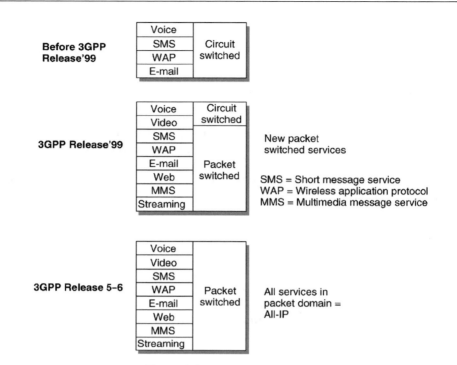

Figure 1.6. Development to all-IP

References

[1] Pirhonen, R., Rautava, T. and Penttinen, J., 'TDMA Convergence for Packet Data Services', *IEEE Personal Communications Magazine*, June 1999, Vol. 6, No. 3, pp. 68–73.

[2] *FCC news, Action by the Commission*, October 16, 2003, by Report and Order (FCC 03-251).

[3] Andermo, P.-G. (ed.), 'UMTS Code Division Testbed (CODIT)', CODIT Final Review Report, September 1995.

[4] Nikula, E., Toskala, A., Dahlman, E., Girard, L. and Klein, A., 'FRAMES Multiple Access for UMTS and IMT-2000', *IEEE Personal Communications Magazine*, April 1998, pp. 16–24.

[5] Ojanperä, T., Rikkinen, K., Häkkinen, H., Pehkonen, K., Hottinen, A. and Lilleberg, J., 'Design of a 3rd Generation Multirate CDMA System with Multiuser Detection, MUD-CDMA', *Proc. ISSSTA'96*, Mainz, Germany, September 1996, pp. 334–338.

[6] Pajukoski, K. and Savusalo, J., 'Wideband CDMA Test System', *Proc. IEEE Int. Conf. on Personal Indoor and Mobile Radio Communications*, PIMRC'97, Helsinki, Finland, 1–4 September 1997, pp. 669–672.

[7] Holma, H., Toskala, A. and Latva-aho, M., 'Asynchronous Wideband CDMA for IMT-2000', *SK Telecom Journal*, South Korea, Vol. 8, No. 6, 1998, pp. 1007–1021.

2

UMTS Services and Applications

Harri Holma, Martin Kristensson, Jouni Salonen and Antti Toskala

2.1 Introduction

2nd generation systems like GSM, were originally designed for efficient delivery of voice services. UMTS networks are, on the contrary, designed from the beginning for flexible delivery of any type of service, where each new service does not require particular network optimisation. In addition to the flexibility, the WCDMA radio solution brings advanced capabilities that enable new services. Such capabilities are:

- High bit rates theoretically up to 2 Mbps in 3GPP Release '99, and beyond 10 Mbps in 3GPP Release 5. Practical bit rates are up to 384 kbps initially, and beyond 2 Mbps with Release 5;

- Low delays with packet round trip times below 200 ms;

- Seamless mobility also for packet data applications;

- Quality of Service differentiation for high efficiency of service delivery;

- Simultaneous voice and data capability;

- Interworking with existing GSM/GPRS networks.

The WCDMA radio capabilities are described in more detail in Chapters 10, 11 and 12. These advanced radio capabilities, combined with the IP Multimedia Sub-system, IMS, allow fast introduction of new services. This chapter presents a few example UMTS services in Sections 2.2–2.5. The services are divided into person-to-person services, content-to-person services and business connectivity. Person-to-person refers to a peer-to-peer or intermediate server based connection between two persons or a group of persons. Content-to-person services are characterised by the access to information or download of content. Business connectivity refers to the laptop access to internet or intranet using WCDMA as the radio modem. The chapter further introduces IP Multimedia Sub-system, IMS, in Section 2.6 and Quality of service differentiation in Section 2.7. The cost of service delivery and the

WCDMA for UMTS, third edition.   Edited by Harri Holma and Antti Toskala
© 2004 John Wiley & Sons, Ltd   ISBN: 0-470-87096-6

maximum system capacity are analysed in Section 2.8. Section 2.9 presents terminal capability classes and Section 2.10 covers location services in WCDMA.

2.2 Person-to-Person Circuit Switched Services

This section considers AMR voice, wideband AMR voice and video. These services are initially provided through the circuit switched core network in WCDMA, but they can later be provided also through the packet switched core network.

2.2.1 AMR Speech Service

The speech codec in UMTS will employ the Adaptive Multirate (AMR) technique. The multirate speech coder is a single integrated speech codec with eight source rates: 12.2 (GSM-EFR), 10.2, 7.95, 7.40 (IS-641), 6.70 (PDC-EFR), 5.90, 5.15 and 4.75 kbps. The AMR bit rates can be controlled by the radio access network. To facilitate interoperability with existing cellular networks, some of the modes are the same as in existing cellular networks. The 12.2 kbps AMR speech codec is equal to the GSM EFR codec, 7.4 kbps is equal to the US-TDMA speech codec, and 6.7 kbps is equal to the Japanese PDC codec. The AMR speech codec is capable of switching its bit rate every 20 ms speech frame upon command. For AMR mode switching, in-band signalling is used.

The AMR coder operates on speech frames of 20 ms corresponding to 160 samples at the sampling frequency of 8000 samples per second. The coding scheme for the multirate coding modes is the so-called Algebraic Code Excited Linear Prediction Coder (ACELP). The multirate ACELP coder is referred to as MR-ACELP. Every 160 speech samples, the speech signal is analysed to extract the parameters of the CELP model. The speech parameter bits delivered by the speech encoder are rearranged according to their subjective importance before they are sent to the network. The rearranged bits are further sorted, based on their sensitivity to errors and are divided into three classes of importance: A, B and C. Class A is the most sensitive, and the strongest channel coding is used for class A bits in the air interface.

During a normal telephone conversation, the participants alternate so that, on the average, each direction of transmission is occupied about 50 % of the time. The AMR has four basic functions to effectively utilise discontinuous activity:

- Voice Activity Detector (VAD) on the TX side;

- Evaluation of the background acoustic noise on the TX side, in order to transmit characteristic parameters to the RX side;

- The transmission of comfort noise information to the RX side is achieved by means of a Silence Descriptor (SID) frame, which is sent at regular intervals;

- Generation of comfort noise on the RX side during periods when no normal speech frames are received.

Discontinuous transmission (DTX) has some obvious positive implications: in the user terminal, battery life will be prolonged or a smaller battery could be used for a given operational duration. From the network point of view, the average required bit rate is reduced, leading to a lower interference level and hence increased capacity.

The AMR specification also contains error concealment. The purpose of frame substitution is to conceal the effect of lost AMR speech frames. The purpose of muting the output in the case of several lost frames is to indicate the breakdown of the channel to the user and to avoid generating possibly annoying sounds as a result of the frame substitution procedure [1] [2]. The AMR speech codec can tolerate about a 1% frame error rate (FER) of class A bits without any deterioration of the speech quality. For class B and C bits a higher FER is allowed. The corresponding bit error rate (BER) of class A bits will be about 10^{-4}.

The bit rate of the AMR speech connection can be controlled by the radio access network depending on the air interface loading and the quality of the speech connections. During high loading, such as during busy hours, it is possible to use lower AMR bit rates to offer higher capacity while providing slightly lower speech quality. Also, if the mobile is running out of the cell coverage area and using its maximum transmission power, a lower AMR bit rate can be used to extend the cell coverage area. The capacity and coverage of the AMR speech codec is discussed in Chapter 12. With the AMR speech codec it is possible to achieve a trade-off between the network's capacity, coverage and speech quality according to the operator's requirements.

2.2.1.1   AMR Source Based Rate Adaptation – Higher Voice Capacity [3]

AMR codec uses voice activity detection (VAD) together with discontinuous transmission (DTX) to optimise the network capacity and the power consumption of the mobile terminal. Active speech is coded by fixed bit rate that is selected by the radio network according to network capacity and radio channel conditions. Although the network capacity is optimised during silence periods using VAD/DTX, it can be further optimised during active speech with source controlled rate adaptation. Thus AMR codec mode is selected for each speech frame depending on the source signal characteristics, see Figure 2.1. The speech codec mode can be updated in every 20 ms frame in WCDMA.

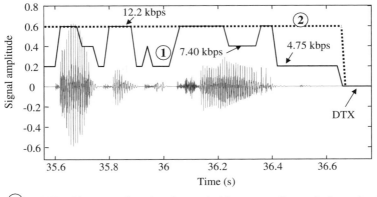

① = AMR with source adaptation changes its bit rate according to the input signal
② = AMR today uses fixed bit rate (+DTX)

Figure 2.1. AMR source based mode selection as a function of time and speech content

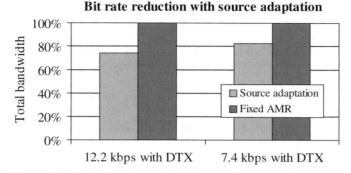

Figure 2.2. Reduction of required bit rate with equal voice quality

AMR source adaptation allows provisioning of the same voice quality with lower average bit rate. The bit rate reduction is typically 20–25% and is illustrated in Figure 2.2. The reduced AMR bit rate can be utilised to lower the required transmission power of the radio link, and it can thus further enhance AMR voice capacity. The WCDMA flexible Layer 1 allows adaptation of the bit rate and the transmission power for each 20 ms frame. The estimated capacity gain is 15–20%. The bit stream format of source adapted AMR is fully compatible with the existing fixed-rate AMR speech codec format, therefore, the decoding part is independent of source based adaptation. The AMR source based adaptation can be added as a simple upgrade to the networks to enhance WCDMA downlink capacity without any changes to the mobiles.

2.2.1.2 Wideband AMR – Better Voice Quality [4]
3GPP Release 5 introduces AMR wideband speech codec, which brings substantial voice quality enhancements compared to AMR narrowband codec or compared to standard fixed telephone line. In the case of packet switched streaming, AMR-WB is already part of 3GPP Release 4. The AMR-WB codec has also been selected by the ITU-T in the standardisation activity for a wideband codec around 16 kbps. This is of significant importance since this is the first time that the same codec is adopted for wireless as well as wireline services. This will eliminate the need for transcoding, and ease the implementation of wideband voice applications and services across a wide range of communications systems. The AMR-WB codec operates on nine speech coding bit rates between 6.6 and 23.85 kbps. The term wideband comes from the sampling rate, which has been increased from 8 kHz to 16 kHz. This allows covering twice the audio bandwidth compared to the classical telephone voice bandwidth of 4 kHz. While all the previous codecs in mobile communication operate on narrow audio bandwidth limited to 200–3400 Hz, AMR-WB extends the audio bandwidth to 50–7000 Hz. Figure 2.3 shows listening test results, where AMR-WB is compared to AMR-NB. The results are presented as mean opinion scores (MOS), where a higher number indicates better experienced voice quality. The MOS results show that AMR-WB is able to improve the voice quality without increasing the required radio bandwidth. For example, AMR-WB 12.65 kbps provides a clearly higher MOS than AMR-NB 12.2 kbps. The improved voice quality can be obtained because of higher sampling frequency.

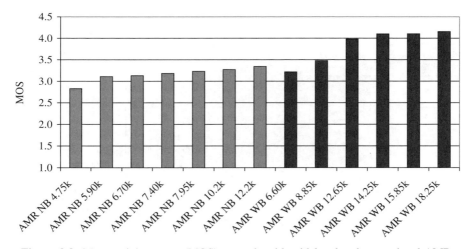

Figure 2.3. Mean opinion score (MOS) example with wideband and narrowband AMR

2.2.2 Video Telephony

Video telephony has similar delay requirements as speech services. Due to the nature of video compression, the BER requirement is more stringent than that of speech. 3GPP has specified that ITU-T Rec. H.324M should be used for video telephony in circuit switched connections and Session Initiation Protocol (SIP) for supporting IP multimedia applications, including video telephony.

2.2.2.1 Multimedia Architecture for Circuit Switched Connections

Originally Rec. H.324 was intended for multimedia communication over a fixed telephone network, i.e. PSTN. It is specified that for PSTN connections, a synchronous V.34 modem is used. Later on, when wireless networks evolved, mobile extensions were added to the specification to make the system more robust against transmission errors. The overall picture of the H.324 system is shown in Figure 2.4 [5].

H.324 consists of the following mandatory elements: H.223 for multiplexing and H.245 for control. Elements that are optional but are typically employed are H.263 video codec, G.723.1 speech codec, and V.8bis. Later, MPEG-4 video and AMR were added as optional codecs into the system. The recommendation defines the seven phases of a call: set-up, speech only, modem training, initialisation, message, end, and clearing. Level 0 of H.223 multiplexing is exactly the same as that of H.324, thus providing backward compatibility with older H.324 terminals. With a standardised negotiation procedure the terminal can adapt to the prevailing radio link conditions by selecting the appropriate error resiliency level.

V.8bis contains procedures for the identification and selection of common modes of operation between data circuit-terminating equipment (DCE) and between data terminal equipment (DTE) over general switched telephone network and leased point-to-point telephone types. The basic features of V.8bis are as follows:

- It allows a desired communication mode to be selected by either the calling or the answering station.

Scope of Rec. H.324 M

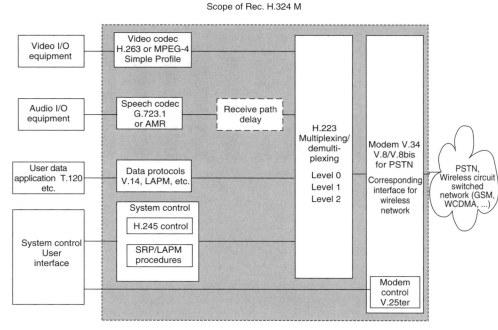

Figure 2.4. Scope of ITU Rec. H.324

- It allows terminals to automatically identify common operating modes (applications).

- It enables automatic selection between multiple terminals that share a common telephone circuit.

- It provides user-friendly switching from normal voice telephony to a modem-based communication mode.

The capabilities exchange feature of V.8bis permits a list of communication modes, as well as software applications, to be exchanged between terminals. Each terminal is therefore able to establish the modes of operation it shares with the remote station. A capability exchange between stations thus ensures, *a priori*, that a selected communication mode is possible. Attempts to establish incompatible modes of operation are thus avoided, which speeds up the application level connection.

As with the mode selection procedure, a capabilities exchange may be performed either at call set-up, automatically under the control of either the calling or the answering station, or during the course of telephony. In the latter case, on completion of the information exchange, the communication link may be configured either to return to voice telephony mode or to adopt immediately one of the common modes of communication.

V.8bis has been designed so that, when a capabilities exchange takes place in telephony mode, and the capabilities exchanged are limited to standard features, the interruption in voice communications is short (less than approximately 2 seconds) and as unobtrusive as possible.

In order to guarantee seamless data services between UMTS and PSTN, the call control mechanism of UMTS should take the V.8bis messages into account. V.8bis messages should be interpreted and converted into UMTS messages and vice versa.

One of the recent developments of H.324 is an operating mode that makes it possible to use an H.324 terminal over ISDN links. This mode of operation is defined in Annex D of the H.324 recommendation and is also referred to as H.324/I. H.324/I terminals use the I.400 series ISDN user-network interface in place of the V.34 modem. The output of the H.223 multiplex is applied directly to each bit of the digital channel, in the order defined by H.223. Operating modes are defined bit rates ranging from 56 kbps to 1920 kbps, so that H.324/I allows the use of several 56 or 64 kbps links at the same time.

H.324/I provides direct interoperability with H.320 terminals, H.324 terminals on the GSTN (using GSTN modems), H.324 terminals operating on ISDN through user substitution of I.400 series ISDN interfaces for V.34 modems, and voice telephones (both GSTN and ISDN). H.324/I terminals support H.324/Annex F ($=$ V.140) which is for establishing communication between two multiprotocol audio-visual terminals using digital channels at a multiple of 64 or 56 kbps [6].

Figure 2.5 shows one of the concept phones for video telephony.

Figure 2.5. 3G concept phone for video telephony

2.3   Person-to-Person Packet Switched Services

2.3.1   Images and Multimedia

It is already common today to send pictures via MMS (multimedia messaging) [7, 8], which, from a user perspective, is perceived almost as an enhanced SMS service. For MMS to be successful it is important that the messages are delivered with a high reliability, while the delivery time is short enough as long as it is roughly below one minute. Since the delivery time is not crucial, it is possible to use a less stringent 3GPP quality of service class for MMS. Another important requirement from an end user point of view is that it should be possible and easy to send MMS messages at the same time as, for example, making a circuit switched call. This requires that both the mobile station and the network are able to handle multiple radio access bearers in parallel.

Although a parallel circuit switched call and MMS transmission is possible, the inter-activity and picture information flow of the MMS service is limited. Another imaging service more powerful than MMS is real time video sharing; see Figure 2.6 for an illustration of this service. From a user point of view real time video sharing is about showing to the

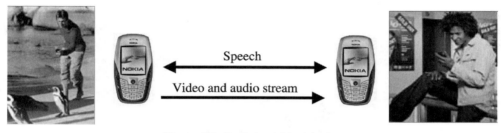

Figure 2.6. Real time video sharing

other end what is going on in your side of the phone connection. A typical usage scenario is that the communication starts out with a normal speech connection and then, when one of the parties has something interesting to show, the one-way video stream is set up. That is, the video stream is only set up when both users feel a clear need to enhance the voice connection with a one-way video connection. This differentiates the one-way video sharing service from full two-way video phone services. The real time video sharing has both professional and private use cases: sharing vacation experiences, showing real estate property for real estate brokers, and explaining what the situation is when there is a need to repair equipment.

The end user performance requirements for the real time video sharing service are that:

- Image quality and update rates should be high enough to enable 'scanning' the environment with the camera.
- Delay between taking a picture and showing it to the other side is low enough to enable true interactivity.
- It is easy and fast to set up the one-way video stream once the voice connection is available.

Note that real time video sharing has different requirements than content-to-person streaming, because in content-to-person streaming there is no or little interactivity and hence no requirements for low delays. For real time video sharing low delays are, on the other hand, crucial. A tolerable delay between taking a picture and showing it to the peer end could be in the order of some seconds (<5 s). When it comes to bit rate requirements it is very much dependent on mobile station display sizes. Based on initial results from video streaming in current networks, a lower bit rate limit for a 3.5 cm times 4 cm mobile phone display is around 40 to 64 kbps. However, note that the required bit rate to use is a non-straightforward function of the tolerable delay, the image update rates and the applied coding schemes.

From a network point of view, the one-way video streaming service has one obvious property that is different from many other proposed services: it requires a fairly high uplink bit rate. The UMTS network must be able to deliver a high and reasonably constant bit rate in order to support the low delay streaming connection as well as the voice connection if the voice connection is mapped over the packet switched domain. These bit rate and delay requirements may be met in a cost efficient way by utilising the QoS differentiation features that are available in UMTS. From a technical perspective, the peer-to-peer connections are in the packet switched domain set up by using the IMS system and by utilising the session initiation protocol (SIP) [9]. From a network and end user perspective this sets requirements on the SIP signalling, that it is fast enough not to disturb the user when setting up the additional video stream connection. Fast SIP signalling may be obtained by supporting one of multiple compression algorithms for SIP.

In Figure 2.7 a simple service evolution path is depicted starting from simple packet switched services like MMS and going towards more demanding services like video telephony. Video telephony has even tighter requirements on the delays than one-way video streaming and a one-way end-to-end delay of less than 400 ms is needed for the connection, while less than 150 ms is preferable [10].

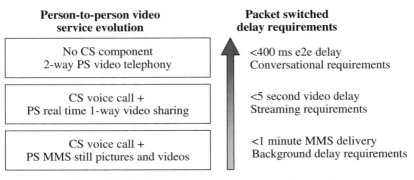

Figure 2.7. Evolution of person-to-person video service

2.3.2   Push-to-Talk over Cellular (PoC)

Push-to-talk over cellular (PoC) service is instant in the sense that the voice connection is established by simply pushing a single button and the receiving user hears the speech without even having to answer the call. While ordinary voice is bi-directional, the PoC service is a one directional service. The basic PoC application may hence be described as a walkie-talkie application over the packet switched domain of the cellular network. In addition to the basic voice communication functionality, the PoC application provides the end user with complementary features like, for example:

- Ad hoc and predefined communication groups;
- Access control so that a user may define who is allowed to make calls to him/her;
- 'Do-not-disturb' in case immediate reception of audio is not desirable.

With ordinary voice calls a bi-directional communication channel is reserved between the end users throughout the duration of the call. In PoC, the channel is only set up to transfer a short speech burst from one to possibly multiple users. Once this speech burst has been transferred, the packet switched communication channel can be released. This difference is highlighted in Figure 2.8.

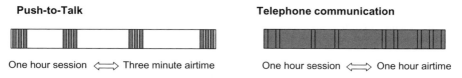

Push-to-Talk   **Telephone communication**

One hour session ⟺ Three minute airtime   One hour session ⟺ One hour airtime

Figure 2.8. Push-to-talk versus ordinary telephone communication

The speech packets are in the PoC solution carried from the sending mobile station to the server by the OPRS/UMTS network. The server then forwards the packets to the receiving mobile stations. In the case of a one-to-many connection, the server multiplies the packets to all the receiving mobile stations. This is illustrated in Figure 2.9. The PoC service is independent of the underlying radio access network. However, as we will see later in this section as well as in Chapter 10, the characteristics of the PoC service also set tight requirements on the underlying radio access network.

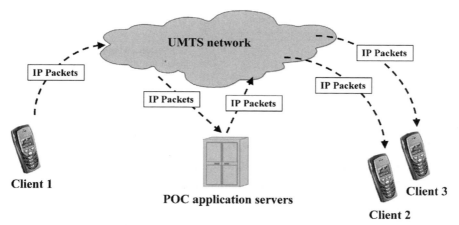

Figure 2.9. Push to talk solution architecture

In order for the PoC service to be well perceived by the end users it must meet multiple requirements. Some examples of end user requirements are:

- Simple user interface, for example, a dedicated push-to-talk button;
- High voice quality and enough sound pressure in the speaker to work also in noisy environments;
- Low delay from pressing the push-to-talk button until it is possible to start talking, called 'start-to-talk time';
- Low delay for the voice packets to receive the peer end, called voice through delay.

The end user is expected to be satisfied with the interactivity of the PoC service if the start-to-talk delay is around or below two seconds, while the speech round trip time should be kept lower than 1.5 seconds. The voice quality is usually evaluated by the mean opinion score (MOS) and is naturally dependent on both the mobile station and the network characteristics. A radio network that hosts PoC connections must, for example, be able to:

- Provide always on packet data connections;
- Reserve and release radio access resources fast in order to keep start-to-talk and speech round trip times low;
- Deliver a constant bit rate with low packet jitter during the duration of one speech burst.

Chapter 10 includes an investigation of the PoC service performance in a WCDMA network.

2.3.3 Voice over IP (VoIP)

The driver for Voice-over-IP, VoIP, in fixed networks has been access to low cost long distance and international voice calls. The driver for VoIP in cellular networks is rather to enable rich calls. A rich call can be defined as a real time communication session between two or more persons which consists of one or more media types. VoIP connection can be complemented with 2-way video, streaming video, images, content sharing, gaming etc., see Figure 2.10. VoIP and rich calls can be carried over WCDMA as the end-to-end network

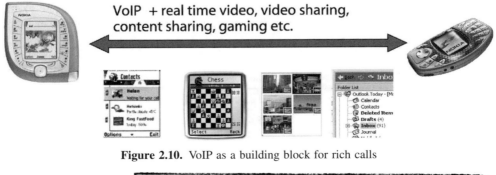

Figure 2.10. VoIP as a building block for rich calls

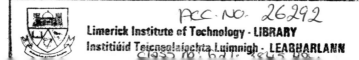

delay is low enough to meet the conversational service requirements. The QoS differentiation and IP header compression are important to make an efficient VoIP service in WCDMA.

2.3.4 Multiplayer Games

We first group the existing multiplayer games into key categories based on their end user
requirements. Three reasonable categories are, according to the study in [11, 12], real time
action games, real time strategy games and turn based strategy games, see Figure 2.11. The

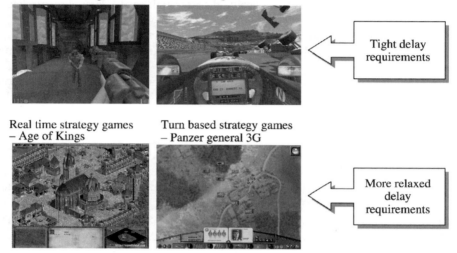

Figure 2.11. Multiplayer game classification

different categories are characterised by the properties and requirements given in Table 2.1.
Note that these requirements have been derived from studies using a fixed network
connection and not a cellular network connection. Although cellular networks behave
somewhat differently than fixed networks, and although mobile station displays are much
smaller than computer displays, the results give indications for what the maximum delay
may be in order to generate a nice gaming experience for the end user.

It can be noted that for experienced players it is an advantage to have significantly lower
end-to-end network delays than what is given by the requirement in Table 2.1; end-to-end
network delays down to as low as 70 to 80 ms are needed to satisfy the most demanding

Table 2.1. Multiplayer game delay and bit rate requirements [11, 12]

Gaming category	End user delay requirements for average player
Real time action games	End to end network delays $< 300\,\text{ms}$
Real time strategy games	End to end network delays $< 900\,\text{ms}$
Turn based strategy games	End to end network delays $< 40\,\text{s}$

users. The end-to-end network delay is particularly noticeable for the users if some users have low delays, like 70 ms, while others have higher delays, like 200 ms. Bearing in mind that today's WCDMA networks provide round trip times of 150–200 ms it is possible to provide real time strategy and turn based strategy games, and even real time action games over WCDMA.

The real time action games are constantly transmitting and receiving packets with typical bit rates of 10–20 kbps. Such bit rates can be easily delivered over cellular networks. However, these packets must be delivered with a very low delay which sets high requirements on the network performance. For real time strategy and turn based strategy games both the requirements on the bit rate and the end-to-end network delays are looser and there is more freedom on how to map these services to radio channels. This mapping is discussed in detail in Chapter 10.

2.4 Content-to-person Services

2.4.1 Browsing

During the early launch and development of WAP for mobile browsing, there were huge expectations that browsing via the mobile station would take off rapidly. Because of several reasons the take off did not happen as fast as expected. However, with better mobile station displays – resolution and colour – and with higher bit rates and increased content, the browsing experience on mobile station devices is increasing rapidly and service usage is going up. There have been several releases of the WAP protocol stack, of which the most important releases are WAP1.1 and WAP2.0; see further Figure 2.12. The WAP version

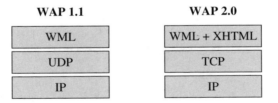

Figure 2.12. Evolution of the WAP protocol stacks

denoted WAP1.1 was approved in June 1999 and the first products based on this version were launched later in the same year. The WAP2.0 version was released in July 2001 by the WAP forum, which is currently part of the Open Mobile Alliance (OMA). The most important difference between WAP1.1 and WAP2.0 is that WAP2.0 is based on the standard Internet transport protocols (TCP/IP, HTTP/XHTML), while the WAP1.1 release utilises WAP1.1 specific transport protocols. From an end user point of view, the TCP/IP protocols provide faster download of large content size.

The focus in WAP1.1 development was to make browsing perform well in systems with large packet round trip times and with limited bit rates. That is, WAP1.1 enables the transmitter to send the packets almost at once, without waiting for connection establishment between the communication peers. This makes WAP1.1 fast for small packets over unreliable links. The weakness is that the link will usually not be fully utilised if the file to transfer is large. The decreased link utilisation lowers the end user bit rate for large files if the air interface bit rate is high.

WAP2.0 introduces standard Internet protocols to the WAP protocol stacks. Because the TCP/IP protocols have well developed link and congestion management algorithms, this makes WAP2.0 more efficient when transferring large files over radio links with high bit rates. To make TCP even more efficient for mobile systems, a particular flavour called wireless TCP (wTCP) has been defined. The wTCP protocol is based on standard TCP features, but in wTCP the support of certain features is mandatory and recommendations for parameter values have been aligned to cope with the higher packet round trip time in wireless networks. The higher link utilisation with TCP/IP for large files is illustrated in Figure 2.13 assuming WCDMA 128 kbps connection. The difference between WAP1.1 and WAP2.0 download times is quite small for small page sizes because of the low round trip time in WCDMA. A low round trip time helps standard Internet protocols perform satisfactorily over WCDMA without special optimisation.

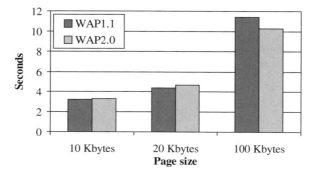

Figure 2.13. Page download time with WAP1.1 and WAP2.0

From a user perspective it is crucial that browsing is easily accessible and fast. Rough performance requirements for browsing are that the first page download time is lower than 10 s and for the second page download, lower than 4 to 7 s is preferred [10]. However, bear in mind that end user service requirements are different from market to market and also in different market segments within the same market. Another user requirement is that it should be possible to use browsing smoothly when travelling by car, train or bus. This requires efficient handling of cell reselections in order to prevent connection breaks. Because WCDMA utilises handover for packet switched data, there are no breaks at cell reselection.

From a network perspective the first page download is different from the second page download. The reason is that the first page download time may include GPRS attach, security procedures, PDP context activation and radio bearer set-up times depending on how the network and the mobile station have been configured. For the second and consecutive pages the download time will be lower because the initial set-up messages have already been sent. The second page download time is mainly limited by the basic packet round trip time, the radio channel bit rate, TCP/IP efficiency, HTTP versions and possibly also the radio bearer set-up time depending on the idle period from the last page download.

2.4.2 Audio and Video Streaming

Multimedia streaming is a technique for transferring data such that it can be processed as a steady and continuous stream. Streaming technologies are becoming increasingly important with the growth of the Internet because most users do not have fast enough access to download large multimedia files quickly. Mobile station memory may also limit the size of the downloads. With streaming, the client browser or plug-in can start displaying the data before the entire file has been transmitted.

For streaming to work, the client side receiving the data must be able to collect the data and send it as a steady stream to the application that is processing the data and converting it to sound or pictures. Streaming applications are very asymmetric and therefore typically withstand more delay than more symmetric conversational services. This also means that they tolerate more jitter in transmission. Jitter can be easily smoothed out by buffering.

Internet video products and the accompanying media industry as a whole are clearly divided into two different target areas: (1) Web broadcast and (2) video streaming on-demand. Web broadcast providers usually target very large audiences that connect to a highly performance-optimised media server (or choose from a multitude of servers) via the actual Internet. The on-demand services are more often used by big corporations that wish to store video clips or lectures to a server connected to a higher bandwidth local intranet – these on-demand lectures are seldom used simultaneously by more than hundreds of people.

Both application types use basically similar core video compression technology, but the coding bandwidths, level of tuning within network protocol use, and robustness of server technology needed for broadcast servers differ from the technology used in on-demand, smaller-scale systems. This has led to a situation where the few major companies developing and marketing video streaming products have specialised their end user products to meet the needs of these two target groups. Basically, they have optimised their core products differently: those directed to the '28.8 kbps market' for bandwidth variation-sensitive streaming over the Internet and those for the 100–7300 kbps intranet market.

At the receiver the streaming data or video clip is played by a suitable independent media player application or a browser plug-in. Plug-ins can be downloaded from the Web, usually free of charge, or may be readily bundled to a browser. This depends largely on the browser and its version in use – new browsers tend to have integrated plug-ins for the most popular streaming video players.

In conclusion, a client player implementation in a mobile system seems to lead to an application-level module that could handle video streams independently (with independent connection and playback activation) or in parallel with the browser application when the service is activated from the browser. The module would interface directly to the socket interface of applied packet network protocol layers, here most likely UDP/IP or TCP/IP.

Example terminals supporting streaming services are shown in Figure 2.14.

2.4.3 Content Download

Content download examples are shown in Figure 2.15: application downloads, ringing tone downloads, video clips and MP3 music. The content size can vary largely from a few kB

Figure 2.14. Example streaming terminals

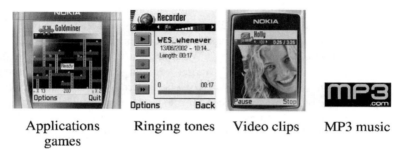

Applications   Ringing tones   Video clips   MP3 music
games

Figure 2.15. Example content download

ringing tones to several MB music files. The download times should preferably be low, which puts high requirements on the radio bit rate, especially for the large downloads with several 100 kB.

2.4.4 Multimedia Broadcast Multicast Service, MBMS

A new service introduced in 3GPP Release 6 specifications is Multimedia Broadcast Multicast Service (MBMS). There are two high level modes of operation in MBMS, as given in [13]

1. Broadcast mode, which allows sending audio and video. The already existing Cell Broadcast Service (CBS) is intended for messaging only. The broadcast mode is expected to be a service without charging and there are no specific activation requirements for this mode.

2. Multicast mode allows sending multimedia data for the end users that are part of a multicast subscription group. End users need to monitor service announcements regarding service availability, and then they can join the currently active service. From the network point of view, the same content can be provided in a point-to-point fashion if there are not enough users to justify the high power transmission. A typical example in

3GPP has been the sport results service where, for example, ice hockey results would be available as well as video clips of the key events in different games of the day. Charging is expected to be applied for the multicast mode.

From the radio point of view, MBMS is considered an application independent way to deliver the MBMS User Services, which are intended to deliver to multiple users simultaneously. The MBMS User Services can be classified into three groups as follows [14]:

1. Streaming services, where a basic example is audio and video stream;

2. File download services;

3. Carousel service, which can be considered as a combination of streaming and file download. In this kind of service, an end user may have an application which is provided data repetitively and updates are then broadcast when there are changes in the content.

For MBMS User Services, an operator controls the distribution of the data. Unlike CBS, the end user needs first to join the service and only users that have joined the service can see the content. The charging can then be based on the subscription or based on the keys which enable an end user to access the data. The MBMS content can be created by the operator itself or by a third party and, as such, all the details of what an MBMS service should look like will not be specified by 3GPP, but left for operators and service providers. One possible MBMS high level architecture is shown in Figure 2.16, where the IP multicast network refers here to any server providing MBMS content over the Internet.

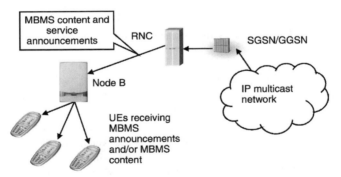

Figure 2.16. Example MBMS high level architecture

The example data rates in [14] range from the 10 kbps text-based information to the 384 kbps video distribution on MBMS. The codecs are expected to be the current ones – such as AMR for voice – to ensure a large interoperability base for different terminals for the services being provided. The MBMS causes changes mostly to Layer 2/3 protocols as described in Chapter 7 in more detail.

2.5   Business Connectivity

Business connectivity considers access to corporate intranet or to Internet services using laptops. We consider shortly two aspects of business connectivity: end-to-end security and the effect of radio latency to the application performance. End-to-end security can be obtained using Virtual Private Networks, VPN, for the encryption of the data. One option is to have a VPN client located on the laptop and the VPN gateway in the corporate premises. Such an approach is often used by large corporates that are able to obtain and maintain required equipment for the remote access service. Another approach uses a VPN connection between the mobile operator core site and the company intranet. The mobile network uses standard UMTS security procedures. In this case the company only needs to subscribe to the operator's VPN service and obtain a VPN gateway. These two approaches are illustrated in Figure 2.17.

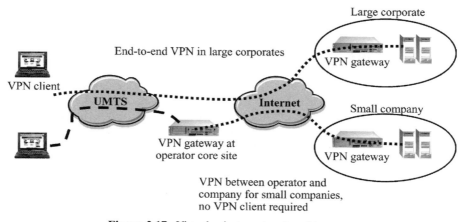

Figure 2.17. Virtual private network architectures

The business connectivity applications can be, for example, web browsing, email access or file download. The application performance should preferably be similar to the performance of DSL or WLAN. The application performance depends on the available bit rate but also on the network latency. The network latency is here measured as the round trip time. The round trip time is the delay of a small IP packet to travel from the mobile to a server and back. The effect of the latency is illustrated using file download over Transmission Control Protocol, TCP, in Figure 2.18. The download process includes TCP connection establishment and file download, including TCP slow start. The end user experienced bit rate is defined here as the download file size divided by the total time. The delay components are illustrated in Figure 2.19.

The user experienced bit rates with round trip times between 0 and 600 ms are shown in Figure 2.20. This figure assumes that a dedicated channel with 384 kbps already exists and no channel allocation is required. The curves show that a low round trip time is beneficial, especially for small file sizes, due to TCP slow start. WCDMA round trip time is analysed in detail in Chapter 10 and it is typically 150–200 ms. Figure 2.21 shows the download time with different round trip times. The download time of a less than 100 kB file is below 3 s as

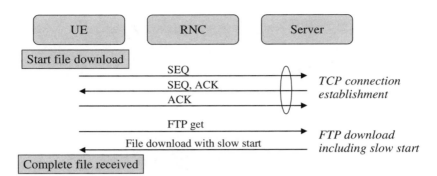

User experienced bit rate = File size / total download time

Figure 2.18. Example signalling flow in file download using TCP

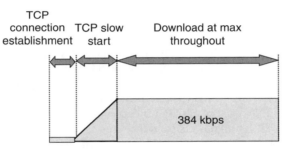

Figure 2.19. Example file download using TCP

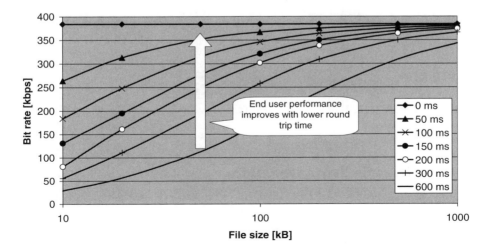

Figure 2.20. Effect of round trip time to the user experienced bit rate with Layer 1 bit rate of 384 kbps

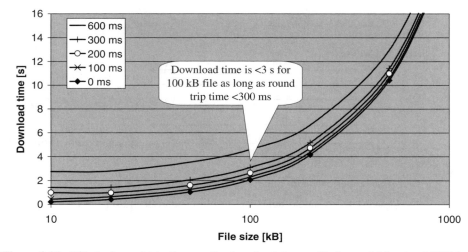

Figure 2.21. Effect of round trip time on the download times with Layer 1 bit rate of 384 kbps

long as the round trip time is below 300 ms. Low round trip time will be more relevant if we need to download several small files using separate TCP sessions.

The application performance is also affected by the application protocol, e.g. HTTP 1.1 vs HTTP 1.0 in web browsing. A web page typically consists of several objects: text and a number of pictures. In the case of HTTP 1.0 each object is downloaded in a separate TCP session, while for HTTP 1.1 all objects from the same server can be downloaded in one TCP connection. The difference is depicted in Figure 2.22. It is beneficial to use HTTP 1.1 to minimise the effect of TCP slow start on the application performance.

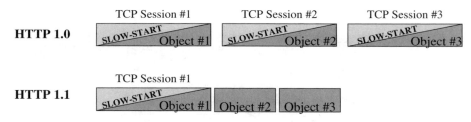

Figure 2.22. HTTP 1.1 uses one TCP connection for multiple objects

2.6 IP Multimedia Sub-system, IMS

IP Multimedia Sub-system, IMS, allows operators to provide their subscribers with multimedia services that are built on Internet applications and protocols [15]. IMS enables IP connectivity between users using the same control and charging mechanisms. The basic session initiation capabilities provided by SIP protocol are utilised to establish peer-to-peer sessions. The IMS concept is shown in Figure 2.23.

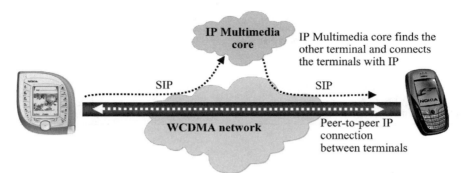

Figure 2.23. Basic principles of the IP Multimedia Sub-system

IMS provides the means for network operators to maintain their role in the value chain by providing new multimedia services and predictable end user performance. The same platform can be used in both real time services, like VoIP, and non-real time services, like content sharing. The IMS network elements are introduced in Chapter 5.

2.7   Quality of Service Differentiation

Chapter 10 covers end-to-end application performance assuming that the system load is reasonably low. When the system load gets higher, it becomes important to prioritise the different services according to their requirements. This prioritisation is called QoS differentiation. 3GPP QoS architecture is designed to provide this differentiation [16]. The terminology is shown in Figure 2.24.

The most relevant parameters of the four UMTS QoS classes are summarised in Table 2.2. The main distinguishing factor between the four traffic classes is how delay-sensitive the traffic

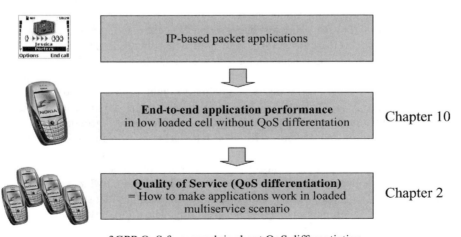

Figure 2.24. Definition of quality of service differentiation

Table 2.2. UMTS QoS classes and their main parameters

	Conversational class	Streaming class	Interactive class	Background class
Transfer delay	80 ms –	250 ms –	—	—
Guaranteed bit rate (kbps)	Up to 2 Mbps	Up to 2 Mbps	—	—
Traffic handling priority	—	—	1,2,3	—
Allocation/retention priority	1,2,3	1,2,3	1,2,3	1,2,3

is: the conversational class is meant for very delay-sensitive traffic, while the background class is the most delay-insensitive. There are, further, three different priority categories, called allocation/retention priority categories, within each QoS class. Interactive has also three traffic handling priorities. Conversational and streaming class parameters also include the guaranteed bit rate and the transfer delay parameters. The guaranteed bit rate defines the minimum bearer bit rate that UTRAN must provide and it can be used in admission control and in resource allocations. The transfer delay defines the required 95th percentile of the delay. It can be used to define the RLC operation mode (acknowledged, non-acknowledged mode) and the number of retransmissions.

The conversational class is characterised by low end-to-end delay and symmetric or nearly symmetric traffic between uplink and downlink in person-to-person communications. The maximum end-to-end delay is given by the human perception of video and audio conversation: subjective evaluations have shown that the end-to-end delay has to be less than 400 ms. The streaming class requires bandwidth to be maintained like conversational class but streaming class tolerates some delay variations that are hidden by dejitter buffer in the receiver. The interactive class is characterised by the request response pattern of the end user. At the message destination there is an entity expecting the message (response) within a certain time. The background class assumes that the destination is not expecting the data within a certain time.

UMTS QoS classes are not mandatory for the introduction of any low delay service. It is possible to support streaming video or conversational Voice over IP from an end-to-end performance point of view by using just background QoS class. QoS differentiation becomes useful for the network efficiency during high load when there are services with different delay requirements. If the radio network has knowledge about the delay requirements of the different services, it will be able to prioritise the services accordingly and improve the efficiency of the network utilisation. The qualitative gain of the QoS differentiation is illustrated in Figure 2.25. Considerable efficiency gains can be obtained in Step 2 just by introducing a few prioritisation classes within interactive or background class by using allocation and retention parameters, ARP. The pure prioritisation in packet scheduling is not alone enough to provide full QoS differentiation gains. Users within the same QoS and ARP class will share the available capacity. If the number of users is simply too high, they will all suffer from bad quality. In that case it would be better to block a few users to guarantee the quality of the existing connections, like streaming videos. That is provided in Step 3 in Figure 2.25 with guaranteed bit rate streaming. The radio network can estimate the available radio capacity and block an incoming user if there is no room to provide the required bandwidth without sacrificing the quality of the existing connections. Finally Step 4 allows further differentiation between guaranteed bit rate services with different delay

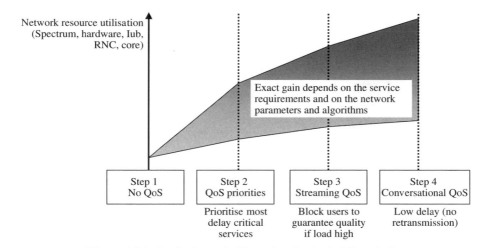

Figure 2.25. Qualitative gain illustration for QoS differentiation

requirements. If the delay requirements are known, the WCDMA RAN can allocate suitable radio parameters – like retransmission parameters – for the new bearer.

An example QoS differentiation scheme is shown in Figure 2.26 with ten different QoS categories: six guaranteed bit rate categories and four non-real time categories. It is assumed in this case that traffic handling priority is equal to allocation and retention priority, and there is no prioritisation within the background class.

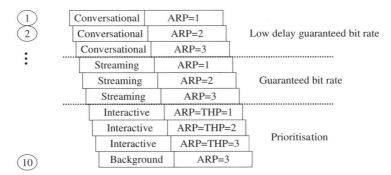

Figure 2.26. Example ten categories by taking a subset of UMTS QoS classes

The next three figures illustrate Steps 1, 2 and 4. Figure 2.27 shows an example where all the services have the same QoS parameters and the same treatment. In this case, all services share the network resources equally: they get the same bit rate and experience the same delay. The network dimensioning must be done so that this bit rate or delay fulfils the most stringent requirements of the services provided in the network. The background type of service, like sending of MMS, will get the same quality, which is unnecessarily good and

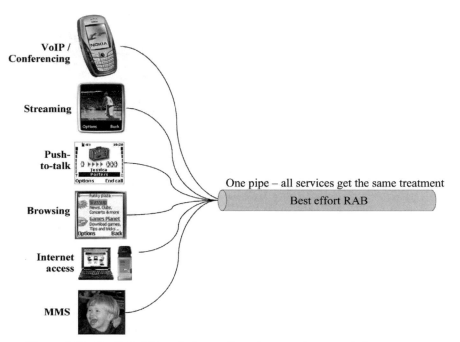

Figure 2.27. No QoS differentiation – all services use the same QoS parameters

wastes network resources. Figure 2.28 shows the case where there are three different pipes with QoS prioritisation in packet scheduling. This approach already provides QoS differentiation and makes the network dimensioning requirements less stringent. Figure 2.29 adds two further pipes with guaranteed bit rates.

The layered architecture of a UMTS bearer service is depicted in Figure 2.30; each bearer service on a specific layer offers its individual services using those provided by the layers below. The QoS parameters are given by the core network to the radio network in radio access bearer set-up.

Figure 2.31 illustrates the mechanisms to define the QoS parameters in radio access bearer set-up.

1. The UE can request QoS parameters. In particular, if the application requires guaranteed bit rate streaming or conversational class, it has to be requested by UE, otherwise, it cannot be given by the network.

2. The access point node, APN, in GGSN can give QoS parameters according to operator settings. Some services may be accessed via certain APNs. That allows the operator to control the QoS parameters for different services and makes it also possible to prioritise operator hosted services compared to accessing other services.

3. The home location register, HLR, may contain subscriber specific limitations for the QoS parameters.

4. The WCDMA radio network must be able to provide the QoS differentiation in packet handling.

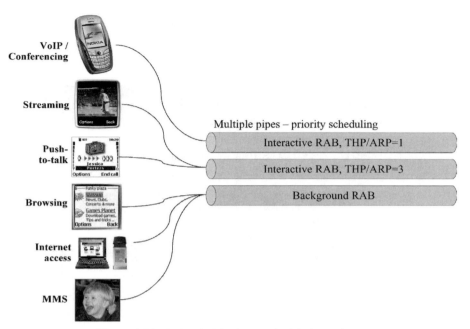

Figure 2.28. QoS prioritisation used with three classes

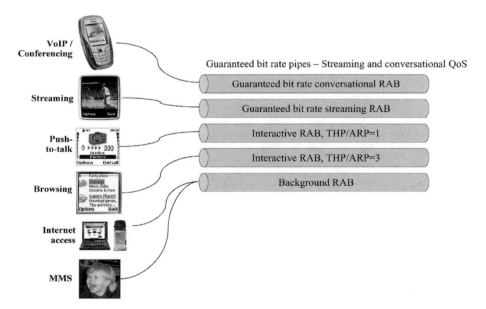

Figure 2.29. QoS differentiation with two guaranteed bit rate classes and three classes for non-real time prioritisation

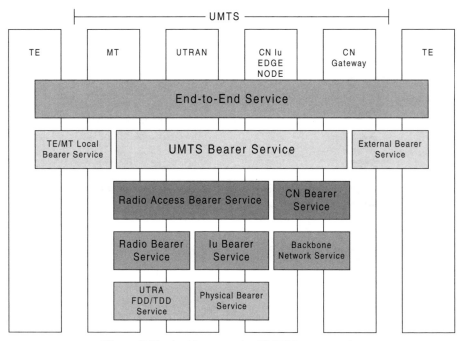

Figure 2.30. Architecture of a UMTS bearer service

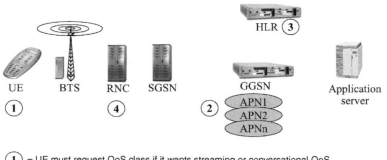

(1) = UE must request QoS class if it wants streaming or conversational QoS

(2) = Different APNs in GGSN can be used to provide different QoS

(3) = User specific QoS limitations can be defined in HLR

(4) = Radio access network must be able to provide QoS differentiation

Figure 2.31. The role of UE, GGSN and HLR in defining QoS class

2.8   Capacity and Cost of Service Delivery

This section considers the maximum capacity of the radio network in delivering new services and estimates the cost of the service delivery from the UMTS network equipment point of view.

2.8.1   Capacity per Subscriber

The maximum capacity per subscriber for data and for voice traffic is presented below. The data traffic is presented as the maximum amount of downloaded megabytes (MB) of data per average subscriber per month. The voice traffic is presented as the maximum mobile-to-mobile voice minutes per average subscriber per month. The following assumptions are used in the calculations:

- WCDMA carrier capacity is 800 kbps/cell or 80 voice channels/cell, see Chapter 12 for more details;

- High-Speed Downlink Packet Access, HSDPA carrier capacity is 2000 kbps/cell, see Chapter 11 for more details;

- Cell capacity utilisation is 80 % during busy hours;

- Busy hour carries 20 % of daily traffic.

- There are 1000 subscribers per site;

- There are three sectors per site;

The percentage of the traffic carried by the busy hour represents how equally the traffic is distributed during the day. That number is affected by the pricing schemes. Attractive evening or weekend pricing schemes can make the traffic distribution more equal (busy hour carries less than 20 % of the traffic) and more traffic can be carried by the same network. The assumption of 1000 subscribers is a typical average figure for large network operators. More subscribers per site can be found in dense areas. The capacity per subscriber per day can be calculated as follows with a 10 MHz WCDMA 2+2+2 site configuration:

$$\frac{0.8\,\text{Mbps/cell}}{8\,\text{bits/byte}} \cdot 6\,\text{cells} \cdot 3600\,\text{s/hour}\frac{80\,\%}{20\,\%}\frac{1}{1000\,\text{subs}} \cdot 30\,\text{days/month}$$
$$= 260\,\text{MB/sub/month} \tag{2.1}$$

The maximum data capacity with WCDMA 10 MHz spectrum allocation is 260 MB/sub/month and with HSDPA 650 MB/sub/month. The voice capacity is up to 1725 minutes/sub/month. The data capacities are shown in Figure 2.32 and voice capacities in Figure 2.33. Since data and voice share the same spectrum, the final capacity depends on the traffic share between data and voice traffic. With a 50/50 split, the capacity could be 325 MB + 862 voice minutes per month. These capacity numbers indicate that there is a lot of traffic growth potential with WCDMA on top of today's traffic numbers. Global average voice traffic is currently 150–200 minutes per month.

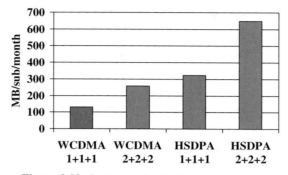

Figure 2.32. Data capacity MB/subscriber/month

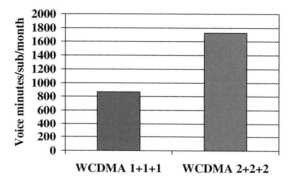

Figure 2.33. Voice capacity minutes/subscriber/month

Most UMTS operators also have GSM900/1800 spectrum that can provide additional capacity for voice and data services. Also, new spectrum for third generation systems will be available at around 2.6 GHz. Chapter 1 gives an overview of the spectrum.

2.8.2 Cost of Capacity Delivery

This section presents cost estimates of delivering megabytes of data or voice minutes over the WCDMA mobile network. The target is to present the calculation methods and to show approximate cost levels. The cost numbers include the depreciation of the radio and the core network capital expenditures (capex) without any implementation costs. The following components are included: base stations, radio transmission, RNC, core network and operations solutions. The price of all this equipment is calculated per transceiver unit (TRX). The assumed per TRX prices are between 15 and 40 k€. The pricing depends on a number of factors and, therefore, a large scale is used. A capex depreciation period of six years is assumed. The mobile network delivery cost per downloaded MB in € can be calculated as follows:

$$\frac{\text{Price per TRX}[\text{€}]}{\dfrac{0.8\,\text{Mbps}}{8\,\text{bits/byte}} \cdot \dfrac{3600\,\text{s/hour} \cdot 80\,\%}{2\,\%} \cdot 365\,\text{days/year} \cdot 6\,\text{years}} \tag{2.2}$$

Delivery cost of downloaded MB

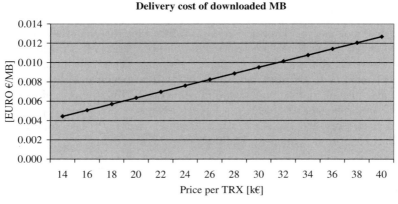

Figure 2.34. Delivery cost of downloaded data MB

The delivery cost for data is shown in Figure 2.34 and for voice in Figure 2.35. The results show that it is possible to push the data delivery capex down to approximately 0.01 €/MB, that means 1 eurocent/MB, and a voice minute below 0.2 eurocent/minute.

The high busy hour utilisation of 80 % can be used for the case when additional capacity is built. If we calculate the delivery cost over the whole network, the busy hour utilisation will be lower on average because parts of the sites are built to provide coverage and they do not collect high traffic volumes.

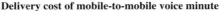

Delivery cost of mobile-to-mobile voice minute

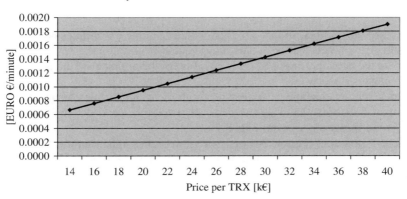

Figure 2.35. Delivery cost of mobile-to-mobile voice minute

The mobile network capex depreciation represents only part of the operator costs. Other costs include, for example, leased line transmission costs, interconnection fees, customer acquisition, advertising and customer care. Therefore, the sales prices cannot be as low as the presented production cost figures which only include capex depreciation. The capacity and the cost calculations above still demonstrate that WCDMA is able to deliver high amounts of traffic with reasonable cost, which are key requirements for enabling new services.

2.9 Service Capabilities with Different Terminal Classes

WCDMA does not use the same principle as GSM with terminal class mark. WCDMA terminals shall tell the network, upon connection set-up, a larger set of parameters indicating the radio access capabilities of the particular terminal. These capabilities determine, for example, the maximum user data rate supported in a particular radio configuration, given independently for the uplink and downlink directions. To provide guidance on which capabilities should be applied together, reference terminal radio access capability combinations have been specified in 3GPP standardisation, see [17]. The following reference combinations have been defined for 3GPP Release '99:

- 32 kbps class. This is intended to provide a basic speech service, including AMR speech as well as some limited data rate capabilities up to 32 kbps.

- 64 kbps class. This is intended to provide a speech and data service, with simultaneous data and AMR speech capability.

- 144 kbps class. This class has the air interface capability to provide, for example, video telephony or various other data services.

- 384 kbps class is being further enhanced from 144 kbps and has, for example, multicode capability, which points toward support of advanced packet data methods provided in WCDMA.

- 768 kbps class has been defined as an intermediate step between 384 kbps and 2 Mbps class.

- 2 Mbps class. This is the state-of-the-art class and has been defined for the downlink direction only.

These classes are defined so that a higher class has all the capabilities covered by a lower class. It should be noted that terminals may deviate from these classes when giving their parameters to the network, thus 2 Mbps is possible for the uplink also, though not covered by any of the classes directly.

3GPP specifications include performance requirements for the bit rates up to 384 kbps, for more details see Section 12.5. Therefore, it is expected that terminals up to 384 kbps will be available in the initial deployment phase.

High-Speed Downlink Packet Access, HSDPA, further enhances the WCDMA bit rate capabilities. HSDPA terminal capabilities are defined in 3GPP Release 5 and extend beyond 10 Mbps. HSDPA is covered in detail in Chapter 11.

2.10 Location Services in WCDMA

2.10.1 Location Services

Location-based services and applications are expected to become one of the new dimensions in UMTS. A location-based service is provided either by a teleoperator or by a third party service provider that utilises available information on the terminal location. The service is either push (e.g. automatic distribution of local information) or pull type (e.g. localisation of emergency calls). Other possible location-based services are discount calls in a certain area,

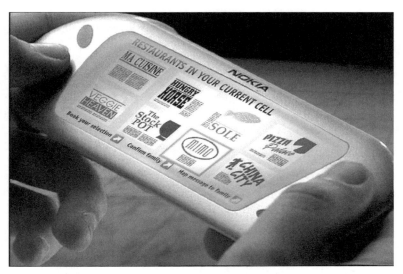

Figure 2.36. 3G concept phone showing location-based service

broadcasting of a service over a limited number of sites (broadcasting video on demand), and retrieval and display of location-based information, such as the location of the nearest gas stations, hotels, restaurants, and so on. Figure 2.36 shows an example. Depending on the service, the data may be retrieved interactively or as background. For instance, before travelling to an unknown city abroad one may request night-time download of certain points of interest from the city. The downloaded information typically contains a map and other data to be displayed on top of the map. By clicking the icon on the map, one gets information from the point. Information to be downloaded background or interactively can be limited by certain criteria and personal interest.

The location information can be input by the user or detected by the network or mobile station. The network architecture of the location services is discussed in Chapter 5. Release '99 of UMTS specifies the following positioning methods:

- the cell coverage based positioning method;
- Observed Time Difference Of Arrival – Idle Period DownLink (OTDOA-IPDL);
- network-assisted GPS methods.

These methods are complementary rather than competing, and are suited for different purposes. These approaches are introduced in the following sections.

2.10.2 Cell Coverage Based Location Calculation

The cell coverage based location method is a network based approach, i.e., it does not require any new functionalities in the mobile. The radio network has the location information with a cell level accuracy when the mobile has been allocated a dedicated channel or when the mobile is in cell_FACH or cell_PCH states. These states are introduced in Chapter 7. If the mobile is in idle state, its location with cell accuracy can be obtained by forcing the mobile to cell_FACH state with a location update as illustrated in Figure 2.37.

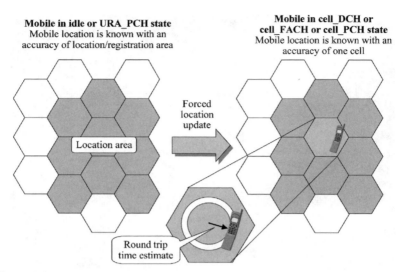

Figure 2.37. Location calculation with cell coverage combined with round trip time

The accuracy of the cell coverage based method depends heavily on the cell size. The typical cell ranges in the urban area are below 1 km and in the dense urban area a few hundred meters providing fairly accurate location information.

The accuracy of the cell coverage based approach can be improved by using the round trip time measurement that can be obtained from the base station. That information is available in cell_DCH state and it gives the distance between the base and the mobile station.

2.10.3   Observed Time Difference Of Arrival, OTDOA

The OTDOA method is based on the mobile measurements of the relative arrival times of the pilot signals from different base stations. At least three base stations must be received by the mobile for the location calculation, as shown in Figure 2.38. A measurement from two base stations defines a hyperbola. With two measurement pairs, i.e. with three base stations, the location can be calculated.

In order to facilitate the OTDOA location measurements and to avoid near–far problems, the WCDMA standard includes idle periods in downlink, IPDL. During those idle periods the mobile is able to receive the pilot signal of the neighbour cells even if the best pilot signal on the same frequency is very strong. A typical frequency of the idle periods is 1 slot every 100 ms, i.e. 0.7 % of the time. The IPDL–OTDOA measurements are shown in Figure 2.39.

The network needs to know the relative transmission times of the pilot signals from different base stations to calculate the mobile location. That relative timing information can be obtained by:

1. OTDOA measurements by the location measurement unit at the base station. The base station measures the relative timing of the adjacent cells. The measurement is similar to the OTDOA measurements by the mobile.

2. The GPS receiver at the base station.

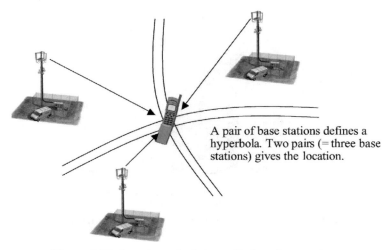

Figure 2.38. Location calculation with three base stations

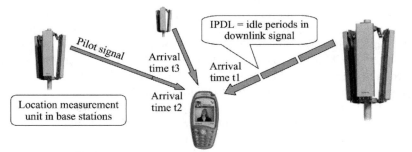

Figure 2.39. IPDL (Idle Period Downlink) – OTDOA (Observed Time Difference of Arrival)

The accuracy of the OTDOA measurements can be in the order of up to tens of meters in very good conditions when several base stations in line-of-sight can be received by the mobile. In practice, such ideal measurement conditions are not typically available in cellular networks. The accuracy depends on the following factors:

1. The number of base stations that the mobile can receive. A minimum of three is required. If more base stations can be received, the accuracy is improved.

2. The relative locations of the base stations. If the base stations are located in different directions from the mobile, the accuracy is improved.

3. The line-of-sight. If there is a line-of-sight between the mobile and the base station, the accuracy is improved.

The requirement of receiving at least three base stations is challenging in the cellular networks. The target of the network planning is to create clear dominance areas of the cells

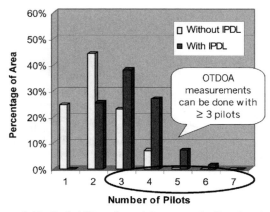

Figure 2.40. Probability of receiving several pilot signals [18]

and to avoid unnecessary overlapping of the cells. That approach maximises the capacity. The clear dominance areas and limited cell overlapping reduce the probability of accurate OTDOA measurements as it is difficult to receive at least three pilot signals. Figure 2.40 shows the probability of a mobile receiving several pilot signals in realistic network scenarios. The probability of receiving at least three pilots is 74 % in Figure 2.40. IPDL allows receipt of the strongest pilot and the second strongest with 100 % probability, but it is challenging to receive at least three pilots with very high probability. The required pilot E_c/I_0 was -18 dB in these simulations and a fully loaded network was assumed. The results show that IPDL greatly improves the performance of OTDOA: without IPDL the probability of receiving at least three pilots would be only 31%. The results also show that it is difficult to obtain a very high probability of OTDOA measurements. The accuracy can be improved by combining OTDOA with the cell coverage based location method.

2.10.4 Assisted GPS

Most accurate location measurements can be obtained with an integrated GPS receiver in the mobile. The network can provide additional information, like visible GPS satellites, reference time and Doppler, to assist the mobile GPS measurements. The assistance data improves the GPS receiver sensitivity for indoor measurements, makes the acquisition times faster and reduces the GPS power consumption. The principle of assisted GPS is shown in Figure 2.41.

A reference GPS receiver in every base station provides the most accurate assistance data and the most accurate GPS measurements by the mobile. The assisted GPS measurements can achieve accuracy of 10 meters outdoors and a few tens of meters indoors. That accuracy also meets the FCC requirements in the USA. If the most stringent measurement probabilities and accuracies are not required, the reference GPS receiver is not needed in every base station, but only a few reference GPS receivers are needed in the radio network. It is also possible to let the mobile GPS make the measurements without any additional assistance data.

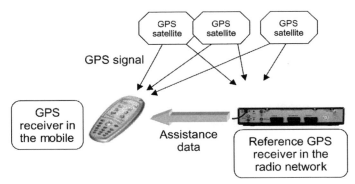

Figure 2.41. Assisted GPS

References

[1] 3GPP, Mandatory Speech Codec Speech Processing Functions, AMR Speech Codec: General Description (3G TS 26.071).

[2] 3GPP, Mandatory Speech Codec Speech Processing Functions, AMR Speech Codec: Frame Structure General Description (3G TS 26.101 version 1.4.0), 1999.

[3] Holma, H., Melero, J., Vainio, J., Halonen, T. and Mäkinen, J. "Performance of Adaptive Multirate (AMR) Voice in GSM and WCDMA", *Proc of IEEE Vehicular Technology Conference* (VTC 2003-Spring), pp. 2177–2181, April 2003.

[4] 3GPP, Technical Specification Group Services and System Aspects, Speech Coded speech processing functions, AMR Wideband Speech Codec, General Decsciption, (3G TS 26.171 version 5.0.0), 2001.

[5] 3GPP, Technical Specification Group Services and System Aspects, Codec for Circuit Switched Multimedia Telephony Service: General Description (3G TS 26.110 version 3.0.1), 1999.

[6] ITU-T H.324, Terminal for Low Bitrate Multimedia Communication, 1998.

[7] 3GPP, Multimedia Messaging Service (3GPP TS 22.140), 2001.

[8] 3GPP, MMS Architecture and functionality (3GPP TS 23.140), 2001.

[9] Handley, M., *et al.*, SIP: Session Initiation Protocol, RFC3261, IETF, June 2002.

[10] 3GPP, *"Service Aspects; Services and Service Capabilities (Release 6)"*, 3GPP TS 22.105, version 6.2.0, June 2003.

[11] Nokia, *"Multiplayer Game Performance over Cellular Networks"*, Forum Nokia, version 1.0, January 20, 2004.

[12] Johanna Anttila, Jani Lakkakorpi, *"On the Effect of Reduced Quality of Service on Multiplayer Online Games"*, IJIGS, Volume 2, Number 2, 2003.

[13] 3GPP Technical Specification, TS 22.146, Multimedia Broadcast/Multicast Service, Stage 1, January 2004, version 6.3.0.

[14] 3GPP Technical Specification, TS 22.246, Multimedia Broadcast/Multicast Service (MBMS) user services; Stage 1, version 6.0.0, January 2004.

[15] 3GPP, IP Multimedia Subsystems (3GPP TS 23.228), 2002.

[16] 3GPP, Technical Specification Group Services and System Aspects, QoS Concept and Architecture (3G TR 23.107).

[17] 3GPP, Technical Specification Group (TSG) RAN, Working Group 2 (WG2), UE Radio Access Capabilities, 3G TS 25.306 version 3.0.0, 2000.

[18] Johnson, C., Joshi, H. and Khalab, J. "WCDMA Radio Network Planning for Location Services and System Capacity", IEE 3G2002 conference in London, 9th May.

3

Introduction to WCDMA

Peter Muszynski and Harri Holma

3.1 Introduction

This chapter introduces the principles of the WCDMA air interface. Special attention is drawn to those features by which WCDMA differs from GSM and IS-95. The main parameters of the WCDMA physical layer are introduced in Section 3.2. The concept of spreading and despreading is described in Section 3.3, followed by a presentation of the multipath radio channel and Rake receiver in Section 3.4. Other key elements of the WCDMA air interface discussed in this chapter are power control and soft and softer handovers. The need for power control and its implementation are described in Section 3.5, and soft and softer handover in Section 3.6.

3.2 Summary of the Main Parameters in WCDMA

We present the main system design parameters of WCDMA in this section and give brief explanations for most of them. Table 3.1 summarises the main parameters related to the WCDMA air interface. Here we highlight some of the items that characterise WCDMA.

- WCDMA is a wideband Direct-Sequence Code Division Multiple Access (DS-CDMA) system, i.e. user information bits are spread over a wide bandwidth by multiplying the user data with quasi-random bits (called chips) derived from CDMA spreading codes. In order to support very high bit rates (up to 2 Mbps), the use of a variable spreading factor and multicode connections is supported. An example of this arrangement is shown in Figure 3.1.

- The chip rate of 3.84 Mcps leads to a carrier bandwidth of approximately 5 MHz. DS-CDMA systems with a bandwidth of about 1 MHz, such as IS-95, are commonly referred to as narrowband CDMA systems. The inherently wide carrier bandwidth of WCDMA supports high user data rates and also has certain performance benefits, such as increased multipath diversity. Subject to his operating licence, the network operator can deploy

WCDMA for UMTS, third edition.   Edited by Harri Holma and Antti Toskala
© 2004 John Wiley & Sons, Ltd   ISBN: 0-470-87096-6

Table 3.1. Main WCDMA parameters

Multiple access method	DS-CDMA
Duplexing method	Frequency division duplex/time division duplex
Base station synchronisation	Asynchronous operation
Chip rate	3.84 Mcps
Frame length	10 ms
Service multiplexing	Multiple services with different quality of service requirements multiplexed on one connection
Multirate concept	Variable spreading factor and multicode
Detection	Coherent using pilot symbols or common pilot
Multiuser detection, smart antennas	Supported by the standard, optional in the implementation

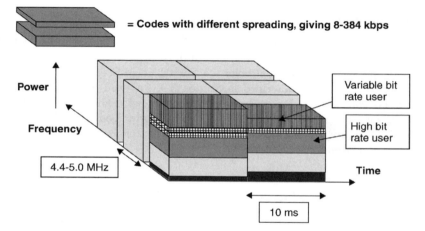

Figure 3.1. Allocation of bandwidth in WCDMA in the time–frequency–code space

multiple 5 MHz carriers to increase capacity, possibly in the form of hierarchical cell layers. Figure 3.1 also shows this feature. The actual carrier spacing can be selected on a 200 kHz grid between approximately 4.4 and 5 MHz, depending on interference between the carriers.

- WCDMA supports highly variable user data rates, in other words the concept of obtaining Bandwidth on Demand (BoD) is well supported. The user data rate is kept constant during each 10 ms frame. However, the data capacity among the users can change from frame to frame. Figure 3.1 also shows an example of this feature. This fast radio capacity allocation will typically be controlled by the network to achieve optimum throughput for packet data services.

- WCDMA supports two basic modes of operation: Frequency Division Duplex (FDD) and Time Division Duplex (TDD). In the FDD mode, separate 5 MHz carrier frequencies are used for the uplink and downlink respectively, whereas in TDD only one 5 MHz is time-shared between the uplink and downlink. Uplink is the connection from the mobile to the base station, and downlink is that from the base station to the mobile.

The TDD mode is based heavily on FDD mode concepts and was added in order to leverage the basic WCDMA system also for the unpaired spectrum allocations of the ITU for the IMT-2000 systems. The TDD mode is described in detail in Chapter 13.

- WCDMA supports the operation of asynchronous base stations, so that, unlike in the synchronous IS-95 system, there is no need for a global time reference such as a GPS. Deployment of indoor and micro base stations is easier when no GPS signal needs to be received.

- WCDMA employs coherent detection on uplink and downlink based on the use of pilot symbols or common pilot. While already used on the downlink in IS-95, the use of coherent detection on the uplink is new for public CDMA systems and will result in an overall increase of coverage and capacity on the uplink.

- The WCDMA air interface has been crafted in such a way that advanced CDMA receiver concepts, such as multiuser detection and smart adaptive antennas, can be deployed by the network operator as a system option to increase capacity and/or coverage. In most second generation systems no provision has been made for such receiver concepts and as a result they are either not applicable or can be applied only under severe constraints with limited increases in performance.

- WCDMA is designed to be deployed in conjunction with GSM. Therefore, handovers between GSM and WCDMA are supported in order to be able to leverage the GSM coverage for the introduction of WCDMA.

In the following sections of this chapter we will briefly review the generic principles of CDMA operation. In the subsequent chapters, the above mentioned aspects specific to the WCDMA standard will be presented and explained in more detail. The basic CDMA principles are also described in references [1], [2], [3] and [4].

3.3   Spreading and Despreading

Figure 3.2 depicts the basic operations of spreading and despreading for a DS-CDMA system.

User data is here assumed to be a BPSK-modulated bit sequence of rate R, the user data bits assuming the values of ± 1. The spreading operation, in this example, is the multiplication of each user data bit with a sequence of 8 code bits, called chips. We assume this also for the BPSK spreading modulation. We see that the resulting spread data is at a rate of $8 \times R$ and has the same random (pseudo-noise-like) appearance as the spreading code. In this case we would say that we used a spreading factor of 8. This wideband signal would then be transmitted across a wireless channel to the receiving end.

During despreading we multiply the spread user data/chip sequence, bit duration by bit duration, with the very same 8 code chips as we used during the spreading of these bits. As shown, the original user bit sequence has been recovered perfectly, provided we have (as shown in Figure 3.2) also perfect synchronisation between the spread user signal and the (de)spreading code.

The increase of the signalling rate by a factor of 8 corresponds to a widening (by a factor of 8) of the occupied spectrum of the spread user data signal. Due to this virtue, CDMA

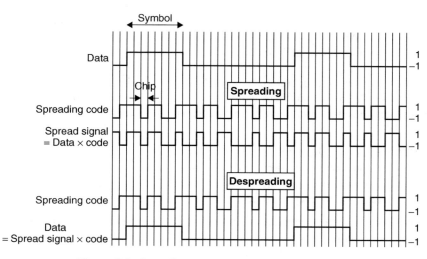

Figure 3.2. Spreading and despreading in DS-CDMA

systems are more generally called spread spectrum systems. Despreading restores a
bandwidth proportional to R for the signal.

The basic operation of the correlation receiver for CDMA is shown in Figure 3.3. The
upper half of the figure shows the reception of the desired own signal. As in Figure 3.2, we
see the despreading operation with a perfectly synchronised code. Then, the correlation
receiver integrates (i.e. sums) the resulting products (data × code) for each user bit.

The lower half of Figure 3.3 shows the effect of the despreading operation when applied to
the CDMA signal of another user whose signal is assumed to have been spread with a

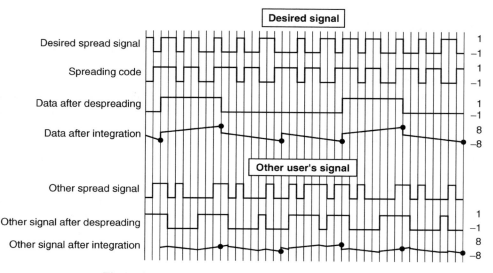

Figure 3.3. Principle of the CDMA correlation receiver

different spreading code. The result of multiplying the interfering signal with the own code and integrating the resulting products leads to interfering signal values lingering around 0.

As can be seen, the amplitude of the own signal increases on average by a factor of 8 relative to that of the user of the other interfering system, i.e. the correlation detection has raised the desired user signal by the spreading factor, here 8, from the interference present in the CDMA system. This effect is termed 'processing gain' and is a fundamental aspect of all CDMA systems, and in general of all spread spectrum systems. Processing gain is what gives CDMA systems the robustness against self-interference that is necessary in order to reuse the available 5 MHz carrier frequencies over geographically close distances. Let's take an example with real WCDMA parameters. Speech service with a bit rate of 12.2 kbps has a processing gain of 25 dB $= 10 \times \log_{10} (3.84e6/12.2e3)$. After despreading, the signal power needs to be typically a few decibels above the interference and noise power. The required power density over the interference power density after despreading is designated as E_b/N_0 in this book, where E_b is the energy, or power density, per user bit and N_0 is the interference and noise power density. For speech service E_b/N_0 is typically in the order of 5.0 dB, and the required wideband signal-to-interference ratio is therefore 5.0 dB minus the processing gain $= -20.0$ dB. In other words, the signal power can be 20 dB under the interference or thermal noise power, and the WCDMA receiver can still detect the signal. The wideband signal-to-interference ratio is also called the carrier-to-interference ratio C/I. Due to spreading and despreading, C/I can be lower in WCDMA than, for example, in GSM. A good quality speech connection in GSM requires $C/I = 9$–12 dB.

Since the wideband signal can be below the thermal noise level, its detection is difficult without knowledge of the spreading sequence. For this reason, spread spectrum systems originated in military applications where the wideband nature of the signal allowed it to be hidden below the omnipresent thermal noise.

Note that within any given channel bandwidth (chip rate) we will have a higher processing gain for lower user data bit rates than for high bit rates. In particular, for user data bit rates of 2 Mbps, the processing gain is less than 2 ($=3.84$ Mcps/2 Mbps $= 1.92$ which corresponds to 2.8 dB) and some of the robustness of the WCDMA waveform against interference is clearly compromised. The performance of high bit rates with WCDMA is presented in Section 12.4.

Both base stations as well as mobiles for WCDMA use essentially this type of correlation receiver. However, due to multipath propagation (and possibly multiple receive antennas), it is necessary to use multiple correlation receivers in order to recover the energy from all paths and/or antennas. Such a collection of correlation receivers, termed 'fingers', is what comprises the CDMA Rake receiver. We will describe the operation of the CDMA Rake receiver in further detail in the following section, but before doing so, we make some final remarks regarding the transformation of spreading/despreading when used for wireless systems.

It is important to understand that spreading/despreading by itself does not provide any signal enhancement for wireless applications. Indeed, the processing gain comes at the price of an increased transmission bandwidth (by the amount of the processing gain).

All the WCDMA benefits come rather 'through the back door' by the wideband properties of the signals when examined at the system level, rather than the level of an individual radio link:

1. The processing gain together with the wideband nature suggest a frequency reuse of 1 between different cells of a wireless system (i.e. a frequency is reused in every cell/ sector). This feature can be used to obtain high spectral efficiency.

2. Having many users share the same wideband carrier for their communications provides interferer diversity, i.e. the multiple access interference from many system users is averaged out, and this again will boost capacity compared to systems where one has to plan for the worst-case interference.

3. However, both the above benefits require the use of tight power control and soft handover to avoid one user's signal blocking the others' communications. Power control and soft handover will be explained later in this chapter.

4. With a wideband signal, the different propagation paths of a wireless radio signal can be resolved at higher accuracy than with signals at a lower bandwidth. This results in a higher diversity content against fading, and thus improved performance.

3.4   Multipath Radio Channels and Rake Reception

Radio propagation in the land mobile channel is characterised by multiple reflections, diffractions and attenuation of the signal energy. These are caused by natural obstacles such as buildings, hills, and so on, resulting in so-called multipath propagation. There are two effects resulting from multipath propagation that we are concerned with in this section:

1. The signal energy (pertaining, for example, to a single chip of a CDMA waveform) may arrive at the receiver across clearly distinguishable time instants. The arriving energy is 'smeared' into a certain multipath delay profile: see Figure 3.4, for example. The delay

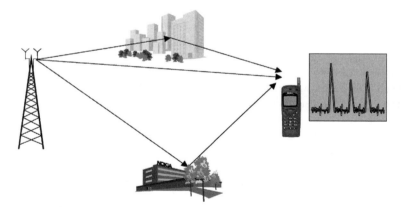

Figure 3.4. Multipath propagation leads to a multipath delay profile

profile extends typically from 1 to 2 μs in urban and suburban areas, although in some cases delays as long as 20 μs or more with significant signal energy have been observed in hilly areas. The chip duration at 3.84 Mcps is 0.26 μs. If the time difference of the multipath components is at least 0.26μs, the WCDMA receiver can separate those multipath components and combine them coherently to obtain multipath diversity. The 0.26 μs delay can be obtained if the difference in path lengths is at least 78 m (= speed of light $\div$ chip rate = $3.0 \cdot 10^8$ ms^{-1} $\div$ 3.84 Mcps). With a chip rate of about 1 Mcps, the difference in the path lengths of the multipath components must be about 300 m, which cannot be obtained in small cells. Therefore, it is easy to see that the 5 MHz WCDMA can provide multipath diversity in small cells, which is not possible with IS-95.

2. Also, for a certain time delay position there are usually many paths nearly equal in length along which the radio signal travels. For example, paths with a length difference of half a wavelength (at 2 GHz this is approximately 7 cm) arrive at virtually the same instant when compared to the duration of a single chip, which is 78 m at 3.84 Mcps. As a result, signal cancellation, called fast fading, takes place as the receiver moves across even short distances. Signal cancellation is best understood as a summation of several weighted phasors that describe the phase shift (usually modulo radio wavelength) and attenuation along a certain path at a certain time instant.

Figure 3.5 shows an exemplary fast fading pattern as would be discerned for the arriving signal energy at a particular delay position as the receiver moves. We see that the received signal power can drop considerably (by 20–30 dB) when phase cancellation of multipath reflections occurs. Because of the underlying geometry causing the fading and dispersion phenomena, signal variations due to fast fading occur several orders of magnitude more frequently than changes in the average multipath delay profile. The statistics of the received signal energy for a short-term average are usually well described by the Rayleigh distribution: see, e.g. [5] and [6]. These fading dips make error-free reception of data bits very difficult, and countermeasures are needed in WCDMA. The countermeasures against fading in WCDMA are shown below.

1. The delay dispersive energy is combined by utilising multiple Rake fingers (correlation receivers) allocated to those delay positions on which significant energy arrives.

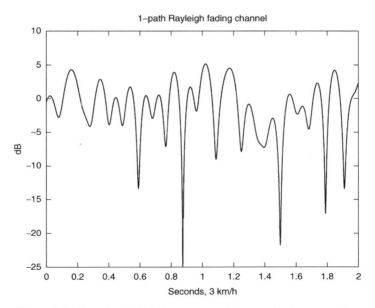

Figure 3.5. Fast Rayleigh fading as caused by multipath propagation

2. Fast power control and the inherent diversity reception of the Rake receiver are used to mitigate the problem of fading signal power.

3. Strong coding and interleaving and retransmission protocols are used to add redundancy and time diversity to the signal and thus help the receiver in recovering the user bits across fades.

The dynamics of the radio propagation suggest the following operating principle for the CDMA signal reception:

1. Identify the time delay positions at which significant energy arrives and allocate correlation receivers, i.e. Rake fingers, to those peaks. The granularity for acquiring the multipath delay profile is in the order of one chip duration (typically within the range of $\frac{1}{4}-\frac{1}{2}$ chip duration) with an update rate in the order of some tens of milliseconds.

2. Within each correlation receiver, track the fast-changing phase and amplitude values originating from the fast fading process and remove them. This tracking process has to be very fast, with an update rate in the order of 1 ms or less.

3. Combine the demodulated and phase-adjusted symbols across all active fingers and present them to the decoder for further processing.

Figure 3.6 illustrates points 2 and 3 by depicting modulation symbols (BPSK or QPSK) as well as the instantaneous channel state as weighted complex phasors. To facilitate point 2, WCDMA uses known pilot symbols that are used to sound the channel and provide an estimate of the momentary channel state (value of the weighted phasor) for a particular finger. Then the received symbol is rotated back, so as to undo the phase rotation caused by the channel. Such channel-compensated symbols can then be simply summed together to recover the energy across all delay positions. This processing is also called Maximal Ratio Combining (MRC).

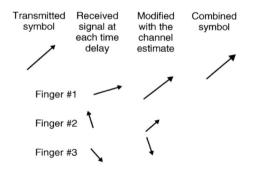

Figure 3.6. The principle of maximal ratio combining within the CDMA Rake receiver

Figure 3.7 shows a block diagram of a Rake receiver with three fingers according to these principles. Digitised input samples are received from the RF front-end circuitry in the form of I and Q branches (i.e. in complex low-pass number format). Code generators and correlator perform the despreading and integration to user data symbols. The channel

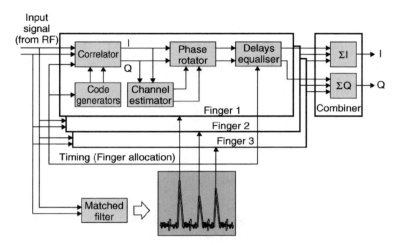

Figure 3.7. Block diagram of the CDMA Rake receiver

estimator uses the pilot symbols for estimating the channel state which will then be removed by the phase rotator from the received symbols. The delay is compensated for the difference in the arrival times of the symbols in each finger. The Rake combiner then sums the channel-compensated symbols, thereby providing multipath diversity against fading. Also shown is a matched filter used for determining and updating the current multipath delay profile of the channel. This measured and possibly averaged multipath delay profile is then used to assign the Rake fingers to the largest peaks.

In typical implementations of the Rake receiver, processing at the chip rate (correlator, code generator, matched filter) is done in ASICs, whereas symbol-level processing (channel estimator, phase rotator, combiner) is implemented by a DSP. Although there are several differences between the WCDMA Rake receiver in the mobile and the base station, all the basic principles presented here are the same.

Finally, we note that multiple receive antennas can be accommodated in the same way as multiple paths received from a single antenna: by just adding additional Rake fingers to the antennas, we can then receive all the energy from multiple paths *and* antennas. From the Rake receiver's perspective, there is essentially no difference between these two forms of diversity reception.

3.5 Power Control

Tight and fast power control is perhaps the most important aspect in WCDMA, in particular on the uplink. Without it, a single overpowered mobile could block a whole cell. Figure 3.8 depicts the problem and the solution in the form of closed loop transmission power control.

Mobile stations MS1 and MS2 operate within the same frequency, separable at the base station only by their respective spreading codes. It may happen that MS1 at the cell edge suffers a path loss, say 70 dB above that of MS2 which is near the base station BS. If there were no mechanism for MS1 and MS2 to be power-controlled to the same level at the base station, MS2 could easily overshout MS1 and thus block a large part of the cell, giving rise to

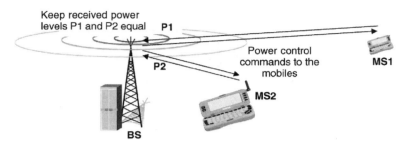

Figure 3.8. Closed loop power control in CDMA

the so-called near–far problem of CDMA. The optimum strategy in the sense of maximising capacity is to equalise the received power per bit of all mobile stations at all times.

While one can conceive open loop power control mechanisms that attempt to make a rough estimate of path loss by means of a downlink beacon signal, such a method would be far too inaccurate. The prime reason for this is that the fast fading is essentially uncorrelated between uplink and downlink, due to the large frequency separation of the uplink and downlink bands of the WCDMA FDD mode. Open loop power control is, however, used in WCDMA, but only to provide a coarse initial power setting of the mobile station at the beginning of a connection.

The solution to power control in WCDMA is fast closed loop power control, also shown in Figure 3.8. In closed loop power control in the uplink, the base station performs frequent estimates of the received Signal-to-Interference Ratio (SIR) and compares it to a target SIR. If the measured SIR is higher than the target SIR, the base station will command the mobile station to lower the power; if it is too low it will command the mobile station to increase its power. This measure–command–react cycle is executed at a rate of 1500 times per second (1.5 kHz) for each mobile station and thus operates faster than any significant change of path loss could possibly happen and, indeed, even faster than the speed of fast Rayleigh fading for low to moderate mobile speeds. Thus, closed loop power control will prevent any power imbalance among all the uplink signals received at the base station.

The same closed loop power control technique is also used on the downlink, though here the motivation is different: on the downlink there is no near–far problem due to the one-to-many scenario. All the signals within one cell originate from the one base station to all mobiles. It is, however, desirable to provide a marginal amount of additional power to mobile stations at the cell edge, as they suffer from increased other-cell interference. Also on the downlink a method of enhancing weak signals caused by Rayleigh fading with additional power is needed at low speeds when other error-correcting methods based on interleaving and error correcting codes do not yet work effectively.

Figure 3.9 shows how uplink closed loop power control works on a fading channel at low speed. Closed loop power control commands the mobile station to use a transmit power proportional to the inverse of the received power (or SIR). Provided the mobile station has enough headroom to ramp the power up, only very little residual fading is left and the channel becomes an essentially non-fading channel as seen from the base station receiver.

While this fading removal is highly desirable from the receiver point of view, it comes at the expense of increased average transmit power at the transmitting end. This means that a

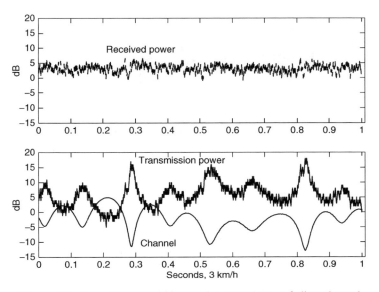

Figure 3.9. Closed-loop power control compensates a fading channel

mobile station in a deep fade, i.e. using a large transmission power, will cause increased interference to other cells. Figure 3.9 illustrates this point. The gain from the fast power control is discussed in more detail in Section 9.2.1.1.

Before leaving the area of closed loop power control, we mention one more related control loop connected with it: outer loop power control. Outer loop power control adjusts the target SIR setpoint in the base station according to the needs of the individual radio link and aims at a constant quality, usually defined as a certain target bit error rate (BER) or block error rate (BLER). Why should there be a need for changing the target SIR setpoint? The required SIR (there exists a proportional E_b/N_0 requirement) for, say, BLER = 1% depends on the mobile speed and the multipath profile. Now, if one were to set the target SIR setpoint for the worst case, i.e. high mobile speeds, one would waste much capacity for those connections at low speeds. Thus, the best strategy is to let the target SIR setpoint float around the minimum value that just fulfils the required target quality. The target SIR setpoint will change over time, as shown in the graph in Figure 3.10, as the speed and propagation environment changes. The gain of outer loop power control is discussed in detail in Section 9.2.2.1.

Outer loop control is typically implemented by having the base station tag each uplink user data frame with a frame reliability indicator, such as a CRC check result obtained during decoding of that particular user data frame. Should the frame quality indicator indicate to the Radio Network Controller (RNC) that the transmission quality is decreasing, the RNC in turn will command the base station to increase the target SIR setpoint by a certain amount. The reason for having outer loop control reside in the RNC is that this function should be performed after a possible soft handover combining. Soft handover will be presented in the next section.

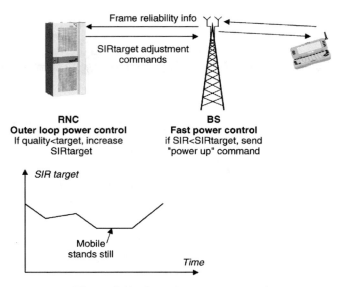

Figure 3.10. Outer loop power control

3.6   Softer and Soft Handovers

During softer handover, a mobile station is in the overlapping cell coverage area of two adjacent sectors of a base station. The communications between mobile station and base station take place concurrently via *two* air interface channels, one for each sector separately. This requires the use of two separate codes in the downlink direction, so that the mobile station can distinguish the signals. The two signals are received in the mobile station by means of Rake processing, very similar to multipath reception, except that the fingers need to generate the respective code for each sector for the appropriate despreading operation. Figure 3.11 shows the softer handover scenario.

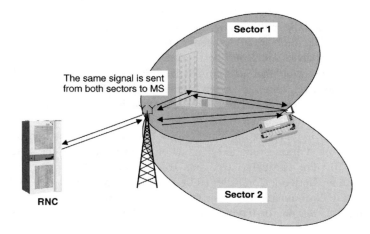

Figure 3.11. Softer handover

In the uplink direction a similar process takes place at the base station: the code channel of the mobile station is received in each sector, then routed to the same baseband Rake receiver and the maximal ratio combined there in the usual way. During softer handover only one power control loop per connection is active. Softer handover typically occurs in about 5–15 % of connections.

Figure 3.12 shows soft handover. During soft handover, a mobile station is in the overlapping cell coverage area of two sectors belonging to different base stations. As in softer handover, the communications between mobile station and base station take place concurrently via two air interface channels from each base station separately. As in softer handover, both channels (signals) are received at the mobile station by maximal ratio combining Rake processing. Seen from the mobile station, there are very few differences between softer and soft handover.

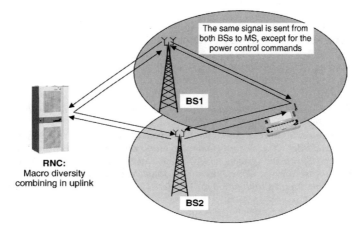

Figure 3.12. Soft handover

However, in the uplink direction soft handover differs significantly from softer handover: the code channel of the mobile station is received from both base stations, but the received data is then routed to the RNC for combining. This is typically done so that the same frame reliability indicator as provided for outer loop power control is used to select the better frame between the two possible candidates within the RNC. This selection takes place after each interleaving period, i.e. every 10–80 ms.

Note that during soft handover two power control loops per connection are active, one for each base station. Power control in soft handover is discussed in Section 9.2.1.3.

Soft handover occurs in about 20–40 % of connections. To cater for soft handover connections, the following additional resources need to be provided by the system and must be considered in the planning phase:

- Additional Rake receiver channels in the base stations;
- Additional transmission links between base station and RNC;
- Additional Rake fingers in the mobile stations.

We also note that soft and softer handover can take place in combination with each other.

Why are these CDMA-specific handover types needed? They are needed for similar reasons as closed loop power control: without soft/softer handover there would be near–far scenarios of a mobile station penetrating from one cell deeply into an adjacent cell without being power-controlled by the latter. Very fast and frequent hard handovers could largely avoid this problem; however, they can be executed only with certain delays during which the near–far problem could develop. So, as with fast power control, soft/softer handovers are an essential interference-mitigating tool in WCDMA. Soft and softer handovers are described in more detail in Section 9.3.

In addition to soft/softer handover, WCDMA provides other handover types:

- Inter-frequency hard handovers that can be used, for example, to hand a mobile over from one WCDMA frequency carrier to another. One application for this is high capacity base stations with several carriers.

- Inter-system hard handovers that take place between the WCDMA FDD system and another system, such as WCDMA TDD or GSM.

References

[1] Ojanperä, T. and Prasad, R., *Wideband CDMA for Third Generation Mobile Communications*, Artech House, 1998.
[2] Viterbi, A., *Principles of Spread Spectrum Communication*, Addison-Wesley, 1997.
[3] Cooper, G. and McGillem, C., *Modern Communications and Spread Spectrum*, McGraw-Hill, 1998.
[4] Dixon, R., *Spread Spectrum Systems with Commercial Applications*, John Wiley & Sons, 1994.
[5] Jakes, W., *Microwave Mobile Communications*, John Wiley & Sons, 1974.
[6] Saunders, S., *Antennas and Propagation for Wireless Communication Systems*, John Wiley & Sons, 1999.

4

Background and Standardisation of WCDMA

Antti Toskala

4.1   Introduction

In the first phase of third generation standardisation, the basic process of selecting the best technology for multiple radio access was conducted in several regions. This chapter describes the selection process that took place in ETSI mainly during 1997, as well as the decisions made by the regional standardisation organisations into early 1998. Other standardisation bodies that carried out WCDMA-related work are also introduced, and then 3GPP, the common standardisation effort to create a global standard for WCDMA, is described, including a description of how a proposal progresses through the WCDMA standardisation process and ends up in the specifications. Finally, the developments in ITU for work on the IMT-2000 recommendations and the relationship of the ITU work to the regional activities are presented.

4.2   Background in Europe

In Europe a long period of research preceded the selection of third generation technology. The RACE I (Research of Advanced Communication Technologies in Europe) programme started the basic third generation research work in 1988. This programme was followed by RACE II, with the development of the CDMA-based CODIT (Code Division Testbed) and TDMA-based ATDMA (Advanced TDMA Mobile Access) air interfaces during 1992–95. In addition, wideband air interface proposals were studied in a number of industrial projects in Europe: see, for example, [1].

   The European research programme ACTS (Advanced Communication Technologies and Services) was launched at the end of 1995 in order to support mobile communications research and development. Within ACTS the FRAMES (Future Radio Wideband Multiple Access System) project [2] was set up with the objective of defining a proposal for a UMTS

WCDMA for UMTS, third edition.   Edited by Harri Holma and Antti Toskala
© 2004 John Wiley & Sons, Ltd   ISBN: 0-470-87096-6

radio access system. The main industrial partners in FRAMES were Nokia, Siemens, Ericsson, France Télécom and CSEM/Pro Telecom, with participation also from several European universities. Based on an initial proposal evaluation phase in FRAMES, a harmonised multiple access platform was defined, consisting of two modes: FMA1, a wideband TDMA [3], and FMA2, a wideband CDMA [4]. The FRAMES wideband CDMA and wideband TDMA proposals were submitted to ETSI as candidates for UMTS air interface and ITU IMT-2000 submission.

The proposals for the UMTS Terrestrial Radio Access (UTRA) air interface received by the milestone were grouped into five concept groups in ETSI in June 1997, after their submission and presentation during 1996 and early 1997. The following groups were formed:

- Wideband CDMA (WCDMA);
- Wideband TDMA (WTDMA);
- TDMA/CDMA;
- OFDMA;
- ODMA.

The concept groups formed in ETSI are introduced briefly in the following section. The evaluation of the proposals was based on the requirements defined in the ITU-R IMT-2000 framework (and in ETSI defined specifically in UMTS 21.01 [5] as well as on the evaluation principles and conditions covered in UMTS 30.03 [6]). The results of the evaluation were collated in UMTS 30.06 [7].

4.2.1   Wideband CDMA

The WCDMA concept group was formed around the WCDMA proposals from FRAMES/ FMA2, Fujitsu, NEC and Panasonic. Several European, Japanese and US companies contributed to the development of the WCDMA concept. The physical layer of the WCDMA uplink was adopted mainly from FRAMES/FMA2, while the downlink solution was modified following the principles of the other proposals made to the WCDMA concept group.

The basic system features consisted of:

- Wideband CDMA operation with 5 MHz;
- Physical layer flexibility for integration of all data rates on a single carrier;
- Reuse 1 operation.

The enhancements covered included:

- Transmit diversity;
- Adaptive antenna operation;
- Support for advanced receiver structures.

The WCDMA concept achieved the greatest support, one of the technical motivating issues being the flexibility of the physical layer for accommodating different service types simultaneously. This was considered to be an advantage, especially with respect to low and medium bit rates. Among the drawbacks of WCDMA, it was recognised that in an unlicensed system in the TDD band, with the continuous transmit and receive operation, pure WCDMA technology does not facilitate interference avoidance techniques in cordless-like operating environments.

4.2.2 Wideband TDMA

The WTDMA concept group was formed by taking the non-spread option from the FRAMES/FMA1 proposal. FRAMES/FMA1 was basically a TDMA-based system concept with 1.6 MHz carrier spacing for wideband service implementation. The concept aimed at high capacity with the aid of interference averaging over the operator bandwidth, with fractional loading and frequency hopping.

The basic system features consisted of:

- Equalisation with training sequences in TDMA bursts;

- Interference averaging with frequency hopping;

- Link adaptation;

- Two basic burst types, 1/16 th and 1/64 th burst lengths for high and low data rates respectively;

- Low reuse sizes.

The enhancements covered included:

- Inter-cell interference suppression;
- Support of adaptive antennas;
- TDD operation;
- Less complex equalisers for large delay spread environments.

The main limitation associated with the system was considered to be the range with respect to low bit rate services. This is due to the fact that in TDMA-based operation the slot duration is, at a minimum, only 1/64 th of the frame timing, which results in either very high peak power or a low average output power level. This means that for large ranges with, for example, speech, the WTDMA concept would not have been competitive on its own, but would have required a narrowband option as a companion.

4.2.3 Wideband TDMA/CDMA

The WTDMA/CDMA group was based on the spreading option in the FRAMES/FMA1 proposal, resulting in the hybrid CDMA/TDMA concept with 1.6 MHz carrier spacing.

The basic system features consisted of:

- TDMA burst structure with midamble for channel estimation;

- CDMA concept applied on top of the TDMA structure for additional flexibility;

- Reduction of intra-cell interference by multiuser detection for users within a times lot on the same carrier;

- Low reuse sizes, down to 3.

Enhancements covered included:

- Frequency hopping;
- Inter-cell interference cancellation;
- Support of adaptive antennas;
- Operation in TDD mode;
- Dynamic Channel Allocation (DCA).

This proposal, especially the issues related to receiver complexity, led to lively discussions during the selection process.

4.2.4   OFDMA

The OFDMA group was based on OFDMA technology with inputs mainly from Telia, Sony and Lucent. The system concept was shaped by the discussions about OFDMA in other forums, such as the Japanese standardisation forum, ARIB.
   The basic concept features included:

- Operation with slow frequency hopping with TDMA and OFDM multiplexing;
- A 100 kHz wide bandslot from the OFDM signal as the basic resource unit;
- Higher rates built by allocating several bandslots, creating a wideband signal;
- Diversity provided by dividing the information among several bandslots over the carrier.

The enhancement techniques covered were:

- Transmit diversity;
- Multiuser detection for interference cancellation;
- Adaptive antenna solutions.

A main technical weakness of the system concept was the uplink transmission direction, where the resulting envelope variations caused concern for power amplifier design.

4.2.5   ODMA

Vodafone proposed Opportunity Driven Multiple Access (ODMA), basically a relaying protocol, not a pure multiple access as such. ODMA was later integrated in the WCDMA and WCDMA/TDMA concept groups and was not considered in the selection process as a concept on its own. ODMA was later considered for a while in 3GPP standardisation but was not eventually included in the specifications.

4.2.6   ETSI Selection

All the proposed technologies were basically able to fulfil the UMTS requirements, although it was difficult to reach a consensus on issues such as system capacity, since the results of simulations can vary greatly depending on the assumptions. However, it soon became evident in the selection process that WCDMA and TDMA/CDMA were the main candidates. Also, issues such as the global potential of a technology naturally had an impact in cases where obvious technical conclusions were very limited; in this respect, the outcome of the ARIB technology selection in Japan gave support to WCDMA.

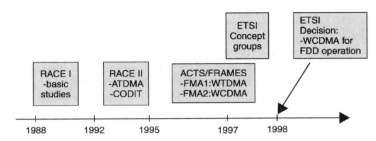

Figure 4.1. European research programmes towards third generation systems and the ETSI decision

ETSI decided between the technologies in January 1998 [8], selecting WCDMA as the standard for the UTRA (UMTS Terrestrial Radio Access) air interface on the paired frequency bands, i.e. for FDD (Frequency Division Duplexing) operation, and WTDMA/ CDMA for operation with unpaired spectrum allocation, i.e. for TDD (Time Division Duplexing) operation. As illustrated in Figure 4.1, it took ten years from the initiation of the European research programs to reach a decision on the UTRA technology. The detailed standardisation of UTRA proceeded within ETSI until the work was handed over to the 3rd Generation Partnership Project (3GPP). The technical work was transferred to 3GPP with the contribution of UTRA in early 1999.

4.3 Background in Japan

In Japan, ARIB (the Association for Radio Industries and Businesses) evaluated possible third generation systems around three different main technologies based on WCDMA, WTDMA and OFDMA.

The WCDMA technology in Japan was very similar to that being considered in Europe in ETSI; indeed, the members of ARIB contributed their technology to ETSI's WCDMA concept group. Details of FRAMES/FMA2 were provided from Europe for consideration in the ARIB process. The other technologies considered, WTDMA and OFDMA, also had many similarities to the candidates in the ETSI selection process.

The outcome of the ARIB selection process in 1997 was WCDMA, with both FDD and TDD modes of operation. Since WCDMA had been chosen in ARIB before the process was completed in ETSI, it carried more weight in the ETSI selection as the global technological alternative. Since the creation of 3GPP for the third generation standardisation framework, ARIB have contributed their WCDMA to 3GPP, in the same way as ETSI have contributed UTRA. In Japan, work on higher layer specifications is the responsibility of TTC (the Telecommunication Technology Committee), which has also shifted the activity to 3GPP.

4.4 Background in Korea

In Korea, the Telecommunications Technology Association (TTA) TTAadopted a two-track approach to the development of third generation CDMA technology. The TTA1 and TTA2 air interface proposals (later renamed Global CDMA 1 and 2 respectively) were based on synchronous and asynchronous wideband CDMA technologies respectively. TTA1 WCDMA was similar to WCDMA technology in ETSI, ARIB and T1P1, while TTA2 was similar to cdma2000 in TR45.5.

Several technical details of the Korean technology that differed from the ETSI and ARIB solutions were submitted to the ETSI and ARIB standardisation processes, leading to a high degree of commonality between the ETSI, ARIB and TTA WCDMA solutions. The Korean standardisation efforts were later moved to 3GPP and 3GPP2 to contribute to WCDMA and cdma2000 standardisation respectively.

4.5   Background in the United States

In the US there exist several second generation technologies, the most widely distributed digital systems being those based on either GSM1900, US-TDMA (D-AMPS) or US-CDMA (IS-95) standards. For all those technologies, a natural path of evolution towards the third generation had been defined. In addition, a third generation CDMA proposal that had no direct relation to second generation systems, namely WIMS W-CDMA, came from the TR46.1 standardisation committee.

4.5.1   W-CDMA N/A

Work on GSM1900 related standardisation was carried out within T1P1, as with GSM standardisation in ETSI, with similar discussions concerning technology selection. As a result, W-CDMA N/A (N/A for North America) was submitted to the ITU-R IMT-2000 process. The proposals had much in common with the ETSI and ARIB WCDMA technologies, since the contributing companies had also been active in the ETSI and ARIB selection processes.

4.5.2   UWC-136

In TR45.3, discussions concerning the evolution of IS-136 (Digital AMPS) technology towards the third generation took place. The resulting selection was a combination of narrowband and wideband TDMA technologies, with the narrowband component identical to the EDGE concept, part of GSM evolution in ETSI and T1P1. The wideband part for indoor service provision up to 2 Mbps was based on the same WTDMA concept as was considered in ETSI. The selection of EDGE technology in TR45.3 will create a clear connection between TDMA technologies within TR45.3, T1P1 and ETSI. The development of TDMA-based technology has accelerated around the common component in the air interface in the form of the EDGE component and thus several existing IS-136 operators have announced the roll-out of GSM/EDGE as part of their radio access evolution.

4.5.3   cdma2000

The cdma2000 air interface proposal to ITU was the result of work in TR45.5 on the evolution of IS-95 towards the third generation. The cdma2000 proposal is based partly on IS-95 principles with respect to synchronous network operation, common pilot channels, and so on, but it is a wideband version with three times the bandwidth of IS-95. The ITU proposal contains further bandwidth options as well as the multicarrier option for downlink. The cdma2000 proposal had a high degree of commonality with the Global CDMA 1 ITU proposal from TTA, Korea.

The cdma2000 multicarrier option is covered in more detail in Chapter 14, as standardised by 3GPP2.

4.5.4   TR46.1

The WIMS W-CDMA was not based on work derived from an existing second generation technology but was a new third generation technology proposal with no direct link to any second generation standardisation. It was based on the constant processing gain principle with a high number of multicodes in use, thus showing some fundamental differences, but also a level of commonality, with WCDMA technology in other forums.

4.5.5   WP-CDMA

WP-CDMA (Wideband Packet CDMA) resulted from the convergence between W-CDMA N/A of T1P1 and WIMS W-CDMA of TR46.1 in the US. The main features of the WIMS W-CDMA proposal were merged with the principles of W-CDMA N/A. The merged proposal was submitted to the ITU-R IMT-2000 process towards the end of 1998, and to the 3GPP process at the beginning of 1999. Its most characteristic feature, compared with the other WCDMA-based proposals, was a common packet mode channel operation for the uplink direction, but there were also a few smaller differences.

4.6   Creation of 3GPP

As similar technologies were being standardised in several regions around the world, it became evident that achieving identical specifications to ensure equipment compatibility globally would be very difficult with work going on in parallel. Also, having to discuss similar issues in several places was naturally a waste of resources for the participating companies. Therefore, initiatives were made to create a single forum for WCDMA standardisation for a common WCDMA specification.

The standardisation organisations involved in the creation of the 3rd Generation Partnership Project (3GPP) [9] were ARIB (Japan), ETSI (Europe), TTA (Korea), TTC (Japan) and T1P1 (USA). The partners agreed on joint efforts for the standardisation of UTRA, now standing for Universal Terrestrial Radio Access, as distinct from UTRA (UMTS Terrestrial Radio Access) from ETSI, also submitted to 3GPP. Companies such as manufacturers and operators are members of 3GPP through the respective standardisation organisation to which they belong, as illustrated in Figure 4.2.

Later during 1999, CWTS CWTS (the China Wireless Telecommunication Standard Group) also joined 3GPP and contributed technology from TDSCDMA, a TDD-based CDMA third generation technology already submitted to ITU-R earlier.

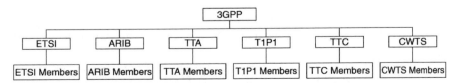

Figure 4.2. 3GPP organisational partners

3GPP also includes market representation partners: GSM Association, UMTS Forum, Global Mobile Suppliers Association, IPv6 Forum and Universal Wireless Communications Consortium (UWCC).

The work was initiated formally at the end of 1998 and the detailed technical work was started in early 1999, with the aim of having the first version of the common specification, called Release '99, ready by the end of 1999.

Within 3GPP, four different technical specification groups (TSGs) were set up as follows:

- Radio Access Network TSG;
- Core Network TSG;
- Service and System Aspects TSG;
- Terminals TSG.

Within these groups the one most relevant to the WCDMA technology is the Radio Access Network TSG (RAN TSG), which has been divided into four different working groups as illustrated in Figure 4.3.

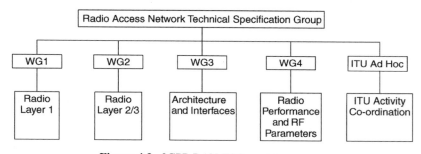

Figure 4.3. 3GPP RAN TSG working groups

The RAN TSG will produce Release '99 of the UTRA air interface specification. The work done within the 3GPP RAN TSG working groups has been the basis of the technical description of the UTRA air interface covered in this book. Without such a global initiative, this book would have been forced to focus on a single regional specification, though with many similarities to those of other regions. Thus, the references throughout this book are to the specification volumes from 3GPP.

During the first half of 1999 the inputs from the various participating organisations were merged in a single standard, leaving the rest of the year to finalise the detailed parameters for the first full release, Release '99, of UTRA from 3GPP. The member organisations have undertaken individually to produce standard publications based on the 3GPP specification. Thus, for example, the Release '99 UMTS specifications from ETSI are identical to the Release '99 specifications produced by 3GPP. The latest specifications can be obtained from 3GPP [9].

During 2000, further work on GSM evolution was moved from ETSI and other forums to 3GPP, including work on GPRS and EDGE. A new TSG, TSG GERAN was set up for this purpose.

4.7   How does 3GPP Operate?

In 3GPP the work is organised around work items, which basically define the justification and objective for a new feature. For a smaller topic there need be only a single work item in one working group if the impacts are limited to that group, or at least are mainly for the specifications under the responsibility of that group. For bigger items, such as HSDPA, there were work tasks done for each of the four RAN working groups and these work tasks were under a common work item, a building block, named HSDPA.

Of the currently on-going items, MBMS is a having a feature-level description, as that is also covering other groups than TSG RAN, and, respectively, TSG RAN is having a work item as a building block for the feature.

The work item sheets also usually contain the specifications to be impacted as well as the expected schedule of the work the latter is usually rather optimistic though. A work item needs to have four supporting companies but also it needs to have justification that can be agreeable at the respective TSG RAN level. (Note that some variations in the way of working exist between TSGs). For a larger topic, quite often a feasibility study (or study item in TSG RAN) is needed before the decision of actually creating a work item. A feasibility study will simply focus on the pain vs. gain ratio of the new feature, comparing the advantages and the resulting impacts on the equipment and existing features (the latter is known as backwards compatibility).

For each work item a raporteur is nominated, who has the responsibility of coordinating the work and reporting the progress from WGs to TSG level. At TSG level, every meeting (called a plenary) monitors progress every three months and makes any necessary synchronisation between working groups and TSGs. Sometimes a work item is determined not to have reached the expected target and it may be altered or removed from the work program. Once the work item is completed in all working groups, Change Requests (CRs) are brought to the plenary for approval. CRs contain the changes needed in each particular specification and once the plenary level approval is obtained, the specification will be updated to a new version with the changes resulting from the new feature. The simplified illustration of the process from feasibility study to specification finalisation is shown in Figure 4.4. The latest work item descriptions can be found from [9].

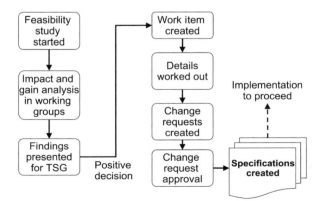

Figure 4.4. Example of 3GPP standardisation process

Creation of the specification does not necessarily mean that everything is 100% completed. Typically, some meeting rounds are then taken for potential corrections, which usually emerge as implementations proceed and details are being verified in implementation and testing. The CRs are used to introduce the corrections, they are agreed in working groups and, once approved by the following TSG plenary meeting, the CRs are then included in the specification.

4.8   Creation of 3GPP2

Work done in TR45.5 and TTA was merged to form 3GPP2, focused on the development of cdma2000 Direct-Sequence (DS) and Multicarrier (MC) mode for the cdma2000 third generation component. This activity has been running in parallel with the 3GPP project, with participation from ARIB, TTC and CWTS as member organisations. The focus shifted to MC-mode after global harmonisation efforts, but later work started to focus more on the narrowband IS-95 evolution, as reflected in the IS-2000 standards series.

4.9   Harmonisation Phase

During 1999, efforts were made to bring further harmonisation and convergence between the CDMA-based third generation solutions. For the 3GPP framework the ETSI, ARIB, TTA and T1P1 concepts had already been merged to a single WCDMA specification, while cdma2000 was still on its own in TR45.5. Eventually, the manufacturers and operators agreed to adopt a harmonised global third generation CDMA standard consisting of three modes: Multicarrier (MC), Direct Spread (DS) and Time Division Duplex (TDD). The MC mode was based on the cdma2000 multicarrier option, the DS mode on WCDMA (UTRA FDD), and the TDD mode on UTRA TDD.

The main technical impacts of these harmonisation activities were the change of UTRA FDD and TDD mode chip rate from 4.096 Mcps to 3.84 Mcps and the inclusion of a common pilot for UTRA FDD. The work in 3GPP2 focused on the MC mode, and the DS mode from cdma2000 was abandoned. Eventually, the work in 3GPP2 resumed on the 1.28 Mcps evolution and development of the MC mode has been stopped. The result is that globally there is only one Direct Spread (DS) wideband CDMA standard, WCDMA.

4.10   IMT-2000 Process in ITU

In the ITU, recommendations have been developed for third generation mobile communications systems, the ITU terminology being called IMT-2000 [10], formerly FPLMTS. In the ITU-R, ITU-R TG8/1 has worked on the radio-dependent aspects, while the radio-independent aspects have been covered in ITU-T SG11.

In the radio aspects, ITU-R TG8/1 received a number of different proposals during the IMT-2000 candidate submission process. In the second phase of the process, evaluation results were received from the proponent organisations as well as from the other evaluation groups that studied the technologies. During the first half of 1999 recommendation IMT.RKEY, which describes the IMT-2000 multimode concept, was created.

The ITU-R IMT-2000 process was finalised at the end of 1999, when the detailed specification (IMT-RSCP) was created and the radio interface specifications were approved by ITU-R [11]. The detailed implementation of IMT-2000 will continue in the regional standards bodies. The ITU-R process has been an important external motivation and timing source for IMT-2000 activities in regional standards bodies. The requirements set by ITU for an IMT-2000 technology have been reflected in the requirements in the regional standards bodies, for example in ETSI UMTS 21.01 [5], in order for the ETSI submission to fulfil the IMT-2000 requirements. The ITU-R interaction between regional standardisation bodies in the IMT-2000 process is reflected in Figure 4.5.

The ITU-R IMT-2000 grouping, with TDMA- and CDMA-based groups, is illustrated in Figure 4.6. The UTRA FDD (WCDMA) and cdma2000 are each part of the CDMA

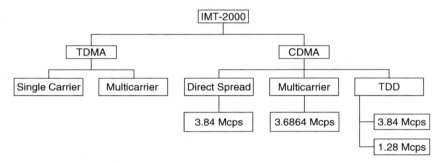

Figure 4.5. ITU-R IMT-2000 grouping

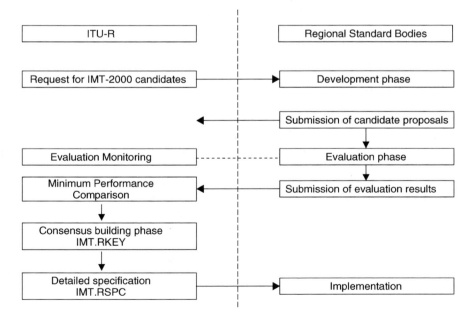

Figure 4.6. Relationship of ITU-R to the regional standards bodies

interface, as CDMA Direct Spread and CDMA Multicarrier respectively. UWC-136 and DECT are part of the TDMA-based interface in the concept, as TDMA Single Carrier and TDMA Multicarrier respectively. The TDD part in CDMA consists of UTRA TDD from 3GPP and TD-SCDMA from CWTS. For the FDD part in the CDMA interface, harmonisation has been completed, and the harmonisation process for the CDMA TDD modes within 3GPP resulted in the 1.28 Mcps TDD being included in the 3GPP Release 4 specifications, completed 03/2001.

4.11   Beyond 3GPP Release '99

Upon completion of the Release '99 specifications, work will concentrate on specifying new features as well as making the necessary corrections to Release 1999. Typically such corrections arise as implementation proceeds and test systems are updated to include the latest changes in the specifications. As experience in various forums has shown, a major step forward in system capabilities with many new features requires a phasing-in period for the specifications. Fortunately, the main functions have been verified in the various test systems in operation since 1995, but only the actual implementation will reveal any errors and inconsistencies in the fine detail of the specifications.

In 3GPP the next version of the specifications was originally considered as Release 2000, but in the meantime the Release naming was adjusted, so that the next release in March 2001 was called Release 4. Release 4 contained only minor adjustments with respect to Release 1999. Bigger items that were included in Release 5 were High-Speed Downlink Packet Access (HSDPA) and IP-based transport layer, see Chapters 11 and 5 respectively. Release 5 was completed 03/2002 for the WCDMA radio aspects. Release '99 specifications have a version number starting with 3 while Release 4 and 5 specifications have version numbers starting logically with 4 and 5 respectively.

On the TDD side, the narrowband (1.28 Mcps) TDD mode originally from CWTS (China) was included in 3GPP Release 4. The 1.28 Mcps UTRA TDD mode, or TD-SCDMA, is covered in Chapter 13.

Besides the IP-based transport option in Release 5 the protocols developed by the Internet Engineering Task Force (IETF) have also influenced the WCDMA specifications. Release 4 specifications contain robust IP header compression suitable for cellular transmission to enable an efficient Voice over IP (VoIP) service.

The next step in the evolution is Release 6, which now has first versions of the specifications available (RAN side), but all features are currently scheduled to be available by 2H/2004. For Release 6, work has been done, e.g., on Multimedia Broadcast Multicast Service (MBMS), feasibility of HSDPA-related enhancements for uplink, radio resource management supporting measurements for beamforming and on other features to enhance the system performance.

Work will then continue for Release 7. The exact timing for Release 7 has not been fixed yet, but it is expected to be towards the end of 2005. Which Release a particular feature will end up in will be determined only once the feature is mature enough to enable a change request to be written for the specifications.

References

[1] Pajukoski, K. and Savusalo, J., 'Wideband CDMA Test System', *Proc. IEEE Int. Conf. on Personal Indoor and Mobile Radio Communications*, PIMRC'97, Helsinki, Finland, 1–4 September 1997, pp. 669–672.

[2] Nikula, E., Toskala, A., Dahlman, E., Girard, L. and Klein, A., 'FRAMES Multiple Access for UMTS and IMT-2000', *IEEE Personal Communications Magazine*, April 1998, pp. 16–24.

[3] Klein, A., Pirhonen, R., Sköld, J. and Suoranta, R., 'FRAMES Multiple Access Mode 1 – Wideband TDMA with and without Spreading', *Proc. IEEE Int. Conf. on Personal Indoor and Mobile Radio Communications*, PIMRC'97, Helsinki, Finland, 1–4 September 1997, pp. 37–41.

[4] Ovesjö, F., Dahlman, E., Ojanperä, T., Toskala, A. and Klein, A., 'FRAMES Multiple Access Mode 2 – Wideband CDMA', *Proc. IEEE Int. Conf. on Personal Indoor and Mobile Radio Communications*, PIMRC'97, Helsinki, Finland, 1–4 September 1997, pp. 42–46.

[5] Universal Mobile Telecommunications System (UMTS), Requirements for the UMTS Terrestrial Radio Access System (UTRA), ETSI Technical Report, UMTS 21.01 version 3.0.1, November 1997.

[6] Universal Mobile Telecommunications System (UMTS), Selection Procedures for the Choice of Radio Transmission Technologies of the UMTS, ETSI Technical Report, UMTS 30.03 version 3.1.0, November 1997.

[7] Universal Mobile Telecommunications System (UMTS), UMTS Terrestrial Radio Access System (UTRA) Concept Evaluation, ETSI Technical Report, UMTS 30.06 version 3.0.0, December 1997.

[8] ETSI Press Release, SMG Tdoc 40/98, 'Agreement Reached on Radio Interface for Third Generation Mobile System, UMTS', Paris, France, January 1998.

[9] *http://www.3GPP.org*

[10] *http://www.itu.int/imt/*

[11] ITU Press Release, ITU/99-22, 'IMT-2000 Radio Interface Specifications Approved in ITU Meeting in Helsinki', 5 November 1999.

5

Radio Access Network Architecture

Fabio Longoni, Atte Länsisalmi and Antti Toskala

5.1 System Architecture

This chapter gives a wide overview of the UMTS system architecture, including an introduction to the logical network elements and the interfaces. The UMTS system utilises the same well-known architecture that has been used by all main second generation systems and even by some first generation systems. The reference list contains the related 3GPP specifications.

The UMTS system consists of a number of logical network elements that each has a defined functionality. In the standards, network elements are defined at the logical level, but this quite often results in a similar physical implementation, especially since there are a number of open interfaces (for an interface to be 'open', the requirement is that it has been defined to such a detailed level that the equipment at the endpoints can be from two different manufacturers). The network elements can be grouped based on similar functionality, or based on which sub-network they belong to.

Functionally the network elements are grouped into the Radio Access Network (RAN, UMTS Terrestrial RAN = UTRAN) that handles all radio-related functionality, and the Core Network, which is responsible for switching and routing calls and data connections to external networks. To complete the system, the User Equipment (UE) that interfaces with the user and the radio interface is defined. The high-level system architecture is shown in Figure 5.1.

From a specification and standardisation point of view, both UE and UTRAN consist of completely new protocols, the design of which is based on the needs of the new WCDMA radio technology. On the contrary, the definition of Core Network (CN) is adopted from GSM. This gives the system with new radio technology a global base of known and rugged CN technology that accelerates and facilitates its introduction, and enables such competitive advantages as global roaming.

Another way to group UMTS network elements is to divide them into sub-networks. The UMTS system is modular in the sense that it is possible to have several network elements of the same type. In principle, the minimum requirement for a fully featured and operational

WCDMA for UMTS, third edition. Edited by Harri Holma and Antti Toskala
© 2004 John Wiley & Sons, Ltd ISBN: 0-470-87096-6

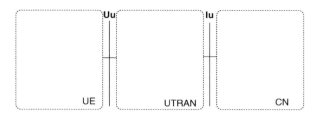

Figure 5.1. UMTS high-level system architecture

network is to have at least one logical network element of each type (note that some features and consequently some network elements are optional). The possibility of having several entities of the same type allows the division of the UMTS system into sub-networks that are operational either on their own or together with other sub-networks, and that are distinguished from each other with unique identities. Such a sub-network is called a UMTS PLMN (Public Land Mobile Network). Typically one PLMN is operated by a single operator, and is connected to other PLMNs as well as to other types of network, such as ISDN, PSTN, the Internet, and so on. Figure 5.2 shows elements in a PLMN and, in order to illustrate the connections, also external networks.

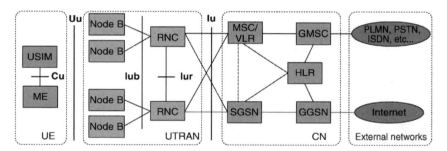

Figure 5.2. Network elements in a PLMN

The UTRAN architecture is presented in Section 5.2. A short introduction to all the elements is given below.

The UE consists of two parts:

- The Mobile Equipment (ME) is the radio terminal used for radio communication over the Uu interface.

- The UMTS Subscriber Identity Module (USIM) is a smartcard that holds the subscriber identity, performs authentication algorithms, and stores authentication and encryption keys and some subscription information that is needed at the terminal.

UTRAN also consists of two distinct elements:

- The Node B converts the data flow between the Iub and Uu interfaces. It also participates in radio resource management. (*Note that the term 'Node B' from the corresponding*

3GPP specifications is used throughout Chapter 5. The more generic term 'Base Station' used elsewhere in this book means exactly the same thing.)

- The Radio Network Controller (RNC) owns and controls the radio resources in its domain (the Node Bs connected to it). RNC is the service access point for all services UTRAN provides the CN, for example, management of connections to the UE.

The main elements of the GSM CN (there are other entities not shown in Figure 5.2, such as those used to provide IN services) are as follows:

- HLR (Home Location Register) is a database located in the user's home system that stores the master copy of the user's service profile. The service profile consists of, for example, information on allowed services, forbidden roaming areas, and supplementary service information such as status of call forwarding and the call forwarding number. It is created when a new user subscribes to the system, and remains stored as long as the subscription is active. For the purpose of routing incoming transactions to the UE (e.g. calls or short messages), the HLR also stores the UE location on the level of MSC/VLR and/or SGSN, i.e. on the level of the serving system.

- MSC/VLR (Mobile Services Switching Centre/Visitor Location Register) is the switch (MSC) and database (VLR) that serves the UE in its current location for Circuit Switched (CS) services. The MSC function is used to switch the CS transactions, and the VLR function holds a copy of the visiting user's service profile, as well as more precise information on the UE's location within the serving system. The part of the network that is accessed via the MSC/VLR is often referred to as the CS domain. MSC also has a role in the early UE handling, as discussed in Chapter 7.

- GMSC (Gateway MSC) is the switch at the point where UMTS PLMN is connected to external CS networks. All incoming and outgoing CS connections go through GMSC.

- SGSN (Serving GPRS (General Packet Radio Service) Support Node) functionality is similar to that of MSC/VLR but is typically used for Packet Switched (PS) services. The part of the network that is accessed via the SGSN is often referred to as the PS domain. Similar to MSC, SGSN support is needed for the early UE handling operation, as covered in Chapter 7.

- GGSN (Gateway GPRS Support Node) functionality is close to that of GMSC but is in relation to PS services.

The external networks can be divided into two groups:

- *CS networks.* These provide circuit-switched connections, like the existing telephony service. ISDN and PSTN are examples of CS networks.

- *PS networks.* These provide connections for packet data services. The Internet is one example of a PS network.

The UMTS standards are structured so that internal functionality of the network elements is not specified in detail. Instead, the interfaces between the logical network elements have been defined. The following main open interfaces are specified:

- *Cu interface.* This is the electrical interface between the USIM smartcard and the ME. The interface follows a standard format for smartcards.

- *Uu interface.* This is the WCDMA radio interface, which is the subject of the main part of this book. The Uu is the interface through which the UE accesses the fixed part of the system, and is therefore probably the most important open interface in UMTS. There are likely to be many more UE manufacturers than manufacturers of fixed network elements.

- *Iu interface.* This connects UTRAN to the CN and is introduced in detail in Section 5.4. Similarly to the corresponding interfaces in GSM, A (Circuit Switched) and Gb (Packet Switched), the open Iu interface gives UMTS operators the possibility of acquiring UTRAN and CN from different manufacturers. The enabled competition in this area has been one of the success factors of GSM.

- *Iur interface.* The open Iur interface allows soft handover between RNCs from different manufacturers, and therefore complements the open Iu interface. Iur is described in more detail in Section 5.5.1.

- *Iub interface.* The Iub connects a Node B and an RNC. UMTS is the first commercial mobile telephony system where the Controller–Base Station interface is standardised as a fully open interface. Like the other open interfaces, open Iub is expected to further motivate competition between manufacturers in this area. It is likely that new manu- facturers concentrating exclusively on Node Bs will enter the market.

5.2 UTRAN Architecture

UTRAN architecture is highlighted in Figure 5.3.

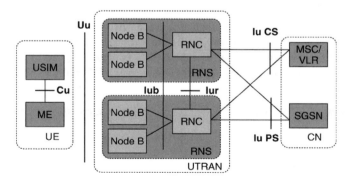

Figure 5.3. UTRAN architecture

UTRAN consists of one or more Radio Network Sub-systems (RNS). An RNS is a sub-network within UTRAN and consists of one Radio Network Controller (RNC) and one or more Node Bs. RNCs may be connected to each other via an Iur interface. RNCs and Node Bs are connected with an Iub interface.

Before entering into a brief description of the UTRAN network elements (in this section) and a more extensive description of UTRAN interfaces (in the following sections), we present the main characteristics of UTRAN that have also been the main requirements for the design of the UTRAN architecture, functions and protocols. These can be summarised in the following points:

- *Support of UTRA* and all the related functionality. In particular, the major impact on the design of UTRAN has been the requirement to support *soft handover* (one terminal connected to the network via two or more active cells) and the WCDMA-specific *Radio Resource Management* algorithms.

- Maximisation of the *commonalities in the handling of packet-switched and circuit-switched data*, with a unique air interface protocol stack and with the use of the same interface for the connection from UTRAN to both the PS and CS domains of the core network.

- Maximisation of the *commonalities with GSM*, when possible.

- Use of the *ATM transport* as the main transport mechanism in UTRAN.

- Use of the IP-based transport as the alternative transport mechanism in UTRAN from Release 5 onwards.

5.2.1 The Radio Network Controller

The RNC (Radio Network Controller) is the network element responsible for the control of the radio resources of UTRAN. It interfaces the CN (normally to one MSC and one SGSN) and also terminates the RRC (Radio Resource Control) protocol that defines the messages and procedures between the mobile and UTRAN. It logically corresponds to the GSM BSC.

5.2.1.1 Logical Role of the RNC

The RNC controlling one Node B (i.e. terminating the Iub interface towards the Node B) is indicated as the *Controlling RNC* (CRNC) of the Node B. The Controlling RNC is responsible for the load and congestion control of its own cells, and also executes the admission control and code allocation for new radio links to be established in those cells.

In case one mobile–UTRAN connection uses resources from more than one RNS (see Figure 5.4), the RNCs involved have two separate logical roles (*with respect to this mobile– UTRAN connection*):

- *Serving RNC*. The SRNC for one mobile is the RNC that terminates both the Iu link for the transport of user data and the corresponding RANAP signalling to/from the core network (this connection is referred to as the RANAP connection). The SRNC also terminates the Radio Resource Control Signalling, that is the signalling protocol between the UE and UTRAN. It performs the L2 processing of the data to/from the radio interface. Basic Radio Resource Management operations, such as the mapping of Radio Access Bearer parameters into air interface transport channel parameters, the handover decision, and outer loop power control, are executed in the SRNC. The SRNC may also (but not always) be the CRNC of some Node B used by the mobile for connection with UTRAN. One UE connected to UTRAN has one and only one SRNC.

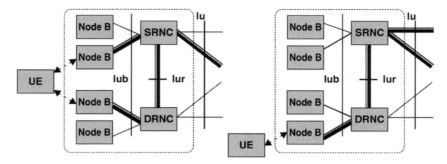

Figure 5.4. Logical role of the RNC for one UE UTRAN connection. The left-hand scenario shows one UE in inter-RNC soft handover (combining is performed in the SRNC). The right-hand scenario represents one UE using resources from one Node B only, controlled by the DRNC

- *Drift RNC.* The DRNC is any RNC, other than the SRNC, that controls cells used by the mobile. If needed, the DRNC may perform macrodiversity combining and splitting. The DRNC does not perform L2 processing of the user plane data, but routes the data transparently between the Iub and Iur interfaces, except when the UE is using a common or shared transport channel. One UE may have zero, one or more DRNCs.

Note that one physical RNC normally contains all the CRNC, SRNC and DRNC functionality.

5.2.2   The Node B (Base Station)

The main function of the Node B is to perform the air interface L1 processing (channel coding and interleaving, rate adaptation, spreading, etc.). It also performs some basic Radio Resource Management operations such as the inner loop power control. It logically corresponds to the GSM Base Station. The enigmatic term 'Node B' was initially adopted as a temporary term during the standardisation process, but then never changed.

The logical model of the Node B is described in Section 5.5.2.

5.3   General Protocol Model for UTRAN Terrestrial Interfaces

5.3.1   General

Protocol structures in UTRAN terrestrial interfaces are designed according to the same general protocol model. This model is shown in Figure 5.5. The structure is based on the principle that the layers and planes are logically independent of each other and, if needed, parts of the protocol structure may be changed in the future while other parts remain intact.

5.3.2   Horizontal Layers

The protocol structure consists of two main layers, the Radio Network Layer and the Transport Network Layer. All UTRAN-related issues are visible only in the Radio Network Layer, and the Transport Network Layer represents standard transport technology that is selected to be used for UTRAN but without any UTRAN-specific changes.

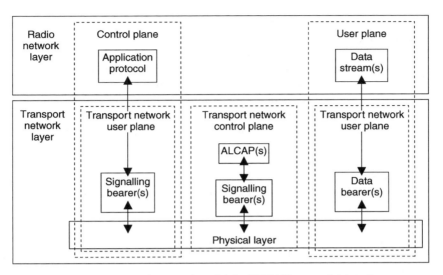

Figure 5.5. General protocol model for UTRAN terrestrial interfaces

5.3.3   Vertical Planes

5.3.3.1   Control Plane

The Control Plane is used for all UMTS-specific control signalling. It includes the Application Protocol (i.e. RANAP in Iu, RNSAP in Iur and NBAP in Iub), and the Signalling Bearer for transporting the Application Protocol messages.

The Application Protocol is used, among other things, for setting up bearers to the UE (i.e. the Radio Access Bearer in Iu and subsequently the Radio Link in Iur and Iub). In the three-plane structure the bearer parameters in the Application Protocol are not directly tied to the User Plane technology, but rather are general bearer parameters.

The Signalling Bearer for the Application Protocol may or may not be of the same type as the Signalling Bearer for the ALCAP. It is always set up by O&M actions.

5.3.3.2   User Plane

All information sent and received by the user, such as the coded voice in a voice call or the packets in an Internet connection, are transported via the User Plane. The User Plane includes the Data Stream(s), and the Data Bearer(s) for the Data Stream(s). Each Data Stream is characterised by one or more frame protocols specified for that interface.

5.3.3.3   Transport Network Control Plane

The Transport Network Control Plane is used for all control signalling within the Transport Layer. It does not include any Radio Network Layer information. It includes the ALCAP protocol that is needed to set up the transport bearers (Data Bearer) for the User Plane. It also includes the Signalling Bearer needed for the ALCAP.

The Transport Network Control Plane is a plane that acts between the Control Plane and the User Plane. The introduction of the Transport Network Control Plane makes it possible for the Application Protocol in the Radio Network Control Plane to be completely independent of the technology selected for the Data Bearer in the User Plane.

When the Transport Network Control Plane is used, the transport bearers for the Data Bearer in the User Plane are set up in the following fashion. First there is a signalling transaction by the Application Protocol in the Control Plane, which triggers the set-up of the Data Bearer by the ALCAP protocol that is specific for the User Plane technology.

The independence of the Control Plane and the User Plane assumes that an ALCAP signalling transaction takes place. It should be noted that ALCAP might not be used for all types of Data Bearer. If there is no ALCAP signalling transaction, the Transport Network Control Plane is not needed at all. This is the case when it is enough to simply select the user plane resources, e.g. selecting end point addresses for IP transport or selecting a preconfigured Data Bearer. It should also be noted that the ALCAP protocol(s) in the Transport Network Control Plane is/are not used for setting up the Signalling Bearer for the Application Protocol or for the ALCAP during real-time operation.

The Signalling Bearer for the ALCAP may or may not be of the same type as that for the Application Protocol. The UMTS specifications assume that the Signalling Bearer for ALCAP is always set up by O&M actions, and do not specify this in detail.

5.3.3.4 Transport Network User Plane

The Data Bearer(s) in the User Plane, and the Signalling Bearer(s) for the Application Protocol, also belong to the Transport Network User Plane. As described in the previous section, the Data Bearers in the Transport Network User Plane are directly controlled by the Transport Network Control Plane during real-time operation, but the control actions required for setting up the Signalling Bearer(s) for the Application Protocol are considered O&M actions.

5.4 Iu, The UTRAN–CN Interface

The Iu interface connects UTRAN to CN. Iu is an open interface that divides the system into radio-specific UTRAN and CN which handles switching, routing and service control. As can be seen from Figure 5.3, the Iu can have two main different instances, which are Iu CS (Iu Circuit Switched) for connecting UTRAN to Circuit Switched (CS) CN, and Iu PS (Iu Packet Switched) for connecting UTRAN to Packet Switched (PS) CN. The additional third instance of Iu, the Iu BC (Iu Broadcast), has been defined to support Cell Broadcast Services (See Section 5.4.5). Iu BC is used to connect UTRAN to the Broadcast domain of the Core Network. The Iu BC interface is not shown in Figure 5.3. The original design goal in the standardisation was to develop only one Iu interface, but then it was realised that fully optimised User Plane transport for CS and PS services can only be achieved if different transport technologies are allowed. Consequently, the Transport Network Control Plane is different. One of the main design guidelines has still been that the Control Plane should be the same for Iu CS and Iu PS, and the differences are minor.

5.4.1 Protocol Structure for Iu CS

The Iu CS overall protocol structure is depicted in Figure 5.6. The three planes in the Iu interface share a common ATM (Asynchronous Transfer Mode) transport which is used for all planes. The physical layer is the interface to the physical medium: optical fibre, radio link

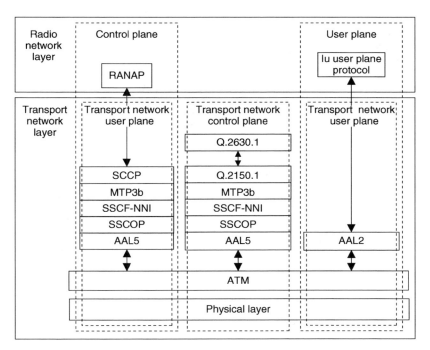

Figure 5.6. Iu CS protocol structure

or copper cable. The physical layer implementation can be selected from a variety of standard off-the-shelf transmission technologies, such as SONET, STM1, or E1.

5.4.1.1 Iu CS Control Plane Protocol Stack

The Control Plane protocol stack consists of RANAP, on top of Broadband (BB) SS7 (Signalling System #7) protocols. The applicable layers are the Signalling Connection Control Part (SCCP), the Message Transfer Part (MTP3-b) and SAAL-NNI (Signalling ATM Adaptation Layer for Network to Network Interfaces). SAAL-NNI is further divided into Service Specific Coordination Function (SSCF), Service Specific Connection Oriented Protocol (SSCOP) and ATM Adaptation Layer 5 (AAL) layers. SSCF and SSCOP layers are specifically designed for signalling transport in ATM networks, and take care of such functions as signalling connection management. AAL5 is used for segmenting the data to ATM cells.

5.4.1.2 Iu CS Transport Network Control Plane Protocol Stack

The Transport Network Control Plane protocol stack consists of the Signalling Protocol for setting up AAL2 connections (Q.2630.1 and adaptation layer Q.2150.1), on top of BB SS7 protocols. The applicable BB SS7 are those described above without the SCCP layer.

5.4.1.3 Iu CS User Plane Protocol Stack

A dedicated AAL2 connection is reserved for each individual CS service. The Iu User Plane Protocol residing directly on top of AAL2 is described in more detail in Section 5.4.4.

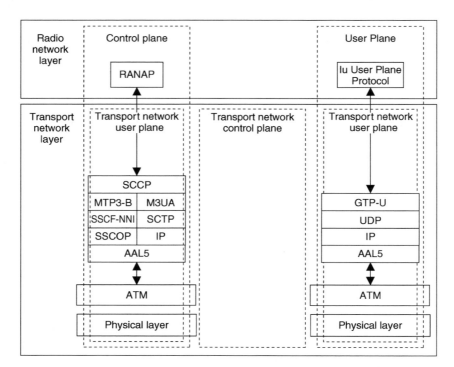

Figure 5.7. Iu PS protocol structure

5.4.2   Protocol Structure for Iu PS

The Iu PS protocol structure is depicted in Figure 5.7. Again, a common ATM transport is applied for both User and Control Plane. Also the physical layer is as specified for Iu CS.

5.4.2.1   Iu PS Control Plane Protocol Stack

The Control Plane protocol stack again consists of RANAP, and the same BB SS7-based signalling bearer as described in Section 5.4.1.1. Also, as an alternative, an IP-based signalling bearer is specified. The SCCP layer is also used commonly for both. The IP-based signalling bearer consists of M3UA (SS7 MTP3 – User Adaptation Layer), SCTP (Simple Control Transmission Protocol), IP (Internet Protocol), and AAL5 which is common to both alternatives. The SCTP layer is specifically designed for signalling transport in the Internet. Specific adaptation layers are specified for different kinds of signalling protocol, such as M3UA for SS7-based signalling.

5.4.2.2   Iu PS Transport Network Control Plane Protocol Stack

The Transport Network Control Plane is not applied to Iu PS. The setting up of the GTP tunnel requires only an identifier for the tunnel, and the IP addresses for both directions, and these are already included in the RANAP RAB Assignment messages. The same information elements that are used in Iu CS for addressing and identifying the AAL2 signalling are used for the User Plane data in Iu CS.

5.4.2.3 Iu PS User Plane Protocol Stack

In the Iu PS User Plane, multiple packet data flows are multiplexed on one or several AAL5 PVCs. The GTP-U (User Plane part of the GPRS Tunnelling Protocol) is the multiplexing layer that provides identities for individual packet data flow. Each flow uses UDP connectionless transport and IP addressing.

5.4.3 RANAP Protocol

RANAP is the signalling protocol in Iu that contains all the control information specified for the Radio Network Layer. The functionality of RANAP is implemented by various RANAP Elementary Procedures. Each RANAP function may require the execution of one or more EPs. Each EP consists of either just the request message (class 2 EP), the request and response message pair (class 1 EP), or one request message and one or more response messages (class 3 EP). The following RANAP functions are defined:

- *Relocation.* This function handles both SRNS Relocation and Hard Handover, including the inter-system case to/from GSM:
  - SRNS Relocation: the serving RNS functionality is relocated from one RNS to another without changing the radio resources and without interrupting the user data flow. The prerequisite for SRNS relocation is that all Radio Links are already in the same DRNC that is the target for the relocation.
  - Inter-RNS Hard Handover: used to relocate the serving RNS functionality from one RNS to another and to change the radio resources correspondingly by a hard handover in the Uu interface. The prerequisite for Hard Handover is that the UE is at the border of the source and target cells.

- *RAB (Radio Access Bearer) Management.* This function combines all RAB handling:
  - RAB Set-up, including the possibility for queuing the set-up;
  - modification of the characteristics of an existing RAB;
  - clearing an existing RAB, including the RAN-initiated case.

- *Iu Release.* Releases all resources (Signalling link and U-Plane) from a given instance of Iu related to the specified UE. Also includes the RAN-initiated case.

- *Reporting Unsuccessfully Transmitted Data.* This function allows the CN to update its charging records with information from UTRAN if part of the data sent was not successfully sent to the UE.

- *Common ID management.* In this function the permanent identification of the UE is sent from the CN to UTRAN to allow paging coordination from possibly two different CN domains.

- *Paging.* This is used by CN to page an idle UE for a UE terminating service request, such as a voice call. A paging message is sent from the CN to UTRAN with the UE common identification (permanent Id) and the paging area. UTRAN will either use an existing signalling connection, if one exists, to send the page to the UE or broadcast the paging in the requested area.

- *Management of tracing.* The CN may, for operation and maintenance purposes, request UTRAN to start recording all activity related to a specific UE–UTRAN connection.

- *UE–CN signalling transfer.* This functionality provides transparent transfer of UE–CN signalling messages that are not interpreted by UTRAN in two cases:

  - Transfer of the first UE message from UTRAN to UE: this may be, for example, a response to paging, a request of a UE-originated call, or just registration to a new area. It also initiates the signalling connection for the Iu.
  - Direct Transfer: used for carrying all consecutive signalling messages over the Iu signalling connection in both the uplink and downlink directions.

- *Security Mode Control.* This is used to set the ciphering or integrity checking on or off. When ciphering is on, the signalling and user data connections in the radio interface are encrypted with a secret key algorithm. When integrity checking is on, an integrity checksum, further secured with a secret key, is added to some or all of the Radio Interface signalling messages. This ensures that the communication partner has not changed, and the content of the information has not been altered.

- *Management of overload.* This is used to control the load over the Iu interface against overload due, for example, to processor overload at the CN or UTRAN. A simple mechanism is applied that allows stepwise reduction of the load and its stepwise resumption, triggered by a timer.

- *Reset.* This is used to reset the CN or the UTRAN side of the Iu interface in error situations. One end of the Iu may indicate to the other end that it is recovering from a restart, and the other end can remove all previously established connections.

- *Location Reporting.* This functionality allows the CN to receive information on the location of a given UE. It includes two elementary procedures, one for controlling the location reporting in the RNC and the other to send the actual report to the CN.

5.4.4   Iu User Plane Protocol

The Iu User Plane protocol is in the Radio Network Layer of the Iu User Plane. It has been defined to be, as much as possible, independent of the CN domain that it is used for. The purpose of the User Plane protocol is to carry user data related to RABs over the Iu interface. Each RAB has its own instance of the protocol. The protocol performs either a fully transparent operation, or framing for the user data segments and some basic control signalling to be used for initialisation and online control. Based on these cases, the protocol has two modes:

- *Transparent Mode.* In this mode of operation the protocol does not perform any framing or control. It is applied for RABs that do not require such features but that assume fully transparent operation.

- *Support Mode for predefined SDU sizes.* In this mode the User Plane performs framing of the user data into segments of predefined size. The SDU sizes typically correspond to AMR (Adaptive Multirate Codec) speech frames, or to the frame sizes derived from the data rate of a CS data call. Also, control procedures for initialisation and rate control are defined, and a functionality is specified for indicating the quality of the frame based, for example, on CRC from the radio interface.

5.4.5   Protocol Structure of Iu BC, and the SABP Protocol

The Iu BC [2] interface connects the RNC in UTRAN with the broadcast domain of the Core
Network, namely with the Cell Broadcast Centre. It is used to define the Cell Broadcast
information that is transmitted to the mobile user via the Cell Broadcast Service (e.g. name
of city/region visualised on the mobile phone display). Note that this shall not be confused
with the UTRAN or Core Network information broadcast on the broadcast common control
channel. Iu BC is a control plane only interface. The protocol structure of Iu BC is shown in
Figure 5.8.

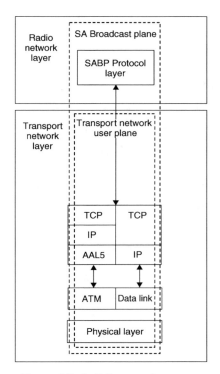

Figure 5.8. Iu BC protocol structure

5.4.5.1   SABP Protocol

The Service Area Broadcast Protocol (SABP) [8] provides the capability for the Cell
Broadcast Centre in the CN to define, modify and remove cell broadcast messages from the
RNC. RNC uses them and the NBAP protocol and RRC signalling to transfer such messages
to the mobile. The SABP has the following functions:

- *Message Handling.* This function is responsible for the broadcast of new messages,
  amendment of existing broadcast messages and prevention of the broadcasting of specific
  messages.

- *Load Handling.* This function is responsible for determining the loading of the broadcast channels at any particular point in time.

- *Reset.* This function permits the CBC to end broadcasting in one or more Service Areas.

5.5   UTRAN Internal Interfaces

5.5.1   RNC–RNC Interface (Iur Interface) and the RNSAP Signalling

The protocol stack of the RNC to RNC interface (Iur interface) is shown in Figure 5.9.

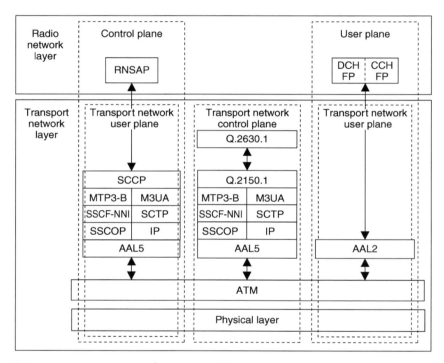

Figure 5.9. Release '99 protocol stack of the Iur interface. As for the Iu interface, two options are possible for the transport of the RNSAP signalling: the SS7 stack (SCCP and MTP3b) and the new SCTP/IP-based transport. Two User Plane protocols are defined (DCH: dedicated channel; CCH: common channel).

Although this interface was initially designed in order to support the inter-RNC soft handover (shown on the left-hand side of Figure 5.4), more features were added during the development of the standard and currently the Iur interface provides four distinct functions:

1. Support of basic inter-RNC mobility;
2. Support of Dedicated Channel traffic;

3. Support of Common Channel traffic;
4. Support of Global Resource Management.

For this reason the Iur signalling protocol itself (RNSAP, *Radio Network System Application Part*) is divided into four different *modules* (intended as groups of procedures). In general, it is possible to implement only part of the four Iur modules between two Radio Network Controllers, according to the operator's need.

5.5.1.1 Iur1: Support of the Basic Inter-RNC Mobility

This functionality requires the *basic* module of RNSAP signalling as described in [12]. This first brick for the construction of the Iur interfaces provides by itself the functionality needed for the mobility of the user between the two RNCs, but does not support the exchange of any user data traffic. If this module is not implemented, the Iur interface as such does not exist, and the only way for a user connected to UTRAN via the RNS1 to utilise a cell in RNS2 is to disconnect itself temporarily from UTRAN (release the RRC connection).

The functions offered by the Iur basic module include:

- *Support of SRNC relocation.*

- *Support of inter-RNC cell and UTRAN registration area update.*

- *Support of inter-RNC packet paging.*

- *Reporting of protocol errors.*

Since this functionality does not involve user data traffic across Iur, the User Plane and the Transport Network Control Plane protocols are not needed.

5.5.1.2 Iur2: Support of Dedicated Channel Traffic

This functionality requires the *Dedicated Channel* module of RNSAP signalling and allows the dedicated and shared channel traffic between two RNCs. Even if the initial need for this functionality is to support the inter-RNC soft handover state, it also allows the anchoring of the SRNC for all the time the user is utilising dedicated channels (dedicated resources in the Node B), commonly for as long as the user has an active connection to the circuit-switched domain.

This functionality requires also the User Plane *Frame Protocol* for the dedicated and shared channel, plus the Transport Network Control Plane protocol (*Q.2630.1*) used for the set-up of the transport connections (AAL2 connections). Each dedicated channel is conveyed over one transport connection, except the coordinated DCH used to obtain unequal error protection in the air interface.

The *Frame Protocol* for dedicated channels, in short DCH FP [16], defines the structure of the *data frames* carrying the user data and the *control frames* used to exchange measurements and control information. For this reason, the Frame Protocol also specifies simple messages and procedures. The user data frames are normally routed transparently through the DRNC; thus the Iur frame protocol is also used in Iub and referred to as Iur/Iub DCH FP. The user plane procedure for the shared channel is described in the Frame Protocol for the common channel in the Iur interface, in short Iur CCH FP [14].

> *The functions offered by the Iur DCH module are:*
>
> * *Establishment, modification and release of the dedicated and shared channel in the DRNC due to handovers in the dedicated channel state.*
> * *Set-up and release of dedicated transport connections across the Iur interface.*
> * *Transfer of DCH Transport Blocks between SRNC and DRNC.*
> * *Management of the radio links in the DRNS via dedicated measurement report procedures, power setting procedures and compress mode control procedures.*

5.5.1.3 Iur3: Support of Common Channel Traffic

This functionality allows the handling of common channel (i.e. RACH, FACH and CPCH) data streams across the Iur interface. It requires the Common Transport Channel module of the RNSAP protocol and the Iur Common Transport Channel Frame Protocol (in short, CCH FP). The Q.2630.1 signalling protocol of the Transport Network Control Plane is also needed if signalled AAL2 connections are used.

If this functionality is not implemented, every inter-RNC cell update always triggers an SRNC relocation, i.e. the serving RNC is always the RNC controlling the cell used for common or shared channel transport.

The identification of the benefits of this feature caused a long debate in the relevant standardisation body. On the one hand, this feature allows the implementation of the total anchor RNC concept, avoiding the SRNC relocation procedure (via the CN); on the other hand, it requires the splitting of the Medium Access Control layer functionality into two network elements, generating inefficiency in the utilisation of the resources and complexity in the Iur interface. The debate could not reach an agreement, thus the feature is supported by the standard but is not essential for the operation of the system.

> *The functions offered by the Iur common transport channel module are:*
>
> * *Set-up and release of the transport connection across the Iur for common channel data streams.*
> * *Splitting of the MAC layer between the SRNC (MAC-d) and the DRNC (MAC-c). The scheduling for DL data transmission is performed in the DRNC.*
> * *Flow control between the MAC-d and MAC-c.*

5.5.1.4 Iur4: Support of Global Resource Management

This functionality provides signalling to support enhanced radio resource management and O&M features across the Iur interface. It is implemented via the *global module* of the RNSAP protocol, and does not require any User Plane protocol, since there is no transmission of user data across the Iur interface. The function is considered optional. This function has been introduced in subsequent releases for the support of common radio resource management between RNCs, advanced positioning methods and Iur optimisation purposes.

The functions offered by the Iur global resource module are:

- *Transfer of cell information and measurements between two RNCs.*
- *Transfer of positioning parameters between controller.*
- *Transfer of Node B timing information between two RNCs.*

5.5.2   RNC–Node B Interface and the NBAP Signalling

The protocol stack of the RNC–Node B interface (Iub interface) is shown, with the typical triple plane notation, in Figure 5.10.

In order to understand the structure of the interface, it is necessary to briefly introduce the logical model of the Node B, depicted in Figure 5.11. This consists of a common control port (a common signalling link) and a set of traffic termination points, each controlled by a dedicated control port (dedicated signalling link). One traffic termination point controls a

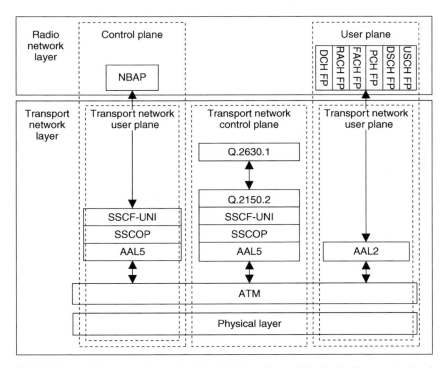

Figure 5.10. Release '99 protocol stack of the Iub interface. This is similar to the Iur interface protocol, the main difference being that in the Radio Network and Transport Network Control Planes the SS7 stack is replaced by the simpler SAAL-UNI as signalling bearer. Note also that the SCTP/IP option is not present here

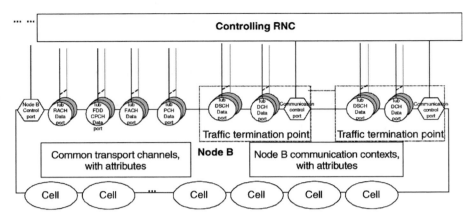

Figure 5.11. Logical model of the Node B for FDD

number of mobiles having dedicated resources in the Node B, and the corresponding traffic is conveyed through dedicated data ports. Common data ports outside the traffic termination points are used to convey RACH, FACH and PCH traffic.

Note that there is no relation between the traffic termination point and the cells, i.e. one traffic termination point can control more than one cell, and one cell can be controlled by more than one traffic termination point.

The Iub interface signalling (NBAP, *Node B Application Part*) is divided into two essential components: the common NBAP, that defines the signalling procedures across the common signalling link, and the dedicated NBAP, used in the dedicated signalling link.

The User Plane Iub frame protocols define the structures of the frames and the basic in-band control procedures for every type of transport channel (i.e. for every type of data port of the model). The Q.2630.1 signalling is used for the dynamic management of the AAL2 connections used in the User Plane.

5.5.2.1 Common NBAP and the Logical O&M

The common NBAP (C-NBAP) procedures are used for the signalling that is not related to one specific UE context already existing in the Node B. In particular, the C-NBAP defines all the procedures for the logical O&M (Operation and Maintenance) of the Node B, such as configuration and fault management.

The main functions of the Common NBAP are:

- *Set-up of the first RL of one UE, and selection of the traffic termination point.*
- *Cell configuration.*
- *Handling of the RACH/FACH/CPCH and PCH channels.*
- *Initialisation and reporting of Cell or Node B specific measurement.*
- *Location Measurement Unit (LMU) control.*
- *Fault management.*

5.5.2.2 Dedicated NBAP

When the RNC requests the first radio link for one UE via the C-NBAP *Radio Link Set-up* procedure, the Node B assigns a traffic termination point for the handling of this UE context, and every subsequent signalling related to this mobile is exchanged with dedicated NBAP (D-NBAP) procedures across the dedicated control port of the given Traffic Termination Point.

The main functions of the Dedicated NBAP are:

- *Addition, release and reconfiguration of radio links for one UE context.*

- *Handling of dedicated and shared channels.*

- *Handling of softer combining.*

- *Initialisation and reporting of radio link specific measurement.*

- *Radio link fault management.*

5.6 UTRAN Enhancements and Evolution

The Release '99 UTRAN Architecture described in the previous chapter defines the basic set of network elements and interface protocols for the support of the Release '99 WCDMA radio interface. Since then, enhancement of the architecture and related specification have been needed in order to support new WCDMA radio interface features, but also as result of the necessity to provide a more efficient, scalable and robust 3GPP system architecture. The four most significant additions to the UTRAN architecture introduced in Release 5 are described in the subsequent sections.

5.6.1 IP Transport in UTRAN

ATM is the transport technology used in the first release of the UTRAN. Even before the completion of the specification it was clear that 3GPP cannot stay immune from the increasing popularity of the IP technology, and a second option for the transport, the 'IP transport' is introduced in the specification in Release 5. Accordingly, user plane FP frames can also be conveyed over UDP/IP protocols on Iur/Iub, and over RTP/UDP/IP protocols in Iu CS interface, in addition to the initially defined option of AAL2/ATM. A second option for the Iub control plane, using SCTP directly below the application part, is also introduced. The protocols to be used to convey IP frames are, in general, left unspecified in order not to limit the use of layer two and physical layer interfaces available in the operator networks. Although the IP transport has required small changes in the specification (and almost none in the control plane application parts), the adoption of the IP technology is a relevant step for both the operator and the vendors, changing the way the network itself is managed and, in some cases, the way the network elements are implemented.

5.6.2 Iu Flex

Release '99 architecture presented in Figure 5.3 is characterised by having only one MSC and one SGSN connected to the RNC, i.e. only one Iu PS and Iu CS interface in the RNC.

This limitation is overcome in the Release 5 specification with the introduction of the *Iu flex* (abbreviation from the word 'flexible') concept, that allows one RNC to have more than one Iu PS and Iu CS interface instances with the core. The main benefits of this feature are to introduce the possibility of load sharing between the core network nodes, and to increase the possibility to anchor the MSC and SGSN in case of SRNS relocation. Iu flex has limited impact on the UTRAN specification, since the core network node to be used is negotiated between the UE and the Core Network.

5.6.3   Stand Alone SMLC and Iupc Interface

Location-based services are expected to be a very important source of revenue for the mobile operators, and a number of different applications are expected to be available and largely used. Following the example of the GSM BSS, UTRAN architecture also includes a stand alone Serving Mobile Location Centre (stand alone SMLC, or, simply, SAS), which is a new network element for the handling of positioning measurements and the calculation of the mobile station position. The SAS is connected to the RNC via the Iupc interface and the Positioning Calculation Application Part (PCAP) is the L3 protocol used for the RNC-SAS signalling. Stand Alone SMLC and Iupc interface are optional elements, since SMLC functionality can be integrated in the RNC as well, thus it is dependent on the individual network implementation whether to use it or not. The first version of the Iupc supported only Assisted GPS but then, for later versions, support for other positioning methods was added.

5.6.4   Interworking between GERAN and UTRAN, and the Iur-g Interface

Iu interface has also been scheduled to be part of the GSM/EDGE Radio Access Network (GERAN) in GERAN Release 5. This allows reusing the 3G Core Network also for the GSM/EDGE radio interface (and frequency band), but also allows a more optimised interworking between the two radio technologies. As an effect of this, the RNSAP basic mobility module (described in Section 5.5.1.1) is enhanced to allow also the mobility to and from GERAN cells in the target and the source, and the RNSAP global module (see 5.5.1.4) is enhanced in order to allow the GERAN cells measurements to be exchanged between controllers. The last feature allows a Common Radio Resource Management (CRRM) between UTRAN and GERAN radios. The term *Iur-g interface* is often used to refer to the above-mentioned set of Iur functionalities that are utilised also by the GERAN.

5.6.5   All IP RAN Concept

The increasing role of IP Technology in modern telecoms and IT networks has been mentioned already in the previous sections to motivate the introduction of the IP Transport option in UTRAN. We will see in the next section how the need to provide an optimised support IP service leads to sensible changes in the architecture of the core network with the introduction of a new sub-system, the IMS, to form what is now commonly referred to as All IP Core Network. Is the introduction of IP Transport in UTRAN enough to provide the radio access network architecture most suitable to be implemented with IP technology, integrated with the always more commonly used IP networks and platforms, and utilised by IP packet services? In 3GPP work has been done to investigate new architecture alternatives with the aim of a more distributed operation than the centralised network structure. However, developments since Release '99, with features like HSDPA in Release 5 or HSUPA, have brought the processing in many cases, especially with packet data, closer to the air interface to improve the system performance, while for basic services, such as speech changing the architecture is not expected to add any real benefit since the current system is already efficient in dealing with that.

The term *All IP RAN* is nowadays often used to refer to this IP Optimised RAN architecture concept and implementation, but is currently not yet associated to any 3GPP standard feature. For this reason, this term is sometimes used to refer to a RAN implementation based on the current architecture but using IP Transport. It remains to be seen whether there will be additional architecture changes in forthcoming 3GPP Releases, other than those introduced or being introduced by HSDPA or by HSUPA, to bring functions such as scheduling or retransmission handling for the Node B closer to the WCDMA air interface.

5.7   UMTS Core Network Architecture and Evolution

While the UMTS radio interface, WCDMA, represented a bigger step in the radio access evolution from GSM networks, the UMTS core network did not experience major changes in the 3GPP Release '99 specification. The Release '99 structure was inherited from the GSM core network and, as stated earlier, both UTRAN and GERAN based radio access network connect to the same core network.

5.7.1   Release '99 Core Network Elements

The Release '99 core network has two domains: Circuit Switched (CS) domain and Packet Switched (PS) domain, to cover the need for different traffic types. The division comes from the different requirements for the data, depending on whether it is real time (circuit switched) or non-real time (packet data). The following sections present the functional split in the core network side, however, it should be understood that several functionalities can be implemented in a single physical entity and all entities are not necessarily existing as separate physical units in real networks. Figure 5.12 illustrates the Release '99 core network structure with both CS and PS domains shown. The Figure also contains registers, as well as the Service Control Point (SCP), to indicate the link for providing a particular service to the end user.

The CS domain has the following elements, as introduced in Section 5.1:

- Mobile Switching Centre (MSC), including Visitor Location Register (VLR);
- Gateway MSC (GMSC).

The PS domain has the following elements, as introduced in Section 5.1:

- Serving GPRS Support Node (SGSN), which covers similar functions as the MSC for the packet data, including VLR type functionality.
- Gateway GPRS Support Node (GGSN) connects PS core network to other networks, for example to the Internet.

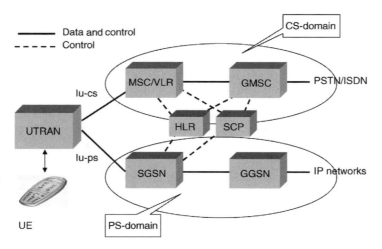

Figure 5.12. Release '99 UMTS core network structure

In addition to the two domains, the network needs various registers for proper operation:

- Home Location Register (HLR) with the functionality as covered in Section 5.1;

- Equipment Identity Register (EIR) contains the information related to the terminal equipment and can be used to, e.g., prevent a specific terminal from accessing the network.

5.7.2 Release 5 Core Network and IP Multimedia Sub-system

The Release 5 core network has many additions compared to Release '99 core networks. Release 4 already included the change in core network CS domain when the MSC was divided into MSC server and Media Gateway (MGW). Also, the GMSC was divided into GMSC server and MGW. Release 5 contains the first phase of IP Multimedia Sub-system (IMS), which will enable a standardised approach for IP-based service provision via PS domain, as discussed in Chapter 2. The capabilities of the IMS will be further enhanced in Release 6. Release 6 IMS will allow the provision of services similar to CS domain services from the PS domain. The following sections summarise the elements in Release 5 based architecture, added to Release '99 and Release 4 architecture. The Release 5 architecture is presented in Figure 5.13, with the simplification that the registers, now part of Home Subscriber Server (HSS), are shown only as an independent item without all the connections to the other elements shown.

From a protocols perspective, the key protocol between the terminal and the IMS is the Session Initiation Protocol (SIP), which is the basis for IMS-related signalling, with the contents as described in Chapter 2.

The following elements have experienced changes in the CS-domain for Release 4.

- The MSC or GMSC server, takes care of the control functionality as MSC or GMSC respectively, but the user data goes via the Media Gateway (MGW). One MSC/GMCS

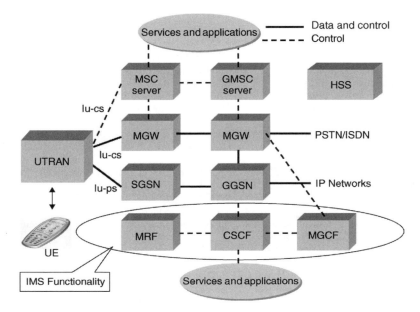

Figure 5.13. Release 5 UMTS core network architecture

server can control multiple MGWs, which allows better scalability of the network when, e.g., the data rates increase with new data services. In that case, only the number of MGWs needs to be increased.

- MGW performs the actual switching for user data and network interworking processing, e.g., echo cancellation or speech decoding/encoding.

In the PS-domain, the SGSN and GGSN are as in Release '99 with some enhancements, but for the IP-based service delivery, the IMS has now the following key elements included:

- Media Resource Function (MRF) which, e.g., controls media stream resources or can mix different media streams. The standard defines further the detailed functional split for MRF.

- Call Session Control Function (CSCF), which acts as the first contact point to the terminal in the IMS (as a proxy). The CSCF covers several functionalities from handling of the session states to being a contact point for all IMS connections intended for a single user and acting as a firewall towards other operator's networks.

- Media Gateway Control Function (MGCF), to handle protocol conversions. This may also control a service coming via the CS domain and perform processing in an MGW, e.g. for echo cancellation.

An overview of the different elements and their interfaces can be found in [24] and further details of the core network protocols in [25].

References

[1] 3GPP Technical Specification 25.401 UTRAN Overall Description.
[2] 3GPP Technical Specification 25.410 UTRAN Iu Interface: General Aspects and Principles.
[3] 3GPP Technical Specification 25.411 UTRAN Iu Interface: Layer 1.
[4] 3GPP Technical Specification 25.412 UTRAN Iu Interface: Signalling Transport.
[5] 3GPP Technical Specification 25.413 UTRAN Iu Interface: RANAP Signalling.
[6] 3GPP Technical Specification 25.414 UTRAN Iu Interface: Data Transport and Transport Signalling.
[7] 3GPP Technical Specification 25.415 UTRAN Iu Interface: CN-RAN User Plane Protocol.
[8] 3GPP Technical Specification 25.419 UTRAN Iu Interface: Service Area Broadcast Protocol (SABP).
[9] 3GPP Technical Specification 25.420 UTRAN Iur Interface: General Aspects and Principles.
[10] 3GPP Technical Specification 25.421 UTRAN Iur Interface: Layer 1.
[11] 3GPP Technical Specification 25.422 UTRAN Iur Interface: Signalling Transport.
[12] 3GPP Technical Specification 25.423 UTRAN Iur Interface: RNSAP Signalling.
[13] 3GPP Technical Specification 25.424 UTRAN Iur Interface: Data Transport and Transport Signalling for CCH Data Streams.
[14] 3GPP Technical Specification 25.425 UTRAN Iur Interface: User Plane Protocols for CCH Data Streams.
[15] 3GPP Technical Specification 25.426 UTRAN Iur and Iub Interface Data Transport and Transport Signalling for DCH Data Streams.
[16] 3GPP Technical Specification 25.427 UTRAN Iur and Iub Interface User Plane Protocols for DCH Data Streams.
[17] 3GPP Technical Specification 25.430 UTRAN Iub Interface: General Aspects and Principles.
[18] 3GPP Technical Specification 25.431 UTRAN Iub Interface: Layer 1.
[19] 3GPP Technical Specification 25.432 UTRAN Iub Interface: Signalling Transport.
[20] 3GPP Technical Specification 25.433 UTRAN Iub Interface: NBAP Signalling.
[21] 3GPP Technical Specification 25.434 UTRAN Iub Interface: Data Transport and Transport Signalling for CCH Data Streams.
[22] 3GPP Technical Specification 25.435 UTRAN Iub Interface: User Plane Protocols for CCH Data Streams.
[23] 3GPP Technical Specification 25.450 UTRAN Iupc Interface: General Aspects and Principles.
[24] 3GPP Technical Specification 23.002 Network Architecture, Version 5.5.0, January 2002.
[25] Kaaranen, H. *et al.*, *UMTS Networks: Architecture, Mobility and Services*, John Wiley & Sons, 2001.

6

Physical Layer

Antti Toskala

6.1 Introduction

In this chapter the WCDMA (UTRA FDD) physical layer is described. The physical layer of the radio interface has been typically the main discussion topic when different cellular systems have been compared against each other. The physical layer structures naturally relate directly to the achievable performance issues when observing a single link between a terminal station and a base station. For the overall system performance the protocols in the other layers, such as handover protocols, also have a great deal of impact. Naturally it is essential to have low Signal-to-Interference Ratio (SIR) requirements for sufficient link performance with various coding and diversity solutions in the physical layer, since the physical layer defines the fundamental capacity limits. The performance of the WCDMA physical layer is described in detail in Chapter 12.

The physical layer has a major impact on equipment complexity with respect to the required baseband processing power in the terminal station and base station equipment. As well as the diversity benefits on the performance side, the wideband nature of WCDMA also offers new challenges in its implementation. As third generation systems are wideband from the service point of view as well, the physical layer cannot be designed around only a single service, such as speech; more flexibility is needed for future service introduction. The new requirements of third generation systems for the air interface are summarised in Section 1.4. This chapter presents the WCDMA physical layer solutions to meet those requirements.

This chapter uses the term 'terminal' for the user equipment. The UTRA FDD physical layer specifications are contained in references [1–5].

This chapter has been divided as follows. First, the transport channels are described, together with their mapping to different physical channels, in Section 6.2. Spreading and modulation for uplink and downlink are presented in Section 6.3, and the physical channels for user data and control data are described in Sections 6.4 and 6.5. In Section 6.6 the key physical layer procedures, such as power control and handover measurements, are covered. The biggest change in Release 5 impacting the physical layer is the addition of the high speed downlink packet access (HSDPA) feature. As there are significant differences in

WCDMA for UMTS, third edition.   Edited by Harri Holma and Antti Toskala
© 2004 John Wiley & Sons, Ltd   ISBN: 0-470-87096-6

HSDPA when compared to Release '99 based operation (which is naturally retained as well), the HSDPA details are covered in a separate section to maintain clear separation between the first phase WCDMA standard and the first evolutionary step of the radio interface development. For further details on HSDPA please refer to Chapter 11. The beamforming on the network side has been made more complete by defining the related measurements on the network side. The introduction of Multimedia Broadcast Multicast Service (MBMS), as discussed in Chapter 7, has some relevance to physical layer operation though it uses the existing physical channels and is thus not discussed separately in this chapter. A work item has been started on the High Speed Uplink Packet Access (HSUPA) operation, with principles covered in Chapter 11, which, once completed, will create changes for the physical layer, but whether there will be something done for Release 6 remains to be seen.

6.2   Transport Channels and their Mapping to the Physical Channels

In UTRA the data generated at higher layers is carried over the air with transport channels, which are mapped in the physical layer to different physical channels. The physical layer is required to support variable bit rate transport channels to offer bandwidth-on-demand services, and to be able to multiplex several services to one connection. This section presents the mapping of the transport channels to the physical channels, and how those two requirements are taken into account in the mapping.

Each transport channel is accompanied by the Transport Format Indicator (TFI) at each time event at which data is expected to arrive for the specific transport channel from the higher layers. The physical layer combines the TFI information from different transport channels to the Transport Format Combination Indicator (TFCI). The TFCI is transmitted in the physical control channel to inform the receiver which transport channels are active for the current frame; the exception to this is the use of Blind Transport Format Detection (BTFD) that will be covered in connection with the downlink dedicated channels. The TFCI is decoded appropriately in the receiver and the resulting TFI is given to higher layers for each of the transport channels that can be active for the connection. In Figure 6.1 two transport channels are mapped to a single physical channel, and also error indication is provided for each transport block. The transport channels may have a different number of blocks and at any moment not all the transport channels are necessarily active.

One physical control channel and one or more physical data channels form a single Coded Composite Transport Channel (CCTrCh). There can be more than one CCTrCh on a given connection but only one physical layer control channel is transmitted in such a case.

The interface between higher layers and the physical layer is less relevant for terminal implementation, since basically everything takes place within the same equipment, thus the interfacing here is rather a tool for specification work. For the network side the division of functions between physical and higher layers is more important, since there the interface between physical and higher layers is represented by the Iub-interface between the base station and Radio Network Controller (RNC) as described in Chapter 5. In the 3GPP specification the interfacing between physical layer and higher layers is covered in [6].

Two types of transport channel exist: dedicated channels and common channels. The main difference between them is that a common channel is a resource divided between all or a group of users in a cell, whereas a dedicated channel resource, identified by a certain code on

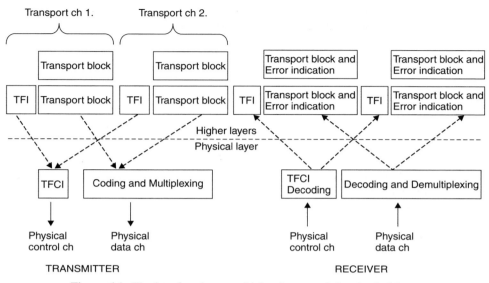

Figure 6.1. The interface between higher layers and the physical layer

a certain frequency, is reserved for a single user only. The transport channels are compared in Section 10.3 for the transmission of packet data.

6.2.1   Dedicated Transport Channel

The only dedicated transport channel is the dedicated channel, for which the term DCH is used in the 25-series of the UTRA specification. The dedicated transport channel carries all the information intended for the given user coming from layers above the physical layer, including data for the actual service as well as higher layer control information. The content of the information carried on the DCH is not visible to the physical layer, thus higher layer control information and user data are treated in the same way. Naturally the physical layer parameters set by UTRAN may vary between control and data.

The familiar GSM channels, the traffic channel (TRCH) or associated control channel (ACCH), do not exist in the UTRA physical layer. The dedicated transport channel carries both the service data, such as speech frames, and higher layer control information, such as handover commands or measurement reports from the terminal. In WCDMA a separate transport channel is not needed because of the support of variable bit rate and service multiplexing.

The dedicated transport channel is characterised by features such as fast power control, fast data rate change on a frame-by-frame basis, and the possibility of transmission to a certain part of the cell or sector with varying antenna weights with adaptive antenna systems. The dedicated channel supports soft handover.

6.2.2   Common Transport Channels

There are six different common transport channel types defined for UTRA in Release '99, which are introduced in the following sections. There are a few differences from second generation systems, for example transmission of packet data on the common channels, and a downlink shared channel for transmitting packet data. Common channels do not have soft handover but some of them can have fast power control. The new transport channel in Release 5, High-speed Downlink Shared Channel (HS-DSCH) is covered in Chapter 11.

6.2.2.1   Broadcast Channel

The Broadcast Channel (BCH) is a transport channel that is used to transmit information specific to the UTRA network or for a given cell. The most typical data needed in every network is the available random access codes and access slots in the cell, or the types of transmit diversity method used with other channels for that cell. As the terminal cannot register to the cell without the possibility of decoding the broadcast channel, this channel is needed for transmission with relatively high power in order to reach all the users within the intended coverage area. From a practical viewpoint, the information rate on the broadcast channel is limited by the ability of low-end terminals to decode the data rate of the broadcast channel, resulting in a low and fixed data rate for the UTRA broadcast channel.

6.2.2.2   Forward Access Channel

The Forward Access Channel (FACH) is a downlink transport channel that carries control information to terminals known to be located in the given cell. This is used, for example, after a random access message has been received by the base station. It is also possible to transmit packet data on the FACH. There can be more than one FACH in a cell. One of the forward access channels must have such a low bit rate that it can be received by all the terminals in the cell area. With more than one FACH, the additional channels can have a higher data rate. The FACH does not use fast power control, and the messages transmitted need to include inband identification information to ensure their correct receipt.

6.2.2.3   Paging Channel

The Paging Channel (PCH) is a downlink transport channel that carries data relevant to the paging procedure, that is, when the network wants to initiate communication with the terminal. The simplest example is a speech call to the terminal: the network transmits the paging message to the terminal on the paging channel of those cells belonging to the location area that the terminal is expected to be in. The identical paging message can be transmitted in a single cell or in up to a few hundred cells, depending on the system configuration. The terminals must be able to receive the paging information in the whole cell area. The design of the paging channel also affects the terminal's power consumption in the standby mode. The less often the terminal has to tune the receiver in to listen for a possible paging message, the longer the terminal's battery will last in standby mode.

6.2.2.4   Random Access Channel

The Random Access Channel (RACH) is an uplink transport channel intended to be used to carry control information from the terminal, such as requests to set up a connection. It can also be used to send small amounts of packet data from the terminal to the network. For proper system operation the random access channel must be heard from the whole desired cell coverage area, which also means that practical data rates have to be rather low, at least for the initial system access and other control procedures.

6.2.2.5 Uplink Common Packet Channel

The uplink common packet channel (CPCH) is an extension to the RACH channel that is intended to carry packet-based user data in the uplink direction. The reciprocal channel providing the data in the downlink direction is the FACH. In the physical layer, the main differences to the RACH are the use of fast power control, a physical layer-based collision detection mechanism and a CPCH status monitoring procedure. The uplink CPCH transmission may last several frames in contrast with one or two frames for the RACH message.

6.2.2.6 Downlink Shared Channel

The downlink shared channel (DSCH) is a transport channel intended to carry dedicated user data and/or control information; it can be shared by several users. In many respects it is similar to the forward access channel, although the shared channel supports the use of fast power control as well as variable bit rate on a frame-by-frame basis. The DSCH does not need to be heard in the whole cell area and can employ the different modes of transmit antenna diversity methods that are used with the associated downlink DCH. The downlink shared channel is always associated with a downlink DCH.

6.2.2.7 Required Transport Channels

The common transport channels needed for basic network operation are RACH, FACH and PCH, while the use of DSCH and CPCH is optional and can be decided by the network.

6.2.3 Mapping of Transport Channels onto the Physical Channels

The different transport channels are mapped to different physical channels, though some of the transport channels are carried by identical (or even the same) physical channel. The transport channel to physical channel mapping is illustrated in Figure 6.2.

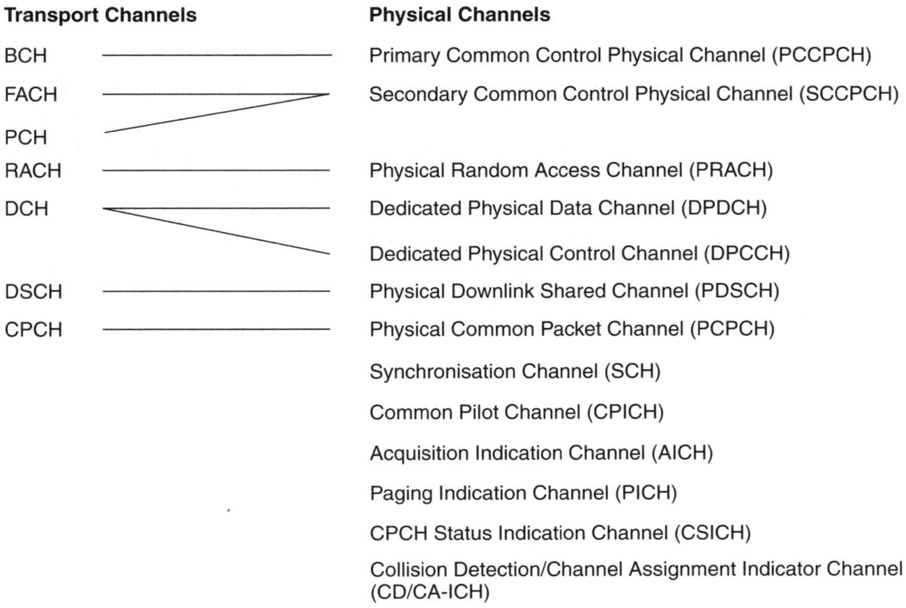

Transport Channels   **Physical Channels**

BCH ——————— Primary Common Control Physical Channel (PCCPCH)

FACH ——————— Secondary Common Control Physical Channel (SCCPCH)

PCH

RACH ——————— Physical Random Access Channel (PRACH)

DCH ——————— Dedicated Physical Data Channel (DPDCH)

Dedicated Physical Control Channel (DPCCH)

DSCH ——————— Physical Downlink Shared Channel (PDSCH)

CPCH ——————— Physical Common Packet Channel (PCPCH)

Synchronisation Channel (SCH)

Common Pilot Channel (CPICH)

Acquisition Indication Channel (AICH)

Paging Indication Channel (PICH)

CPCH Status Indication Channel (CSICH)

Collision Detection/Channel Assignment Indicator Channel (CD/CA-ICH)

Figure 6.2. Transport channel to physical channel mapping

In addition to the transport channels introduced earlier, there exist physical channels to carry only information relevant to physical layer procedures. The Synchronisation Channel (SCH), the Common Pilot Channel (CPICH) and the Acquisition Indication Channel (AICH) are not directly visible to higher layers and are mandatory from the system function point of view, to be transmitted from every base station. The CPCH Status Indication Channel (CSICH) and the Collision Detection/Channel Assignment Indication Channel (CD/CA-ICH) are needed if CPCH is used.

A dedicated channel (DCH) is mapped onto two physical channels. The Dedicated Physical Data Channel (DPDCH) carries higher layer information, including user data, while the Dedicated Physical Control Channel (DPCCH) carries the necessary physical layer control information. These two dedicated physical channels are needed to support efficiently the variable bit rate in the physical layer. The bit rate of the DPCCH is constant, while the bit rate of DPDCH can change from frame to frame.

6.2.4   Frame Structure of Transport Channels

UTRA channels use a 10 ms radio frame structure. The frame structure also employs a longer period, called the system frame period. The System Frame Number (SFN) is a 12-bit number and is used by procedures that span more than a single frame. Physical layer procedures, such as the paging procedure or random access procedure, are examples of procedures that need a longer period than 10 ms for correct definition.

6.3   Spreading and Modulation

6.3.1   Scrambling

The concept of spreading the information in a CDMA system is introduced in Chapter 3. In addition to spreading, part of the process in the transmitter is the scrambling operation. This is needed to separate terminals or base stations from each other. Scrambling is used on top of spreading, so it does not change the signal bandwidth but only makes the signals from different sources separable from each other. With scrambling, it would not matter if the actual spreading were performed with identical codes for several transmitters. Figure 6.3 shows the relationship of the chip rate in the channel to spreading and scrambling in UTRA. As the chip rate is already achieved in spreading by the channelisation codes, the symbol rate is not affected by the scrambling. The concept of channelisation codes is covered in the following section.

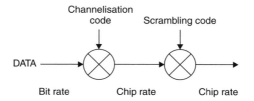

Figure 6.3. Relationship between spreading and scrambling

6.3.2 Channelisation Codes

Transmissions from a single source are separated by channelisation codes, i.e. downlink connections within one sector and the dedicated physical channel in the uplink from one terminal. The spreading/channelisation codes of UTRA are based on the Orthogonal Variable Spreading Factor (OVSF) technique, which was originally proposed in [7].

The use of OVSF codes allows the spreading factor to be changed and orthogonality between different spreading codes of different lengths to be maintained. The codes are picked from the code tree, which is illustrated in Figure 6.4. In case the connection uses a variable spreading factor, the proper use of the code tree also allows despreading according to the smallest spreading factor. This requires only that channelisation codes are used from the branch indicated by the code used for the smallest spreading factor.

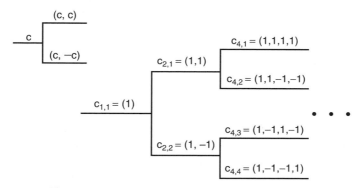

Figure 6.4. Beginning of the channelisation code tree

There are certain restrictions as to which of the channelisation codes can be used for a transmission from a single source. Another physical channel may use a certain code in the tree if no other physical channel to be transmitted using the same code tree is using a code that is on an underlying branch, i.e. using a higher spreading factor code generated from the intended spreading code to be used. Neither can a smaller spreading factor code on the path to the root of the tree be used. The downlink orthogonal codes within each base station are managed by the radio network controller (RNC) in the network.

The functionality and characteristics of the scrambling and channelisation codes are summarised in Table 6.1. Their usage will be described in more detail in Section 6.3.3.

The definition for the same code tree means that for transmission from a single source, from either a terminal or a base station, one code tree is used with one scrambling code on top of the tree. This means that different terminals and different base stations may operate their code trees totally independently of each other; there is no need to coordinate the code tree resource usage between different base stations or terminals.

6.3.3 Uplink Spreading and Modulation

6.3.3.1 Uplink Modulation

In the uplink direction there are basically two additional terminal-oriented criteria that need to be taken into account in the definition of the modulation and spreading methods. The

Table 6.1. Functionality of the channelisation and scrambling codes

	Channelisation code	Scrambling code
Usage	Uplink: Separation of physical data (DPDCH) and control channels (DPCCH) from same terminal Downlink: Separation of downlink connections to different users within one cell	Uplink: Separation of terminal Downlink: Separation of sectors (cells)
Length	4–256 chips (1.0–66.7 μs) Downlink also 512 chips	Uplink: (1) 10 ms = 38 400 chips or (2) 66.7 μs = 256 chips. Option (2) can be used with advanced base station receivers Downlink: 10 ms = 38 400 chips
Number of codes	Number of codes under one scrambling code = spreading factor	Uplink: Several millions Downlink: 512
Code family	Orthogonal Variable Spreading Factor	Long 10 ms code: Gold code Short code: Extended S(2) code family
Spreading	Yes, increases transmission bandwidth	No, does not affect transmission bandwidth

uplink modulation should be designed so that the terminal amplifier efficiency is maximised and/or the audible interference from the terminal transmission is minimised.

Discontinuous uplink transmission can cause audible interference to audio equipment that is very close to the terminal, such as hearing aids. This is a completely separate issue from the interference in the air interface. The audible interference is only a nuisance for the user and does not affect network performance, such as its capacity. With GSM operation we are familiar with the occasional audible interference with audio equipment that is not properly protected. The interference from GSM has a frequency of 217 Hz, which is determined by the GSM frame frequency. This interference falls into the band that can be heard by the human ear. With a CDMA system, the same issues arise when discontinuous uplink transmission is used, for example with a speech service. During the silent periods no information bits need to be transmitted, only the information for link maintenance purposes, such as power control with a 1.5 kHz command rate. With such a rate the transmission of the pilot and the power control symbols with time multiplexing in the uplink direction would cause audible interference in the middle of the telephony voice frequency band. Therefore, in a WCDMA uplink the two dedicated physical channels are not time multiplexed but I-Q/code multiplexing is used.

The continuous transmission achieved with an I-Q/code multiplexed control channel is shown in Figure 6.5. Now, as the pilot and the power control signalling are maintained on a

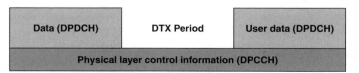

Figure 6.5. Parallel transmission of DPDCH and DPCCH when data is present/absent (DTX)

separate continuous channel, no pulsed transmission occurs. The only pulse occurs when the data channel DPDCH is switched on and off, but such switching happens quite seldom. The average interference to other users and the cellular capacity remain the same as in the time-multiplexed solution. In addition, the link level performance is the same in both schemes if the energy allocated to the pilot and the power control signalling is the same.

For the best possible power amplifier efficiency, the terminal transmission should have as low peak-to-average (PAR) ratio as possible to allow the terminal to operate with a minimal amplifier back-off requirement, mapping directly to the amplifier power conversion efficiency, which in turn is directly proportional to the terminal talk time. With the I-Q/code multiplexing, also called dual-channel QPSK modulation, the power levels of the DPDCH and DPCCH are typically different, especially as data rates increase, and would lead in extreme cases to BPSK-type transmission when transmitting the branches independently. This has been avoided by using a complex-valued scrambling operation after the spreading with channelisation codes.

The signal constellation of the I-Q/code multiplexing before complex scrambling is shown in Figure 6.6. The same constellation is obtained after descrambling in the receiver for the data detection.

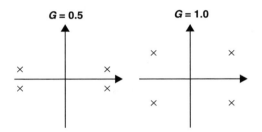

Figure 6.6. Constellation of I-Q/code multiplexing before complex scrambling. G denotes the relative gain factor between DPCCH and DPDCH branches

The transmission of two parallel channels, DPDCH and DPCCH, leads to multicode transmission, which increases the peak-to-average power ratio (crest factor). In Figure 6.6 the peak-to-average ratio changes when G (the relative strengths of the DPDCH and DPCCH) is changed. By using the spreading modulation solution shown in Figure 6.7 the transmitter power amplifier efficiency remains the same as for normal balanced QPSK transmission in general. The complex scrambling codes are formed in such a way that the rotations between consecutive chips within one symbol period are limited to $\pm 90°$. The full $180°$ rotation can happen only between consecutive symbols. This method further reduces the peak-to-average ratio of the transmitted signal from the normal QPSK transmission.

The efficiency of the power amplifier remains constant irrespective of the power difference G between DPDCH and DPCCH. This can be explained with Figure 6.8, which shows the signal constellation for the I-Q/code multiplexed control channel with complex spreading. In the middle constellation with $G = 0.5$ the possible constellation points are only circles or only crosses during one symbol period. Their constellation is the same as for rotated QPSK. Thus, the signal envelope variations with complex spreading are

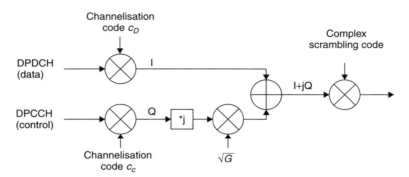

Figure 6.7. I-Q/code multiplexing with complex scrambling

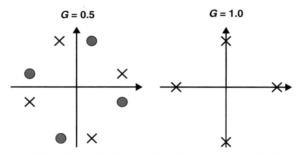

Figure 6.8. Signal constellation for I-Q/code multiplexed control channel with complex scrambling. G denotes the power difference between DPCCH and DPDCH

very similar to QPSK transmission for all values of G. The I-Q/code multiplexing solution with complex scrambling results in power amplifier output back-off requirements that remain constant as a function of the power difference between DPDCH and DPCCH.

The power difference between DPDCH and DPCCH has been quantified in UTRA physical layer specifications to 4-bit words, i.e. 16 different values. At a given point in time the gain value for either DPDCH or DPCCH is set to 1 and then for the other channel a value between 0 and 1 is applied to reflect the desired power difference between the channels. Limiting the number of possible values to 4-bit representation is necessary to make the terminal transmitter implementation simple. The power differences can have 15 different values between −23.5 dB and 0.0 dB and one bit combination for no DPDCH when there is no data to be transmitted.

UTRA will face challenges in amplifier efficiency when compared to GSM. The GSM modulation is GMSK (Gaussian Minimum Shift Keying) which has a constant envelope and is thus optimised for amplifier peak-to-average ratio. As a narrowband system, the GSM signal can be spread relatively more widely in the frequency domain. This allows the use of a less linear amplifier with better power conversion efficiency. Narrowband amplifiers are also easier to linearise if necessary. In practice, the efficiency of the WCDMA power amplifier is slightly lower than that of the GSM power amplifier. On the other hand, WCDMA uses fast power control in the uplink, which reduces the average required transmission power.

Instead of applying combined I-Q and code multiplexing with complex scrambling, it would be possible to use pure code multiplexing. With code multiplexing, multicode transmission occurs with parallel control and data channels. This approach increases transmitted signal envelope variations and sets higher requirements for power amplifier linearity. Especially for low bit rates, as for speech, the control channel can have an amplitude more than 50 % of the data channel, which causes more envelope variations than the combined I-Q/code multiplexing solution.

6.3.3.2 Uplink Spreading

For the uplink DPCCH spreading code, there is an additional restriction. The same code cannot be used by any other code channel even on a different I or Q branch. The reason for this restriction is that physical channels transmitted with the same channelisation codes on I and Q branches with the dual channel QPSK principle cannot be separated before the DPCCH has been detected and channel phase estimates are available. This causes the restriction that, with multicode transmission for DPDCH, the number of parallel spreading codes possible to allocate to DPDCH is six and not eight, when considering the spreading factor of 4 (which would be used in the case of DPDCH multicode transmission).

In the uplink direction the spreading factor on the DPDCH may vary on a frame-by-frame basis. The spreading codes are always taken from the earlier described code tree. When the channelisation code used for spreading is always taken from the same branch of the code tree, the despreading operation can take advantage of the code tree structure and avoid chip-level buffering. The terminal provides data rate information, or more precisely the Transport Format Combination Indicator (TFCI), on the DPCCH, to allow data detection with a variable spreading factor on the DPDCH.

6.3.3.3 Uplink Scrambling Codes

The transmissions from different sources are separated by the scrambling codes. In the uplink direction there are two alternatives: short and long scrambling codes. The long codes with 25 degree generator polynomials are truncated to the 10 ms frame length, resulting in 38 400 chips with 3.84 Mcps. The short scrambling code length is 256 chips. The long scrambling codes are used if the base station uses a Rake receiver. The Rake receiver is described in Section 3.4. If advanced multiuser detectors or interference cancellation receivers are used in the base station, short scrambling codes can be used to make the implementation of the advanced receiver structures easier. The base station multiuser detection algorithms are introduced in Section 11.5.2. Both of the two scrambling code families contain millions of scrambling codes, thus, in the uplink direction, code planning is not needed.

The short scrambling codes have been chosen from the extended S(2) code family. The long codes are Gold codes. The complex-valued scrambling sequence is formed in the case of short codes by combining two codes, and in the case of long codes from a single sequence where the other sequence is the delayed version of the first one.

The complex-valued scrambling code can be formed from two real-valued codes c_1 and c_2 with the decimation principle as:

$$c_{scrambling} = c_1(w_0 + jc_2)(2k)w_1), \quad k = 0, 1, 2 \ldots \tag{6.1}$$

with sequences w_0 and w_1 given as chip rate sequences:

$$w_0 = \{1\,1\}, w_1 = \{1 - 1\} \tag{6.2}$$

The decimation factor with the second code is 2. This way of creating the scrambling codes will reduce the zero crossings in the constellation and will further reduce the amplitude variations in the modulation process.

6.3.3.4 Spreading and Modulation on Uplink Common Channels

The Random Access Channel (RACH) contains preambles that are sent using the same scrambling code sequence as with the uplink transmission, the difference being that only 4096 chips from the beginning of the code period are needed and the modulation state transitions are limited in a different way. The spreading and scrambling process on the RACH is BPSK-valued, thus only one sequence is used to spread and scramble both the in-phase and quadrature branches. This has been chosen to reduce the complexity of the required matched filter in the base station receivers for RACH reception.

The RACH message part spreading and modulation, including scrambling, is identical to that for the dedicated channel. The codes available for RACH scrambling use are transmitted on the BCH of each cell.

For the peak-to-average reduction, an additional rotation function is used on the RACH preamble, given as:

$$b(k) = a(k)e^{j\left(\frac{\pi}{4}+\frac{\pi}{2}k\right)}, \qquad k = 0, 1, 2, \ldots, 4095 \tag{6.3}$$

where $a(k)$ is the binary preamble and $b(k)$ is the resulting complex-valued preamble with limited 90° phase transition between chips. The autocorrelation properties are not affected by this operation.

The RACH preambles have a modulation pattern on top of them, called signature sequences. These have been defined by taking the higher Doppler frequencies as well as frequency errors into account. The sequences have been generated from 16 symbols, which have additionally been interleaved over the preamble duration to avoid large inter-sequence cross-correlations in case of large frequency errors that could otherwise severely degrade the cross-correlation properties between the signature sequences. The 16 signature sequences have been specified for RACH use, but there can be multiple scrambling codes each using the same set of signatures.

The CPCH spreading and modulation are identical to those of the RACH in order to maximise the commonality for both terminal and base station implementation when supporting CPCH. RACH and CPCH processes will be described in more detail in connection with the physical layer procedures.

6.3.4 Downlink Spreading and Modulation

6.3.4.1 Downlink Modulation

In the downlink direction normal QPSK modulation has been chosen with time-multiplexed control and data streams. The time-multiplexed solution is not used in the uplink because it would generate audible interference during discontinuous transmission. The audible interference generated with DTX is not a relevant issue in the downlink since the common channels have continuous transmission in any case. Also, as there exist several parallel code transmissions in the downlink, similar optimisation for peak-to-average (PAR) ratio as with single code (pair) transmission is not relevant. Also, reserving a channelisation code just for DPCCH purposes results in slightly worse code resource utilisation when sending several transmissions from a single source.

Since the I and Q branches have equal power, the scrambling operation does not provide a similar difference to the envelope variations as in the uplink. The discontinuous transmission is implemented by gating the transmission on and off.

6.3.4.2 Downlink Spreading

The spreading in the downlink is based on the channelisation codes, as in the uplink. The code tree under a single scrambling code is shared by several users; typically only one scrambling code and thus only one code tree is used per sector in the base station. The common channels and dedicated channels share the same code tree resource. There is one exception for the physical channels: the synchronisation channel (SCH), which is not under a downlink scrambling code. The SCH spreading codes are covered in a later section.

In the downlink, the dedicated channel spreading factor does not vary on a frame-by-frame basis; the data rate variation is taken care of with either a rate matching operation or with discontinuous transmission, where the transmission is off during part of the slot.

In the case of multicode transmission for a single user, the parallel code channels have different channelisation codes and are under the same scrambling code as normally are all the code channels transmitted from the base station. The spreading factor is the same for all the codes with multicode transmission. Each coded composite transport channel (CCTrCh) may have a different spreading factor, even if received by the same terminal. As in the downlink, normal QPSK modulation is used, the number of spreading codes available (under the same scrambling code) is equal to the spreading factor. If we consider the smallest spreading factor of 4, then at most four of those codes would be available but, due to the common channel requirements for code space, then at most three codes could be allocated for a particular terminal. The number of bits then would be roughly equal to the six codes possible for the uplink, as each QPSK symbol carries two bits.

The special case in the downlink direction is the downlink shared channel (DSCH) which may use a variable spreading factor on a frame-by-frame basis. In this case the channelisation codes taking care of the spreading are allocated from the same branch of the code tree to ease the terminal implementation. The restriction specified is illustrated in Figure 6.9 which shows the spreading factor for maximum data rate and the part of the code tree that may be used by the network to allocate codes when the lower data rate is needed. In such a frame-by-frame operation, the DPCCH of the dedicated channel contains the TFCI information,

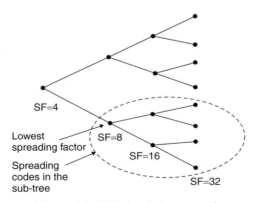

Figure 6.9. DSCH code tree example

which informs the receiver of the spreading code used, as well as other transport format parameters for the DSCH.

6.3.4.3 Downlink Scrambling

The downlink scrambling uses long codes, the same Gold codes as in the uplink. The complex-valued scrambling code is formed from a single code by simply having a delay between the I and Q branches. The code period is truncated to 10 ms; no short codes are used in the downlink direction. The downlink set of the (primary) scrambling codes is limited to 512 codes, otherwise the cell search procedure described in the physical layer procedures section would become excessive. The scrambling codes must be allocated to the sectors in the network planning. Because the number of scrambling codes is so high, the scrambling code planning is a trivial task and can be done automatically by the network planning tool. The 512 primary scrambling codes are expected to be enough from the cell planning perspective, especially as the secondary scrambling codes can be used in the case of beam steering, as used on dedicated channels. This allows the capacity to evolve with adaptive antenna techniques without consuming extra primary scrambling codes and causing problems for downlink code planning.

The actual code period is very long with the 18-degree code generator, but only the first 38 400 chips are used. Limiting the code period was necessary from the system perspective; the terminals would have difficulty in finding the correct code phase with a code period spanning several frames and 512 different codes to choose from.

The secondary downlink scrambling codes can be applied with the exception of those common channels that need to be heard in the whole cell and/or prior to the initial registration. Only one scrambling code should be used per cell or sector to maintain the orthogonality between different downlink code channels. With adaptive antennas the beams provide additional spatial isolation and the orthogonality between different code channels is less important. However, in all cases the best strategy is still to keep as many users as possible under a single scrambling code to minimise downlink interference. If a secondary scrambling code needs to be introduced in the cell, then only those users not fitting under the primary scrambling code should use the secondary code. The biggest loss in orthogonality occurs when the users are shared evenly between two different scrambling codes.

6.3.4.4 Synchronisation Channel Spreading and Modulation

The downlink synchronisation channel (SCH) is a special type of physical channel that is not visible above the physical layer. It contains two channels, primary and secondary SCHs. These channels are utilised by the terminal to find the cells, and are not under the cell-specific primary scrambling code. The terminal must be able to synchronise to the cell before knowing the downlink scrambling code.

The primary SCH contains a code word with 256 chips, with an identical code word in every cell. The primary SCH code word is sent without modulation on top. The code word is constructed from shorter 16-chip sequences in order to optimise the required hardware at the terminal. When detecting this sequence there is normally no prior timing information available and typically a matched filter is needed for detection. Therefore, for terminal complexity and power consumption reasons, it was important to optimise this synchronisation sequence for low-complexity matched filter implementation.

The secondary SCH code words are similar sequences but vary from one base station to another, with a total of 16 sequences in use. These 16 sequences are used to generate a total of 64 different code words which identify to which of the 64 code groups a base station belongs. Like the primary SCH, the secondary SCH is not under the base station-specific scrambling code, but the code sequences are sent without scrambling on top. The SCH code words contain modulation to indicate the use of open loop transmit diversity on the BCH. The SCH itself can use time-switched transmit antenna diversity (TSTD) and is the only channel in UTRA FDD that uses TSTD.

6.3.5   Transmitter Characteristics

The pulse shaping method applied to the transmitted symbols is root-raised cosine filtering with a roll-off factor of 0.22. The same roll-off is valid for both the terminals and the base stations. There are a few other key RF parameters that are introduced here and that have an essential impact on the implementation, as well as on system behaviour.

The nominal carrier spacing in WCDMA is 5 MHz but the carrier frequency in WCDMA can be adjusted with a 200 kHz raster. The central frequency of each WCDMA carrier is indicated with an accuracy of 200 kHz. The target of this adjustment is to provide more flexibility for channel spacing within the operator's band.

The Adjacent Channel Leakage Ratio (ACLR) determines how much of the transmitted power is allowed to leak into the first or second neighbouring carrier. The concept of ACLR is illustrated in Figure 6.10, where $ACLR_1$ and $ACLR_2$ correspond to the power level integrated over the first and second adjacent carriers, with 5 MHz and 10 MHz carrier separation respectively. No separate values are specified for other values of carrier spacing.

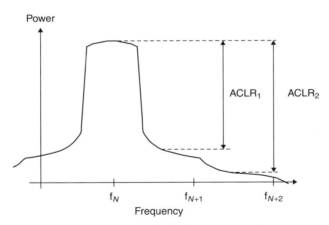

Figure 6.10. Adjacent Channel Leakage Ratio for the first and second adjacent carriers

On the terminal side the ACLR values for the power classes of 21 dBm and 24 dBm have been set to 33 dB and 43 dB for $ACLR_1$ and $ACLR_2$ respectively. On the base station side the corresponding values are 45 dB and 50 dB. In the first phase of network deployment it is also likely that most terminals will belong to the 21 dBm power class and the network needs to be planned accordingly.

The higher the ACLR requirement, the more linearity is required from the power amplifier and the lower is the efficiency of the amplifier. The terminal needs to have a value that allows a power-efficient amplifier. The impact of the ACLR on system performance is studied in Section 8.5.

The frequency accuracy requirements are also directly related to the implementation cost, especially on the terminal side. The terminal frequency accuracy has been defined to be ±0.1 ppm when compared to the received carrier frequency. On the base station side the requirement is tighter: ±0.05 ppm. The baseband timing is tied to the same timing reference as RF. The base station value needs to be tighter than the terminal value, since the base station carrier frequency is the reference for the terminal accuracy. The terminal needs also to be able to search the total frequency uncertainty area caused by the base station frequency error tolerance on top of the terminal tolerances and the error caused by terminal movement. With the 200 kHz carrier raster the looser base station frequency accuracy would start to cause problems. In 3GPP the RF parameters for terminals are specified in [8] and for base stations in [9].

6.4   User Data Transmission

For user data transmission in second generation systems, such as the first versions of GSM, typically only one service has been active at a time, either voice or low-rate data. From the beginning, the technology base has required that the physical layer implementation be defined to the last detail without real flexibility. For example, puncturing patterns in GSM have been defined bit by bit, whereas such a definition for all possible service combinations and data rates is simply not possible for UTRA. Instead, algorithms for generating such patterns are defined. Signal processing technology has also evolved greatly, thus there is no longer a need to have items like puncturing on hardware as in the early days of GSM hardware development. For the circuit switched (CS) traffic (e.g. speech and video), a transmission dedicated channel needs to be used, while for packet data there are additional choices available, RACH and CPCH for the uplink and FACH and DSCH for the downlink.

6.4.1   Uplink Dedicated Channel

As described earlier, the uplink direction uses I-Q/code multiplexing for user data and physical layer control information. The physical layer control information is carried by the Dedicated Physical Control Channel (DPCCH) with a fixed spreading factor of 256. The higher layer information, including user data, is carried on one or more Dedicated Physical Data Channels (DPDCHs), with a possible spreading factor ranging from 256 down to 4. The uplink transmission may consist of one or more Dedicated Physical Data Channels (DPDCH) with a variable spreading factor, and a single Dedicated Physical Control Channel (DPCCH) with a fixed spreading factor.

The DPDCH data rate may vary on a frame-by-frame basis. Typically with a variable rate service the DPDCH data rate is informed on the DPCCH. The DPCCH is transmitted continuously and rate information is sent with the Transport Format Combination Indicator (TFCI), the DPCCH information on the data rate on the current DPDCH frame. If the TFCI is not decoded correctly, the whole data frame is lost. Because the TFCI indicates the transport format of the same frame, the loss of the TFCI does not affect any other frames. The reliability of the TFCI is higher than the reliability of the user data detection on the DPDCH. Therefore, the loss of the TFCI is a rare event. Figure 6.11 illustrates the uplink dedicated channel structure in more detail.

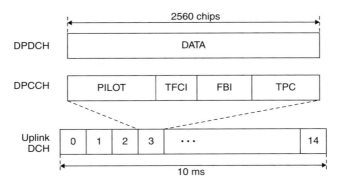

Figure 6.11. Uplink dedicated channel structure

The uplink DPCCH uses a slot structure with 15 slots over the 10 ms radio frame. This results in a slot duration of 2560 chips or about 666 μs. This is actually rather close to the GSM burst duration of 577 μs. Each slot has four fields to be used for pilot bits, TFCI, Transmission Power Control (TPC) bits and Feedback Information (FBI) bits. The pilot bits are used for the channel estimation in the receiver, and the TPC bits carry the power control commands for the downlink power control. The FBI bits are used when closed loop transmission diversity is used in the downlink. The use of FBI bits is covered in the physical layer procedures section. There exist a total of six slot structures for uplink DPCCH. The different options are 0, 1 or 2 bits for FBI bits and these same alternatives with and without TFCI bits. The TPC and pilot bits are always present and their number varies in such a way that the DPCCH slot is always fully used.

It is beneficial to transmit with a single DPDCH for as long as possible, for reasons of terminal amplifier efficiency, because multicode transmission increases the peak-to-average ratio of the transmission, which reduces the efficiency of the terminal power amplifier. The maximum user data rate on a single code is derived from the maximum channel bit rate, which is 960 kbps with spreading factor 4. With channel coding the practical maximum user data rate for the single code case is in the order of 400–500 kbps.

When higher data rates are needed, parallel code channels are used. This allows up to six parallel codes to be used (as explained in Section 6.3.3.2), raising the channel bit rate for data transmission up to 5740 kbps, which can accommodate 2 Mbps user data or an even higher data rate if the coding rate is $\frac{1}{2}$. Therefore, it is possible to offer a user data rate of 2 Mbps even after retransmission. The achievable data rates with different spreading factors are presented in Table 6.2. The rates given assume $\frac{1}{2}$-rate coding and do not include bits taken for coder tail bits or the Cyclic Redundancy Check (CRC). The relative overhead due to tail bits and CRC bits has significance only with low data rates.

The uplink receiver in the base station needs to perform typically the following tasks when receiving the transmission from a terminal:

- The receiver starts receiving the frame and despreading the DPCCH and buffering the DPDCH according to the maximum bit rate, corresponding to the smallest spreading factor.

Table 6.2. Uplink DPDCH data rates

DPDCH spreading factor	DPDCH channel bit rate (kbps)	Maximum user data rate with $\frac{1}{2}$-rate coding (approx.)
256	15	7.5 kbps
128	30	15 kbps
64	60	30 kbps
32	120	60 kbps
16	240	120 kbps
8	480	240 kbps
4	960	480 kbps
4, with 6 parallel codes	5740	2.8 Mbps

- For every slot:

  - obtain the channel estimates from the pilot bits on the DPCCH;
  - estimate the SIR from the pilot bits for each slot;
  - send the TPC command in the downlink direction to the terminal to control its uplink transmission power;
  - decode the TPC bit in each slot and adjust the downlink power of that connection accordingly.

- For every second or fourth slot:

  - decode the FBI bits, if present, over two or four slots and adjust the diversity antenna phases, or phases and amplitudes, depending on the transmission diversity mode.

- For every 10 ms frame:

  - decode the TFCI information from the DPCCH frame to obtain the bit rate and channel decoding parameters for DPDCH.

- For Transmission Time Interval (TTI, interleaving period) of 10, 20, 40 or 80 ms:

  - decode the DPDCH data.

  The same functions are valid for the downlink as well, with the following exceptions:

- In the downlink the dedicated channel spreading factor is constant, as well as with the common channels. The only exception is the Downlink Shared Channel (DSCH) which also has a varying spreading factor.

- The FBI bits are not in use in the downlink direction.

- There is a common pilot channel available in addition to the pilot bits on DPCCH. The common pilot can be used to aid the channel estimation.

- In the downlink transmission may occur from two antennas in the case of transmission diversity. The receiver does the channel estimation from the pilot patterns sent from two antennas and consequently accommodates the despread data sent from two different antennas. The overall impact on the complexity is small, however.

6.4.2 Uplink Multiplexing

In the uplink direction the services are multiplexed dynamically so that the data stream is continuous with the exception of zero rate. The symbols on the DPDCH are sent with equal power level for all services. This means in practice that the service coding and channel multiplexing needs, in some cases, to adjust the relative symbol rates for different services in order to balance the power level requirements for the channel symbols. The rate matching function in the multiplexing chain in Figure 6.12 can be used for such quality balancing operations between services on a single DPDCH. For the uplink DPDCH there do not exist fixed positions for different services, but the frame is filled according to the outcome of the rate matching and interleaving operation(s). The uplink multiplexing is done in 11 steps, as illustrated in Figure 6.12.

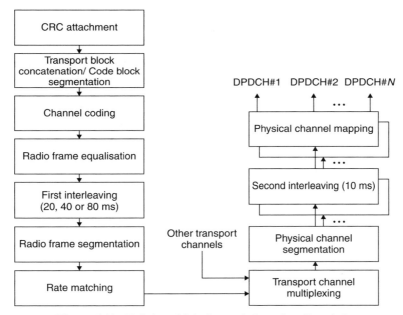

Figure 6.12. Uplink multiplexing and channel coding chain

After receiving a transport block from higher layers, the first operation is CRC attachment. The CRC (Cyclic Redundancy Check) is used for error checking of the transport blocks at the receiving end. The CRC length that can be inserted has four different values: 0, 8, 12, 16 and 24 bits. The more bits the CRC contains, the lower is the probability of an undetected error in the transport block in the receiver. The physical layer provides the transport block to higher layers together with the error indication from the CRC check.

After the CRC attachment, the transport blocks are either concatenated together or segmented to different coding blocks. This depends on whether the transport block fits the available code block size as defined for the channel coding method. The benefit of the concatenation is better performance in terms of lower overhead due to encoder tail bits and, in some cases, due to better channel coding performance because of the larger block size. On the other hand, code block segmentation allows the avoidance of excessively large code blocks that could also be a complexity issue. If the transport block with CRC attached does not fit into the maximum available code block, it will be divided into several code blocks.

The channel encoding is performed on the coding blocks after the concatenation or segmentation operation. Originally it was considered to have the possibility to send data without any channel coding, as is done with AMR class C bits in GSM, but that was removed at a later stage as there was no real need identified.

The function of radio frame equalisation is to ensure that data can be divided into equal-sized blocks when transmitted over more than a single 10 ms radio frame. This is done by padding the necessary number of bits until the data can be in equal-sized blocks per frame.

The first interleaving or inter-frame interleaving is used when the delay budget allows more than 10 ms of interleaving. The interlayer length of the first interleaving has been defined to be 20, 40 and 80 ms. The interleaving period is directly related to the Transmission Time Interval (TTI), which indicates how often data arrives from higher layers to the physical layer. The start positions of the TTIs for different transport channels multiplexed together for a single connection are time aligned. The TTIs have a common starting point, i.e. a 40 ms TTI goes in twice, even for an 80 ms TTI on the same connection. This is necessary to limit the possible transport format combinations from the signalling perspective. The timing relation with different TTIs is illustrated in Figure 6.13. If the first interleaving is used, the frame segmentation will distribute the data coming from the first interleaving over two, four or eight consecutive frames in line with the interleaving length.

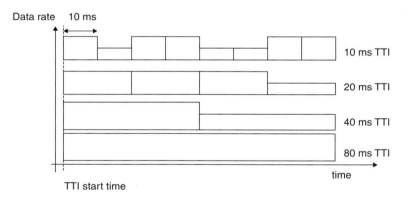

Figure 6.13. TTI start time relationship with different TTIs on a single connection

Rate matching is used to match the number of bits to be transmitted to the number available on a single frame. This is achieved either by puncturing or by repetition. In the uplink direction, repetition is preferred, and basically the only reason why puncturing is used is when facing the limitations of the terminal transmitter or base station receiver. Another reason for puncturing is to avoid multicode transmission. The rate matching operation in Figure 6.12 needs to take into account the number of bits coming from the other transport channels that are active in that frame. The uplink rate matching is a dynamic operation that may vary on a frame-by-frame basis. When the data rate of the service with lowest TTI varies, as in Figure 6.13, the dynamic rate matching adjusts the rate matching parameters for other transport channels as well, so that all the symbols in the radio frame are used. For example, if with two transport channels the other has momentarily zero rate, rate matching increases the symbol rate for the other service sufficiently so that all uplink channel symbols are used, assuming that the spreading factor would stay the same.

The higher layers provide a semi-static parameter, the rate matching attribute, to control the relative rate matching between different transport channels. This is used to calculate the rate matching value when multiplexing several transport channels for the same frame. When this rule is applied as specified, with the aid of the rate matching attribute and TFCI the receiver can calculate backwards the rate matching parameters used and perform the inverse operation. By adjusting the rate matching attribute, the quality of different services can be fine-tuned to reach an equal or near-equal symbol power level requirement.

The different transport channels are multiplexed together by the transport channel multiplexing operation. This is a simple serial multiplexing on a frame-by-frame basis. Each transport channel provides data in 10 ms blocks for this multiplexing. In case more than one physical channel (spreading code) is used, physical channel segmentation is used. This operation simply divides the data evenly on the available spreading codes, as currently no cases have been specified where the spreading factors would be different in multicode transmissions. The use of serial multiplexing also means that with multicode transmission the lower rates can be implemented by sending fewer codes than with the full rate.

The second interleaving performs 10 ms radio frame interleaving, sometimes called intra-frame interleaving. This is a block interleaver with inter-column permutations applied to the 30 columns of the interleaver. It is worth noting that the second interleaving is applied separately for each physical channel, in case more than a single code channel is used. From the output of the second interleaver the bits are mapped on the physical channels. The number of bits given for a physical channel at this stage is exactly the number that the spreading factor of that frame can transmit. Alternatively, the number of bits to transmit is zero and the physical channel is not transmitted at all.

6.4.3   User Data Transmission with the Random Access Channel

In addition to the uplink dedicated channel, user data can be sent on the Random Access Channel (RACH), mapped on the Physical Random Access Channel (PRACH). This is intended for low data rate operation with packet data where continuous connection is not maintained. In the RACH message it will be possible to transmit with a limited set of data rates based on prior negotiations with the UTRA network. The RACH operation does not include power control; thus the validity of the power level obtained with the PRACH power ramping procedure will be valid only for a short period, over one or two frames at most, depending on the environment.

The PRACH has, as a specific feature, preambles that are sent prior to data transmission. These use a spreading factor of 256 and contain a signature sequence of 16 symbols, resulting in a total length of 4096 chips for the preamble. Once the preamble has been detected and acknowledged with the Acquisition Indicator Channel (AICH), the 10 ms (or 20 ms) message part is transmitted. The spreading factor for the message part may vary from 256 up to 32 depending on the transmission needs, but is subject to prior agreement with the UTRA network. Additionally, the 20 ms message length has been defined for range improvement reasons. The AICH structure is covered in the signalling part, while the RACH procedure is covered in detail in the physical layer procedures section.

6.4.4 Uplink Common Packet Channel

As well as the previously covered user data transmission methods, an extension for RACH has been defined. The main differences in the uplink from RACH data transmission are the reservation of the channel for several frames and the use of fast power control, which is not needed with RACH when sending only one or two frames. The uplink Common Packet Channel (CPCH) has, as a pair, the DPCCH in the downlink direction, providing fast power control information. Also the network has an option to tell the terminals to send an 8-slot power control preamble before the actual message transmission. This is beneficial in some cases as it allows the power control to converge before the actual data transmission starts.

The higher layer downlink signalling to a terminal using uplink CPCH is provided by the Forward Access Channel (FACH). The main reason for not using the DPDCH of the dedicated channel carrying the DPCCH for that is that the CPCH is a fast set-up and fast release channel, handled similarly to RACH reception by the physical layer at the base station site. The DPDCH content is taken care of by the higher layer signalling protocols, which are located in a Radio Network Controller (RNC). In case the RNC wants to send a signalling message for a terminal as a response to CPCH activity, an ARQ message for example, the CPCH connection might have already been terminated by the base station. The differences in uplink CPCH operation from the RACH procedure are covered in the physical layer procedures section in more detail.

6.4.5 Downlink Dedicated Channel

The downlink dedicated channel is transmitted on the Downlink Dedicated Physical Channel (Downlink DPCH). The Downlink DPCH applies time multiplexing for physical control information and user data transmission, as shown in Figure 6.14. As in the uplink, the terms Dedicated Physical Data Channel (DPDCH) and Dedicated Physical Control Channel (DPCCH) are used in the 3GPP specification for the downlink dedicated channels.

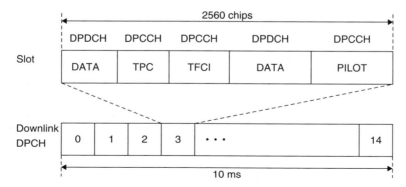

Figure 6.14. Downlink Dedicated Physical Channel (Downlink DPCH) control/data multiplexing

The spreading factor for the highest transmission rate determines the channelisation code to be reserved from the given code tree. The variable data rate transmission may be implemented in two ways:

- In case TFCI is not present, the positions for the DPDCH bits in the frame are fixed. As the spreading factor is also always fixed in the Downlink DPCH, the lower rates are implemented with Discontinuous Transmission (DTX) by gating the transmission on/off. Since this is done on the slot interval, the resulting gating rate is 1500 Hz. As in the uplink, there are 15 slots per 10 ms radio frame; this determines the gating rate. The data rate, in case of more than one alternative, is determined with Blind Transport Format Detection (BTFD) which is based on the use of a guiding transport channel or channels that have different CRC positions for different Transport Format Combinations (TFCs). For a terminal it is mandatory to have BTFD capability with relatively low rates only, such as with AMR speech service. With higher data rates also the benefits from avoiding the TFCI overhead are insignificant and the complexity of BTFD rates starts to increase.

- With TFCI available it is also possible to use flexible positions, and it is up to the network to select which mode of operation is used. With flexible positions it is possible to keep continuous transmission and implement the DTX with repetition of the bits. In such a case the frame is always filled as in the uplink direction.

The downlink multiplexing chain in Figure 6.16 (Section 6.4.6) is also impacted by the DTX, the DTX indication having been inserted before the first interleaving.

In the downlink the spreading factors range from 4 to 512, with some restrictions on the use of spreading factor 512 in the case of soft handover. The restrictions are due to the timing adjustment step of 256 chips in soft handover operation, but in any case the use of a spreading factor of 512 for soft handover is not expected to occur very often. Typically, such a spreading factor is used to provide information on power control, etc. when providing services with minimal downlink activity, as with file uploading and so on. This is also the case with the CPCH where power control information for the limited duration uplink transmission is provided with a DPCCH with spreading factor 512. In such a case soft handover is not needed either.

Modulation causes some differences between the uplink and downlink data rates. While the uplink DPDCH consists of BPSK symbols, the downlink DPDCH consists of QPSK symbols, each carrying two bits. As the BPSK symbols carry only one bit per symbol, use of the same spreading ratio in uplink and downlink DPDCH gives a double data rate in the downlink direction, especially at higher data rates where time multiplexed DPCCH overhead is very small. These downlink data rates are given in Table 6.3 with raw bit rates calculated from the QPSK-valued symbols in the downlink reserved for data use.

The Downlink DPCH can use either open loop or closed loop transmit diversity to improve performance. The use of such enhancements is not required from the network side but is mandatory in terminals. It was made mandatory as it was felt that this kind of feature had a strong relation to such issues as network planning and system capacity, so it was made a baseline implementation capability. The open loop transmit diversity coding principle is shown in Figure 6.15, where the information is coded to be sent from two antennas. The method is also denoted in the 3GPP specification as space time block coding based transmit diversity (STTD). Another possibility is to use feedback mode transmit

Table 6.3. Downlink Dedicated Channel symbol and bit rates

Spreading factor	Channel symbol rate (kbps)	Channel bit rate (kbps)	DPDCH channel bit rate range (kbps)	Maximum user data rate with $\frac{1}{2}$-rate coding (approx.)
512	7.5	15	3–6	1–3 kbps
256	15	30	12–24	6–12 kbps
128	30	60	42–51	20–24 kbps
64	60	120	90	45 kbps
32	120	240	210	105 kbps
16	240	480	432	215 kbps
8	480	960	912	456 kbps
4	960	1920	1872	936 kbps
4, with 3 parallel codes	2880	5760	5616	2.8 Mbps

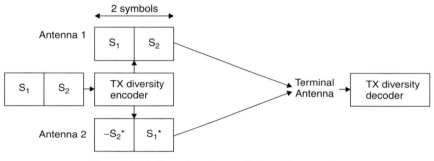

Figure 6.15. Open loop transmit diversity encoding

diversity, where the signal is sent from two antennas based on the feedback information from the terminal. The feedback mode uses phase, and in some cases also amplitude, offsets between the antennas. The feedback mode of transmit diversity is covered in the physical layer procedures section.

6.4.6 Downlink Multiplexing

The multiplexing chain in the downlink is mainly similar to that in the uplink but there are also some functions that are done differently.

As in the uplink, the interleaving is implemented in two parts, covering both intra-frame and inter-frame interleaving. Also the rate matching allows one to balance the required channel symbol energy for different service qualities. The services can be mapped to more than one code as well, which is necessary if the single code capability in either the terminal or base station is exceeded.

There are differences in the order in which rate matching and segmentation functions are performed, as shown in Figure 6.16. Whether fixed or flexible bit positions are used determines the DTX indication insertion point. The DTX indication bits are not transmitted over the air; they are just inserted to inform the transmitter at which bit positions the

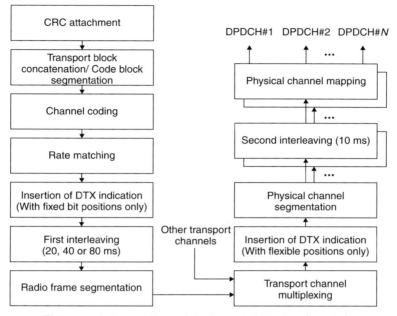

Figure 6.16. Downlink multiplexing and channel coding chain

transmission should be turned off. They were not needed in the uplink where the rate matching was done in a more dynamic way, always filling the frame when there was something to transmit on the DPDCH.

The use of fixed positions means that for a given transport channel, the same symbols are always used. If the transmission rate is below the maximum, then DTX indication bits are used for those symbols. The different transport channels do not have a dynamic impact on the rate matching values applied for another channel, and all transport channels can use the maximum rate simultaneously as well. The use of fixed positions is partly related to the possible use of blind rate detection. When a transport channel always has the same position regardless of the data rate, the channel decoding can be done with a single decoding 'run' and the only thing that needs to be tested is which position of the output block is matched with the CRC check results. This naturally requires that different rates have different numbers of symbols.

With flexible positions the situation is different since now the channel bits unused by one service may be utilised by another service. This is useful when it is possible to have such a transport channel combination that they do not all need to be able to reach the full data rate simultaneously, but can alternate with the need for full rate transmission. This allows the necessary spreading code occupancy in the downlink to be reduced. The concept of flexible versus fixed positions in the downlink is illustrated in Figure 6.17. The use of blind rate detection is also possible in principle with flexible positions, but is not required by the specifications. If the data rate is not too high and number of possible data rates is not very high, the terminal can run channel decoding for all the combinations and check which of the cases comes out with the correct CRC result.

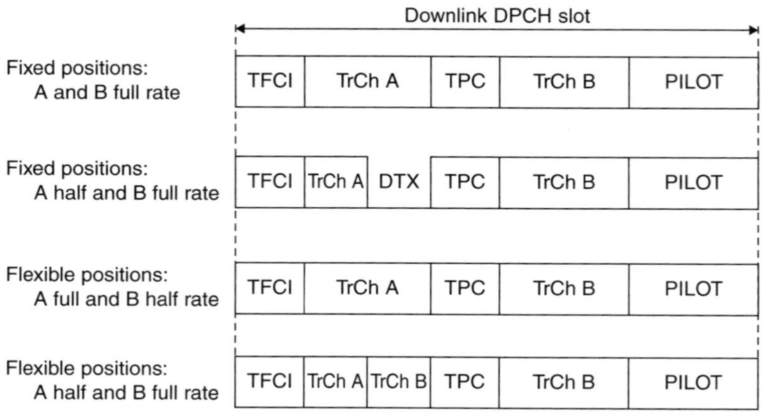

Figure 6.17. Flexible and fixed transport channel slot positions in the downlink

6.4.7 Downlink Shared Channel

Transmitting data with high peak rate and low activity cycle in the downlink quickly causes the channelisation codes under a single scrambling code to start to run out. To avoid this problem, basically two alternatives exist: use of either additional scrambling codes or common channels. The additional scrambling code approach loses the advantage of the transmissions being orthogonal from a single source, and thus should be avoided. Using a shared channel resource maintains this advantage and at the same time reduces the downlink code resource consumption. As such, resource sharing cannot provide a 100 % guarantee of available physical channel resource at all times, its applicability in practice is limited to packet-based services.

As in a CDMA system one has to ensure the availability of power control and other information continuously, the Downlink Shared Channel (DSCH) has been defined to be always associated with a Downlink Dedicated Channel (Downlink DCH). The DCH provides, in addition to the power control information, an indication to the terminal when it has to decode the DSCH and which spreading code from the DSCH it has to despread. For this indication two alternatives have been specified: either TFCI based on a frame-by-frame basis or higher layer signalling based on a longer allocation period. Thus, the DSCH data rate without coding is directly the channel bit rate indicated in Table 6.3 for the Downlink DCH. The small difference from the downlink DCH spreading codes is that spreading factor 512 is not supported by DSCH. The DSCH also allows the mixing of terminals with different data rate capabilities under a single branch from the code resource, making the configuration manageable with evolving terminal capabilities. The DSCH code tree was illustrated in Figure 6.9 in connection with the downlink spreading section.

With DSCH the user may be allocated different data rates, for example 384 kbps with spreading factor 8 and then 192 kbps with spreading factor 16. The DSCH code tree definition allows sharing the DSCH capacity on a frame-by-frame basis, for example with either a single user active with a high data rate or with several lower-rate users active in parallel. The DSCH may be mapped to a multicode case as well; for example, three channelisation codes with spreading factor 4 provide a DSCH with 2 Mbps capability.

In the uplink direction, such concerns for code resource usage do not exist, but there is the question of how to manage the total interference level and in some cases the resource usage on the receiver side. Thus, an operation similar to DSCH is not specified in the uplink in UTRA FDD.

The physical channel carrying the DSCH is the Physical Downlink Shared Channel (PDSCH). The timing relation of the PDSCH to the associated downlink Dedicated Physical Channel (DPCH) is shown in Figure 6.18. The PDSCH frame may not start before three slots

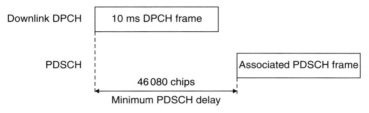

Figure 6.18. PDSCH timing relationship to DPCH

after the end of the associated dedicated channel frame. This ensures that buffering requirements for DSCH reception do not increase compared to the other buffering needs in the receiver.

6.4.8   *Forward Access Channel for User Data Transmission*

The Forward Access Channel (FACH) can be used for transmission of user (packet) data. The channel is typically multiplexed with the paging channel to the same physical channel, but can exist as a stand alone channel as well. The main difference with the dedicated and shared channels is that FACH does not allow the use of fast power control and applies either slow power control or no power control at all. Slow power control is possible if a lot of data is transmitted between the base station and the terminal and the latter provides feedback on the quality of the received packets. This type of power control cannot combat the effect of the fading channel but rather the longer-term changes in the propagation environment. For less frequent transmission, FACH needs to use more or less the full power level. The power control for FACH is also typically very slow, since the FACH data transmission is controlled by RNC, which means rather a large delay for any feedback information from the base station.

Whether the FACH contains pilot symbols or not depends on whether it applies beam forming techniques. Normally FACH does not contain pilot symbols and the receiver uses the common pilot channel as phase reference.

As FACH needs to be received by all terminals, the primary FACH cannot use high data rates. If higher data rates were desired of FACH, this would require a separate physical channel where only the capabilities in terms of maximum data rates of those terminals allocated to that channel need to be taken into account. The necessary configuration would become rather complicated when terminals with different capabilities are included. The FACH has a fixed spreading factor, and reserving FACH for very high data rates is not optimised from the code resource point of view, especially if not all the terminals can decode the high data rate FACH.

Messages on FACH normally need in-band signalling to tell for which user the data was intended. In order to read such information, the terminal must decode FACH messages first. Running such decoding continuously is not desirable due to power consumption, especially with higher FACH rates.

6.4.9   Channel Coding for User Data

In UTRA two channel coding methods have been defined. Half-rate and $\frac{1}{3}$-rate convolutional coding are intended to be used with relatively low data rates, equivalent to the data rates provided by second generation cellular networks today, though an upper limit has not been specified. For higher data rates, $\frac{1}{3}$-rate turbo coding can be applied and typically brings performance benefits when large enough block sizes are achieved. It has been estimated that roughly 300 bits should be available per TTI in order to give turbo coding some gain over convolutional coding. This also depends on the required quality level and operational environment.

The convolutional coding is based on constraint length 9 coding with the use of tail bits. The selected turbo encoding/decoding method is 8-state PCCC (parallel concatenated convolutional code). The main motivation for turbo coding for higher bit rates has been performance, while for low rates the main reason not to use it has been both low rate or low block length performance as well as the desire to allow the use of simple blind rate detection with low rate services such as speech. Blind rate detection with turbo coding typically requires detection of all transmission rates, while with convolutional coding trial methods can allow only a single Viterbi pass for determining which transmission rate was used. This is performed together with the help of CRC and applying a proper interleaving technique.

Turbo coding has specific interleaving which has been designed with a large variety of data rates in mind. The maximum turbo coding block size has been limited to 5114 information bits, since after that block size only memory requirements increase but no significant effect on the performance side can be observed. For the higher amount of data per interleaving period, several blocks are used, with a block size as equal as possible at or below 5114 bits. The actual block size for data is a little smaller, since the tail bits as well as CRC bits are to be accommodated in the block size.

The minimum block size for turbo coding was initially defined to be 320 bits, which corresponds to 32 kbps with 10 ms interleaving or down to 4 kbps with 80 ms interleaving. The possible range of block sizes was, however, extended down to 40 bits, since with variable rate connection it is not desirable to change the codec 'on the fly' when coming down from the maximum rate. Nor may a transport channel change the channel coding method on a frame-by-frame basis. Data rates below 40 bits can be transmitted with turbo coding as well, but in such a case padding with dummy bits is used to fill the 40 bits minimum size interleaver.

With speech service, AMR coding uses an unequal error protection scheme. This means that the three different classes of bits have different protection. Class A bits – those that contribute the most to voice quality – have the strongest protection, while class C bits are sent with weaker channel coding. This gives around 1 dB gain in E_b/N_0 compared to the equal error protection scheme. The coding methods usable by different channels are summarised in Table 6.4. Although the FACH has two options given, the cell access use of FACH is based on convolutional coding, as not all terminals support turbo coding.

Table 6.4. Channel coding options with different channels

DCH	Turbo coding or convolutional coding
CPCH	Turbo coding or convolutional coding
DSCH	Turbo coding or convolutional coding
FACH	Turbo coding or convolutional coding
Other common channels	$\frac{1}{2}$-rate convolutional coding

6.4.10   Coding for TFCI Information

The Transport Format Combination Indicator (TFCI) may carry from 1 to 10 bits of transport format information. As well as the normal mode of operation, there is also 'split' mode where the TFCI code word is sent with two different code words and not every cell necessarily sends both code words. In this case both code words are capable of carrying 5 bits. The typical split mode operation would be that an RNC for a downlink dedicated channel would be different from an RNC for controlling a DSCH. The split mode is valid for the downlink direction only. For Release 5 the split mode is enhanced to allow, in addition to 5:5 split, also 6:4 and 7:3 splits with small enhancements to the TFCI coding scheme.

The coding in the normal mode is second-order Reed–Muller code punctured from 32 bits to 30 bits, carrying up to 10 bits of information. The TFCI coding is illustrated in Figure 6.19. The coding with split mode (with 5:5 split) is biorthogonal (16,5) block code.

Figure 6.19. TFCI information coding

6.5   Signalling

For signalling purposes a lot of information needs to be transmitted between the network and the terminals. The following sections describe the methods used for transmitting signalling messages generated above the physical layer, as well as the required physical layer control channels needed for system operation but not necessarily visible for higher layer functionality.

6.5.1   Common Pilot Channel (CPICH)

The common pilot channel is an unmodulated code channel, which is scrambled with the cell-specific primary scrambling code. The function of the CPICH is to aid the channel estimation at the terminal for the dedicated channel and to provide the channel estimation reference for the common channels when they are not associated with the dedicated channels or not involved in the adaptive antenna techniques.

UTRA has two types of common pilot channel, primary and secondary. The difference is that the Primary CPICH is always under the primary scrambling code with a fixed channelisation code allocation and there is only one such channel for a cell or sector. The

Secondary CPICH may have any channelisation code of length 256 and may be under a secondary scrambling code as well. The typical area of Secondary CPICH usage would be operations with narrow antenna beams intended for service provision at specific 'hot spots' or places with high traffic density.

An important area for the primary common pilot channel is the measurements for the handover and cell selection/reselection. The use of CPICH reception level at the terminal for handover measurements has the consequence that, by adjusting the CPICH power level, the cell load can be balanced between different cells. Reducing the CPICH power causes part of the terminals to hand over to other cells, while increasing it invites more terminals to hand over to the cell, as well as to make their initial access to the network in that cell.

The CPICH does not carry any higher layer information, neither is there any transport channel mapped to it. The CPICH uses the spreading factor of 256. It may be sent from two antennas in case transmission diversity methods are used in the base station. In this case, the transmissions from the two antennas are separated by a simple modulation pattern on the CPICH transmitted from the diversity antenna, called diversity CPICH. The diversity pilot is used with both open loop and closed loop transmission diversity schemes.

6.5.2 Synchronisation Channel (SCH)

The Synchronisation Channel (SCH) is needed for the cell search. It consists of two channels, the primary and secondary synchronisation channels.

The Primary SCH uses a 256-chip spreading sequence identical in every cell. The system-wide sequence has been optimised for matched filter implementations, as described in connection with SCH spreading and modulation in Section 6.3.4.4.

The Secondary SCH uses sequences with different code word combination possibilities representing different code groups. Once the terminal has identified the secondary synchronisation channel, it has obtained frame and slot synchronisation as well as information on the group the cell belongs to. There are 64 different code groups in use, pointed out by the 256 chip sequences sent on the secondary SCHs. Such a full cell search process with a need to search for all groups is needed naturally only at the initial search upon terminal power-on or when entering a coverage area, otherwise a terminal has more information available on the neighbouring cells and not all the steps are always necessary.

As with the CPICH, no transport channel is mapped on the SCH, as the code words are transmitted for cell search purposes only. The SCH is time multiplexed with the Primary Common Control Physical Channel. For the SCH there are always 256 chips out of 2560 chips from each slot. The Primary and Secondary SCH are sent in parallel, as illustrated in Figure 6.20. Further details on the cell search procedure are covered in Section 6.6.6.

6.5.3 Primary Common Control Physical Channel (Primary CCPCH)

The Primary Common Control Physical Channel (Primary CCPCH) is the physical channel carrying the Broadcast Channel (BCH). It needs to be demodulated by all the terminals in the system. As a result, the parameters with respect to, for example, the channel coding and spreading code contain no flexibility, as they need to be known by all terminals made since the publication of the Release '99 specifications. The contents of the signalling messages have room for flexibility as long as the new message structures are such that they do not cause unwanted or unpredictable behaviour in the terminals deployed in the network.

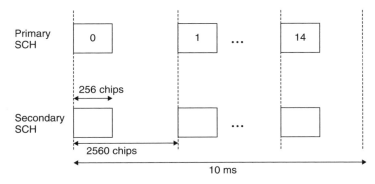

Figure 6.20. Primary and secondary synchronisation channel principle

The Primary CCPCH contains no Layer 1 control information as it is fixed rate and does not carry power control information for any of the terminals. The pilot symbols are not used, since the Primary CCPCH needs to be available over the whole cell area and does not use specific antenna techniques but is sent with the same antenna radiation pattern as the common pilot channel. This allows the common pilot channel to be used for channel estimation with coherent detection in connection with the Primary CCPCH.

The channel bit rate is 30 kbps with spreading ratio of the permanently allocated channelisation code of 256. The total bit rate is reduced further as the Primary CCPCH alternates with the Synchronisation Channel (SCH), reducing the bit rate without coding available for system information to 27 kbps. This is illustrated in Figure 6.21, where the 256-chip idle period on the Primary CCPCH is shown.

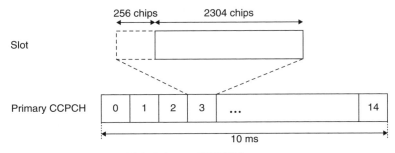

Figure 6.21. Primary CCPCH frame structure

The channel coding with the Primary CCPCH is $\frac{1}{2}$-rate convolutional coding with 20 ms interleaving over two consecutive frames. It is important to keep the data rate with the Primary CCPCH low, as in practice it will be transmitted with very high power from the base station to reach all terminals, having a direct impact on system capacity. If Primary CCPCH decoding fails, the terminals cannot access the system if they are unable to obtain the critical system parameters such as random access codes or code channels used for other common channels.

As a performance improvement method, the Primary CCPCH may apply open loop transmission diversity. In such a case the use of open transmission diversity on the Primary CCPCH is indicated in the modulation of the Secondary SCH. This allows the terminals to have the information before attempting to decode the BCH with the initial cell search.

6.5.4   Secondary Common Control Physical Channel (Secondary CCPCH)

The Secondary Common Control Physical Channel (Secondary CCPCH) carries two different common transport channels: the Forward Access Channel (FACH) and the Paging Channel (PCH). The two channels can share a single Secondary CCPCH or can use different physical channels. This means that in the minimum configuration each cell has at least one Secondary CCPCH. In case of a single Secondary CCPCH fewer degrees of freedom exist in terms of data rates, and so on, since again all the terminals in the network need to be able to detect the FACH and PCH. Since there can be more than one FACH or PCH, however, for the additional Secondary CCPCHs the data rates can vary more, as long as the terminals are not capable of demodulating higher data rates using another, lower data rate Secondary CCPCH.

The spreading factor used in a Secondary CCPCH is fixed and determined according to the maximum data rate. The data rate may vary with DTX or rate matching parameters, but the channelisation code is always reserved according to the maximum data rate. The maximum data rate usable is naturally dependent on the terminal capabilities. As with the Primary CCPCH, the channel coding method is $\frac{1}{2}$-rate convolutional coding when carrying the channels used for cell access, FACH or PCH. When used to carry PCH, the interleaving period is always 10 ms. For data transmission with FACH, turbo coding or $\frac{1}{3}$-rate convolutional coding may also be applied.

The Secondary CCPCH does not contain power control information, and for other layer 1 control information the following combinations can be used:

- Neither pilot symbols nor rate information (TFCI). Used with PCH and FACH when no adaptive antennas are in use and a channel needs to be detected by all terminals.

- No pilot symbols, but rate information with TFCI. Used typically with FACH when it is desired to use FACH for data transmission with variable transport format and data rate. In such a case, variable transmission rates are implemented by DTX or repetition.

- Pilot symbol with or without rate information (TFCI). Typical for the case when an uplink channel is used to derive information for adaptive antenna processing purposes and user-specific antenna radiation patterns or beams are used.

The FACH and PCH can be multiplexed to a single Secondary CCPCH, as the paging indicators used together with the PCH are multiplexed to a different physical channel, called the Paging Indicator Channel (PICH). The motivation for multiplexing the channels together is base station power budget. Since both of the channels need to be transmitted at full power for all the terminals to receive, avoiding the need to send them simultaneously obviously reduces base station power level variations. In order to enable this multiplexing, it has been necessary to terminate both FACH and PCH at RNC.

As a performance improvement method, open loop transmission diversity can be used with a Secondary CCPCH as well. The performance improvement of such a method is higher for common channels in general, as neither Primary nor Secondary CCPCH can use fast power control. Also, since they are often sent with full power to reach the cell edge, reducing the required transmission power level improves downlink system capacity.

6.5.5 Random Access Channel (RACH) for Signalling Transmission

The Random Access Channel (RACH) is typically used for signalling purposes, to register the terminal after power-on to the network or to perform location update after moving from one location area to another or to initiate a call. The structure of the physical RACH for signalling purposes is the same as when using the RACH for user data transmission, as described in connection with the user data transmission. With signalling use the major difference is that the data rate needs to be kept relatively low, otherwise the range achievable with RACH signalling starts to limit the system coverage. This is more critical the lower the data rates used as a basis for network coverage planning. The detailed RACH procedure will be covered in connection with the physical layer procedures.

The RACH that can be used for initial access has a relatively low payload size, since it needs to be usable by all terminals. The ability to support 16 kbps data rate on RACH is a mandatory requirement for all terminals regardless of what kind of services they provide.

6.5.6 Acquisition Indicator Channel (AICH)

In connection with the Random Access Channel, the Acquisition Indicator Channel (AICH) is used to indicate from the base station the reception of the random access channel signature sequence. The AICH uses an identical signature sequence as the RACH on one of the downlink channelisation codes of the base station to which the RACH belongs. Once the base station has detected the preamble with the random access attempt, then the same signature sequence that has been used on the preamble will be echoed back on AICH. As the structure of AICH is the same as with the RACH preamble, it also uses a spreading factor of 256 and 16 symbols as the signature sequence. There can be up to 16 signatures, acknowledged on the AICH at the same time. Both signature sets can be used with AICH. The procedure with AICH and RACH is described in the physical layer procedures section.

For the detection of AICH the terminal needs to obtain the phase reference from the common pilot channel. The AICH also needs to be heard by all terminals and needs to be sent typically at high power level without power control.

The AICH is not visible to higher layers but is controlled directly by the physical layer in the base station, as operation via a radio network controller would make the response time too slow for a RACH preamble. There are only a few timeslots to detect the RACH preamble and to transmit the response to the terminal on AICH. The AICH access slot structure is shown in Figure 6.22.

6.5.7 Paging Indicator Channel (PICH)

The Paging Channel (PCH) is operated together with the Paging Indicator Channel (PICH) to provide terminals with efficient sleep mode operation. The paging indicators use a channelisation code of length 256. The paging indicators occur once per slot on the corresponding physical channel, the Paging Indicator Channel (PICH). Each PICH frame carries 288 bits to be used by the paging indicator bit, and 12 bits are left idle. Depending on

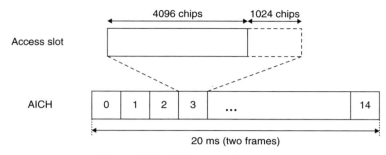

Figure 6.22. AICH access slot structure

the paging indicator repetition ratio, there can be 18, 36, 72 or 144 paging indicators per PICH frame. How often a terminal needs to listen to the PICH is parameterised, and the exact moment depends on running the system frame number (SFN).

For detection of the PICH the terminal needs to obtain the phase reference from the CPICH, and as with the AICH, the PICH needs to be heard by all terminals in the cell and thus needs to be sent at high power level without power control. The PICH frame structure with different PI repetition factors is illustrated in Figure 6.23.

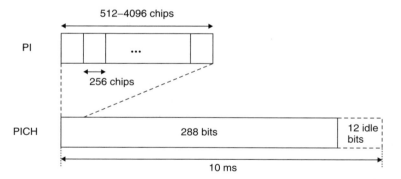

Figure 6.23. PICH structure with different PI repetition rates

6.5.8 *Physical Channels for the CPCH Access Procedure*

For the CPCH access procedure, a set of CPCH specific physical channels has been specified. These channels carry no transport channels, but only information needed in the CPCH access procedure. The channels are:

- CPCH Status Indication Channel (CSICH);
- CPCH Collision Detection Indicator Channel (CD-ICH);
- CPCH Channel Assignment Indicator Channel (CA-ICH);
- CPCH Access Preamble Acquisition Channel (AP-AICH).

The CSICH uses the part of the AICH channel that is defined as unused, as shown in Figure 6.22. The CSICH bits indicate the availability of each physical CPCH channel and are used to tell the terminal to initiate access only on a free channel but, on the other hand, to accept a channel assignment command to an unused channel. The CSICH shares the downlink channelisation code resource with the AP-AICH.

The CD-ICH carries the collision detection information to the terminal. When the CA-ICH channel is used, the CD-ICH and CA-ICH are sent in parallel to the terminal. Both have 16 different bit patterns specified.

The AP-AICH is identical to the AICH used with RACH and may share the same channelisation code when sharing access resources with RACH. In this case CSICH uses also the same channelisation code as the CPCH and RACH AICH channels.

6.6   Physical Layer Procedures

In the physical layer of a CDMA system there are many procedures essential for system operation. Examples include the fast power control and random access procedures. Other important physical layer procedures are paging, handover measurements and operation with transmit diversity. These procedures have been naturally shaped by the CDMA-specific properties of the UTRA FDD physical layer.

6.6.1   Fast Closed Loop Power Control Procedure

The fast closed loop power control procedure is denoted in the UTRA specifications as inner loop power control. It is known to be essential in a CDMA-based system due to the uplink near–far problem illustrated in Chapter 3. The fast power control operation operates on a basis of one command per slot, resulting in a 1500 Hz command rate. The basic step size is 1 dB. Additionally, multiples of that step size can be used and smaller step sizes can be emulated. The emulated step size means that the 1 dB step is used, for example, only every second slot, thus emulating the 0.5 dB step size. 'True' step sizes below 1 dB are difficult to implement with reasonable complexity, as the achievable accuracy over the large dynamic range is difficult to ensure. The specifications define the relative accuracy for a 1 dB power control step to be ±0.5 dB. The other 'true' step size specified is 2 dB.

Fast power control operation has two special cases: operation with soft handover and with compressed mode in connection with handover measurements. Soft handover needs special concern as there are several base stations sending commands to a single terminal, while with compressed mode operation breaks in the command stream are periodically provided to the terminal.

In soft handover the main issue for terminals is how to react to multiple power control commands from several sources. This has been solved by specifying the operation such that the terminal combines the commands but also takes the reliability of each individual command decision into account in deciding whether to increase or decrease the power.

In the compressed mode case, the fast power control uses a larger step size for a short period after a compressed frame. This allows the power level to converge more quickly to the correct value after a break in the control stream. The need for this method depends heavily on the environment and it is not relevant for the lower terminal or very short transmission gap lengths.

The SIR target for closed loop power control is set by the outer loop power control. The latter power control is introduced in Section 3.5 and described in detail in Section 9.2.2.

On the terminal side, what is expected to be done inside a terminal in terms of (fast) power control operation is specified rather strictly. On the network side there is much greater freedom to decide how a base station should behave upon reception of a power control command, as well as the basis on which the base station should tell a terminal to increase or decrease the power.

6.6.2 Open Loop Power Control

In UTRA FDD there is also open loop power control, which is applied only prior to initiating the transmission on the RACH or CPCH. Open loop power control is not very accurate, since it is difficult to measure large power dynamics accurately in the terminal equipment. The mapping of the actual received absolute power to the absolute power to be transmitted shows large deviations, due to variation in the component properties as well as to the impact of environmental conditions, mainly temperature. Also, the transmission and reception occur at different frequencies, but the internal accuracy inside the terminal is the main source of uncertainty. The requirement for open loop power control accuracy is specified to be within ± 9 dB in normal conditions.

Open loop power control was used in earlier CDMA systems, such as IS-95, being active in parallel with closed loop power control. The motivation for such usage was to allow corner effects or other sudden environmental changes to be covered. As the UTRA fast power control has almost double the command rate, it was concluded that a 15 dB adjustment range does not need open loop power control to be operated simultaneously. Additionally, the fast power control step size can be increased from 1 dB to 2 dB, which would allow a 30 dB correction range during a 10 ms frame.

The use of open loop power control while in active mode also has some impact on link quality. The large inaccuracy of open loop power control can cause it to make adjustments to the transmitted power level even when they are not needed. As such, behaviour depends on terminal unit tolerances and on various environmental variables, running open loop power control makes it more difficult from the network side to predict how a terminal will behave in different conditions.

6.6.3 Paging Procedure

The Paging Channel (PCH) operation is organised as follows. A terminal, once registered to a network, has been allocated a paging group. For the paging group there are Paging Indicators (PI) which appear periodically on the Paging Indicator Channel (PICH) when there are paging messages for any of the terminals belonging to that paging group.

Once a PI has been detected, the terminal decodes the next PCH frame transmitted on the Secondary CCPCH to see whether there was a paging message intended for it. The terminal may also need to decode the PCH in case the PI reception indicates low reliability of the decision. The paging interval is illustrated in Figure 6.24.

The less often the PIs appear, the less often the terminal needs to wake up from the sleep mode and the longer the battery life becomes. The trade-off is obviously the response time to the network-originated call. An infinite paging indicator interval does not lead to infinite battery duration, as there are other tasks the terminal needs to perform during idle mode as well.

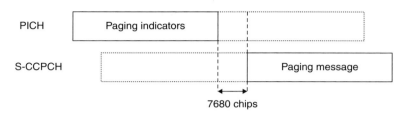

Figure 6.24. PICH relationship to PCH

6.6.4   RACH Procedure

The Random Access procedure in a CDMA system has to cope with the near–far problem, as when initiating the transmission there is no exact knowledge of the required transmission power. The open loop power control has a large uncertainty in terms of absolute power values from the received power measurement to the transmitter power level setting value, as stated in connection with the open loop description. In UTRA the RACH procedure has the following phases:

- The terminal decodes the BCH to find out the available RACH sub-channels and their scrambling codes and signatures.

- The terminal selects randomly one of the RACH sub-channels from the group its access class allows it to use. Furthermore, the signature is also selected randomly from among the available signatures.

- The downlink power level is measured and the initial RACH power level is set with the proper margin due to the open loop inaccuracy.

- A 1 ms RACH preamble is sent with the selected signature.

- The terminal decodes AICH to see whether the base station has detected the preamble.

- In case no AICH is detected, the terminal increases the preamble transmission power by a step given by the base station, as multiples of 1 dB. The preamble is retransmitted in the next available access slot.

- When an AICH transmission is detected from the base station, the terminal transmits the 10 ms or 20 ms message part of the RACH transmission.

The RACH procedure is illustrated in Figure 6.25, where the terminal transmits the preamble until acknowledgement is received on AICH, and then the message part follows.

In the case of data transmission on RACH, the spreading factor and thus the data rate may vary; this is indicated with the TFCI on the DPCCH on PRACH. Spreading factors from 256 to 32 have been defined to be possible, thus a single frame on RACH may contain up to 1200 channel symbols which, depending on the channel coding, maps to around 600 or 400 bits. For the maximum number of bits the achievable range is naturally less than what can be achieved with the lowest rates, especially as RACH messages do not use methods such as macro-diversity as in the dedicated channel.

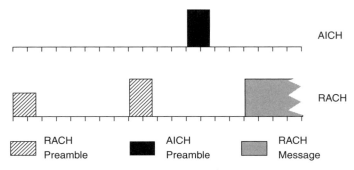

Figure 6.25. PRACH ramping and message transmission

6.6.5 CPCH Operation

Uplink Common Packet Channel (CPCH) operation, as illustrated in Figure 6.26, is rather similar to RACH operation. The main difference is the Layer 1 Collision Detection (CD) based on a signal structure similar to that of the RACH preamble. The operation follows the RACH procedure until the terminal detects AICH. After that a CD preamble with the same power level is still sent back with another signature, randomly chosen from a given set. Then the base station is expected to echo this signature back to the terminal on the CD Indication Channel (CD-ICH) and in this way to create a method of reducing the collision probability on Layer 1. After the correct preamble has been sent by the base station on the collision detection procedure, the terminal starts the transmission, which may last over several frames. The longer duration of the transmission highlights the need for the physical layer-based collision detection mechanism. In RACH operation only one RACH message may end up lost due to collision, whereas with CPCH operation an undetected collision may cause several frames to be sent and cause only extra interference.

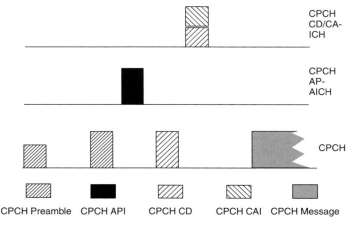

Figure 6.26. CPCH access procedure operation

The fast power control on CPCH helps to reduce the interference due to the data transmission while it also highlights the importance of the added collision detection to RACH. A terminal transmitting data over several frames and following a power control command stream intended for another terminal would create a severe interference problem in the cell, especially when high data rates are involved. At the beginning of the CPCH transmission, an optional power control preamble can be sent before actual data transmission is initiated. This is to allow power control to converge, as there is a longer delay with CPCH than with RACH between the acknowledged preamble and actual data frame transmission. The 8-slot power control preamble also uses a 2 dB step size for faster power control convergence.

A CPCH transmission needs to have a restriction on maximum duration, since CPCH supports neither soft handover nor compressed mode to allow inter-frequency or inter-system measurements. UTRAN sets the maximum CPCH transmission during service negotiations.

The latest addition to CPCH operation is the status monitoring and channel assignment functionality. The CPCH Status Indication Channel (CSICH) is a separate physical channel, sent from the base station, that has indicator bits to indicate the status of different CPCH channels. This avoids unnecessary access attempts when all CPCH channels are busy, so it will also improve CPCH throughput. The Channel Assignment functionality is a system option, in the form of a CA message that may direct the terminal to a CPCH channel other than the one used for the access procedure. The CA message is sent in parallel with the collision detection message.

6.6.6 Cell Search Procedure

The cell search procedure or synchronisation procedure in an asynchronous CDMA system differs greatly from the procedure in a synchronous system like IS-95. Since the cells in an asynchronous UTRA CDMA system use different scrambling codes and not just different code phase shifts, terminals with today's technology cannot search for 512 codes of 10 ms duration without any prior knowledge. There would be too many comparisons to make and users would experience too long an interval from power-on to the service availability indication in the terminal.

The cell search procedure using the synchronisation channel has basically three steps, though from the standards point of view there will be no requirements as to which steps to perform and when. Rather the standard will set requirements for performance in terms of maximum search duration in reference test conditions. The basic steps for the initial cell search are typically as follows:

1. The terminal searches the 256-chip primary synchronisation code, being identical for all cells. As the primary synchronisation code is the same in every slot, the peak detected corresponds to the slot boundary.

2. Based on the peaks detected for the primary synchronisation code, the terminal seeks the largest peak from the Secondary SCH code word. There are 64 possibilities for the secondary synchronisation code word. The terminal needs to check all 15 positions, as the frame boundary is not available before Secondary SCH code word detection.

3. Once the Secondary SCH code word has been detected, the frame timing is known. The terminal then seeks the primary scrambling codes that belong to that particular code

group. Each group consists of eight primary scrambling codes. These need to be tested for a single position only, as the starting point is known already.

When setting the network parameters, the properties of the synchronisation scheme need to be taken into account for optimum performance. For the initial cell search there is no practical impact, but the target cell search in connection with handover can be optimised. Basically, since there is rather a large number of code groups, in a practical planning situation one can, in most cases, implement the neighbouring cell list so that all the cells in the list for one cell belong to a different code group. Thus, the terminal can search for the target cell and skip step 3 totally, just confirming detection without needing to compare the different primary scrambling codes for that step. Alternatively one can, as has been shown in practical networks resulting in similar performance, aim to have all neighbouring cells under one code group as well. Initially, having more groups was expected not to be optimal but, as experience from the field has shown, using one group only is the preferred alternative.

Further ways of improving cell search performance include the possibility of providing information on the relative timing between cells. This kind of information, which is being measured by the terminals for soft handover purposes in any case, can be used to improve especially the step 2 performance. The more accurate the relative timing information, the fewer slot positions need to be tested for the Secondary SCH code word, and the better is the probability of correct detection.

6.6.7 Transmit Diversity Procedure

As was mentioned in connection with the downlink channels, UTRA uses two types of transmit diversity transmission for user data performance improvement, as studied in Chapter 11. These methods are classified as open loop and closed loop methods. In this section the feedback procedure for closed loop transmit diversity is described. The open loop method was covered in connection with the downlink dedicated channel description.

In the case of closed loop transmit diversity, the base station uses two antennas to transmit the user information. The use of these two antennas is based on the feedback from the terminal, transmitted in the Feedback (FB) bits in the uplink DPCCH. The closed loop transmit diversity itself has two modes of operation.

In mode 1, the terminal feedback commands control the phase adjustments that are expected to maximise the power received by the terminal. The base station thus maintains the phase with antenna 1 and then adjusts the phase of antenna 2 based on the sliding averaging over two consecutive feedback commands. With this method, four different phase settings are applied to antenna 2.

In mode 2, the amplitude is adjusted in addition to the phase adjustment. The same signalling rate is used, but now the command is spread over four bits in four uplink DPCCH slots, with a single bit for amplitude and three bits for phase adjustment. This gives a total of eight different phase and two different amplitude combinations, thus a total of 16 combinations for signal transmission from the base station. The amplitude values have been defined to be 0.2 and 0.8, while the phase values are naturally distributed evenly for the antenna phase offsets, from $-135°$ to $+180°$ phase offset. In this mode the last three slots of the frame contain only phase information, while amplitude information is taken from the previous four slots. This allows the command period to go even with 15 slots as with mode 1, where the average at the frame boundary is slightly modified by averaging the commands from slot 13 and slot 0 to avoid discontinuities in the adjustment process.

The closed loop method may be applied only on the dedicated channels or with a DSCH together with a dedicated channel. The open loop method may be used on both the common and dedicated channels. With HSDPA open loop and closed loop mode 1 are applicable.

6.6.8   Handover Measurements Procedure

Within the UTRA FDD the possible handovers are as follows:

- Intra-mode handover, which can be soft handover, softer handover or hard handover. Hard handover may take place as intra- or inter-frequency handover.

- Inter-mode handover as handover to the UTRA TDD mode.

- Inter-system handover, which in Release '99 means only GSM handover. The GSM handover may take place to a GSM system operating at 850 MHz, 900 MHz, 1800 MHz and 1900 MHz.

The main relevance of the handover to the physical layer is what to measure for handover criteria and how to obtain the measurements.

6.6.8.1   Intra-Mode Handover

The UTRA FDD intra-mode handover relies on the E_c/N_0 measurement performed from the common pilot channel (CPICH). The quantities defined that can be measured by the terminal from the CPICH are as follows:

- Received Signal Code Power (RSCP), which is the received power on one code after despreading, defined on the pilot symbols.

- Received Signal Strength Indicator (RSSI), which is the wideband received power within the channel bandwidth.

- E_c/N_0, representing the received signal code power divided by the total received power in the channel bandwidth, which is defined as RSCP/RSSI.

There are also other items that can be used as a basis for handover decisions in UTRAN, as the actual handover algorithm decisions are left as an implementation issue. One such parameter mentioned in the standardisation discussions has been the dedicated channel SIR, giving information on the cell orthogonality and being measured in any case for power control purposes.

Additional essential information for soft handover purposes is the relative timing information between the cells. As in an asynchronous network, there is a need to adjust the transmission timing in soft handover to allow coherent combining in the Rake receiver, otherwise the transmissions from the different base stations would be difficult to combine, and especially the power control operation in soft handover would suffer additional delay. The timing measurement in connection with the soft handover operation is illustrated in Figure 6.27. The new base station adjusts the downlink timing in steps of 256 chips based on the information it receives from the RNC.

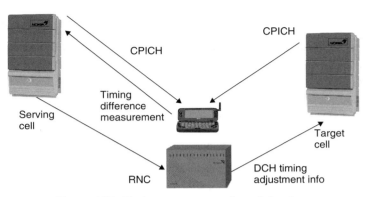

Figure 6.27. Timing measurement for soft handover

When the cells are within the 10 ms window, the relative timing can be found from the primary scrambling code phase, since the code period used is 10 ms. If the timing uncertainty is larger, the terminal needs to decode the System Frame Number (SFN) from the Primary CCPCH. This always takes time and may suffer from errors, which requires also a CRC check to be made on the SFN. The 10 ms window has no relevance when the timing information is provided in the neighbouring cell list. In such a case only the phase difference of the scrambling codes needs to be considered, unless the base stations are synchronised to chip level.

For the hard handover between frequencies such accurate timing information on chip level is not needed. Obtaining the other measurements is slightly more challenging as the terminal must make the measurements on a different frequency. This is typically done with the aid of compressed mode, which is described later in this chapter.

6.6.8.2 Inter-Mode Handover

On request from UTRAN, the dual-mode FDD–TDD terminals operating in FDD measure the power level from the TDD cells available in the area. The TDD CCPCH bursts sent twice during the 10 ms TDD frame can be used for measurement, since they are guaranteed to always exist in the downlink. The TDD cells in the same coverage area are synchronised, thus finding one slot with the reference midamble means that other TDD cells have roughly the same timing for their burst with reference power. UTRA TDD is covered in further detail in Chapter 13.

6.6.8.3 Inter-System Handover

For UTRA–GSM handover, basically similar requirements are valid as for GSM–GSM handover. Normally the terminal receives the GSM Synchronisation Channel (GSM SCH) during compressed frames in UTRA FDD to allow measurements from other frequencies. GSM1800 set special requirements for compressed mode and required that compressed mode was specified for the uplink also. This was also needed for TDD measurements.

6.6.9 Compressed Mode Measurement Procedure

The compressed mode, often referred to as the slotted mode, is needed when making measurements from another frequency in a CDMA system without a full dual receiver terminal. The compressed mode means that transmission and reception are halted for a short time, in the order of a few milliseconds, in order to perform measurements on the other frequencies. The intention is not to lose data but to compress the data transmission in the time domain. Frame compression can be achieved with three different methods:

- Lowering the data rate from higher layers, as higher layers have knowledge of the compressed mode schedule for the terminal.

- Increasing the data rate by changing the spreading factor. For example, using spreading factor 64 instead of spreading factor 128 doubles the number of available symbols and makes it very straightforward to achieve the desired compression ratio for the frame.

- Reducing the symbol rate by puncturing at the physical layer multiplexing chain. In practice, this is limited to the rather short Transmission Gap Lengths (TGL), since puncturing has some practical limits. The benefit is obviously in keeping the existing spreading factor and not causing new requirements for channelisation code usage.

The compressed frames are provided normally in the downlink and in some cases in the uplink as well. If they appear in the uplink, they need to be simultaneous with the downlink frames, as illustrated in Figure 6.28.

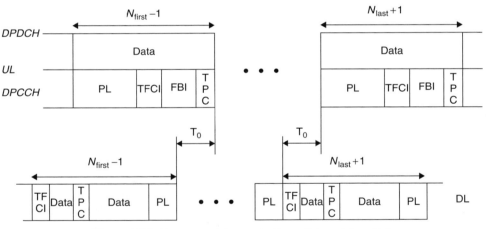

Figure 6.28. Compressed frames in the uplink and downlink

The specified TGL lengths are 3, 4, 5, 7, 10 and 14 slots. TGL lengths of 3, 4 and 7 can be obtained with both single- and double-frame methods. For TGL lengths of 10 or 14 only the double-frame method can be used. An example of the double-frame method is illustrated in Figure 6.29, where the idle slots are divided between two frames. This allows minimising the impact during a single frame and keeping, for example, the required increment in the transmission power lower than with the single-frame method.

The case when uplink compressed frames are always needed with UTRA is the GSM1800 measurements, where the close proximity of the GSM1800 downlink frequency band to the

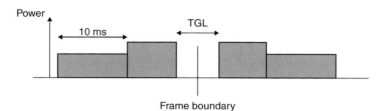

Figure 6.29. Compressed mode with the double-frame method

core UTRA FDD uplink frequency band at 1920 MHz and upwards is too close to allow simultaneous transmission and reception.

Use of the compressed mode in the uplink with GSM900 measurements or UTRA inter-frequency handover depends on terminal capability. For maintaining the continuous uplink, the terminal needs to have a means of generating the additional frequency parallel while maintaining the existing frequency. In practice, this means additional oscillators for frequency generation as well as some other duplicated components, which add to terminal power consumption.

The use of compressed mode has an inevitable impact on link performance, as studied in [10] for the uplink compressed mode and in [11] for the downlink. Link performance does not deteriorate very much if the terminal is not at the cell edge, since there is room to compensate the momentary performance loss with fast power control. The impact is largest at the cell edge; the difference in uplink performance between compressed mode and non-compressed mode is very slight until headroom is less than 4 dB. At 0 dB headroom the difference from normal transmission is between 2 and 4 dB, depending on the transmission gap duration with compressed frames. The 0B headroom corresponds to terminal operation at full power at the cell edge with no possibility of (soft) handover and with no room to run fast power control any more. The use of soft handover (or handover in general) will improve the situation, since low headroom values are less likely to occur, as with typical planning there is some overlap in the cell coverage area and the 0 dB headroom case should occur only when leaving the coverage area. The compressed mode performance is analysed in Section 9.3.2.

The actual time available for sampling on another frequency is reduced from the above values, due to the time taken by the hardware to switch the frequency; thus very short values of 1 or 2 slots have been excluded, since there is no really practical time available for measurements. The smallest value used in the specifications is 3, which itself allows only a very short measurement time window and should be considered for use only in specific cases.

6.6.10   Other Measurements

In the base station other measurements are needed to give RNC sufficient information on uplink status and base station transmission power resource usage. The following have been specified for the base station, to be supported by signalling between base station and RNC:

- RSSI, to give information on the uplink load;

- Uplink SIR on the DPCCH;

- Total transmission power on a single carrier at a base station transmitter, giving information on the available power resources at the base station;

- The transmission code on a single code for one terminal. This is used, for example, in balancing power between radio links in soft handover;

- Block Error Rate (BLER) and Bit Error Rate (BER) estimates for different physical channels.

The BLER measurement is to be supported by the terminals as well. The main function of terminal BLER measurement is to provide feedback for outer loop power control operation in setting the SIR target for fast power control operation.

Support of position location functionality needs measurements from the physical layer. For that purpose a second type of timing measurement has been specified that gives the timing difference between the primary scrambling codes of different cells with $\frac{1}{4}$-chip resolution for improved position location accuracy. The achievable position accuracy in theory can thus be estimated from the fact that a single chip corresponds to roughly 70 m in distance. In a cellular environment there are obviously further factors contributing to the achievable accuracy. To alleviate the impact of the near–far problem for a terminal that is very close to a base station, the specifications contain also a method of introducing idle periods in base station transmission. This enables timing measurements from base stations that would otherwise be too weak due to close proximity of the serving base station.

6.6.11   Operation with Adaptive Antennas

UTRA has been designed to allow the use of adaptive antennas, also known as beamforming, in both the uplink and downlink directions. Basically there are two types of beamforming one may use. Either a beam may use the secondary common pilot channel (S-CPICH) or then a beam may use only the dedicated pilot symbols. From the physical layer point of view the use of adaptive antennas is fully covered with Release '99 but the exact performance requirements for the terminals in different operation scenarios are covered in Release 5.

What kind of beamforming may be applied with different channels depends on whether the channel contains dedicated pilot symbols or not. For the S-CCPCH, beamforming could be used in theory, but in reality it is not practical as the channel is intended to be received by several terminals and thus the slot formats with pilots are not supported by the terminals. Also, the typically short duration communication would be difficult as it takes some time to adapt the receiver at the terminal for a different delay profile than the P-CPICH transmission. A summary of the use of beamforming on different downlink channel types is given in Table 6.5.

If it is desired to use beamforming together with any of the transmit diversity modes, then S-CPICH needs to be transmitted, including the diversity pilot, in the same antenna beam.

In the UTRAN side the full support for beamforming parameterisation was completed for Release 5. This includes phase reference change signalling from RNC to Node B. To further improve the radio resource management of the beamforming, Release 6 contains added functionality. The uplink SIR measurement has been extended to be possible for all the 'beams' in the uplink direction. The practical example of having e.g., four beams with S-PCICH per sector allows the Node B to report for each terminal from which beam there is the best received signal and thus RNC can make decisions to reconfigure phase reference

Table 6.5. Application of beamforming concepts on downlink physical channel types

Physical channel type	Beamforming with S-CPICH	Beamforming without S-CPICH
P-CCPCH	No	No
SCH	No	No
S-CCPCH	No	No
DPCH	Yes	Yes
PICH	No	No
PDSCH, HS-PDSCH and HS-SCCH (with associated DPCH)	Yes	Yes[1]
AICH	No	No
CSICH	No	No

[1]UE capability with HSDPA

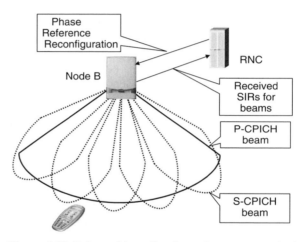

Figure 6.30 Release 6 beamforming enhancement method

(i.e. change the fixed beam) of the terminal. The enhancement is illustrated in Figure 6.30. The physical layer modifications are covered in the Release 6 version of [5].

6.6.12   Site Selection Diversity Transmission

The Site Selection Diversity Transmission (SSDT) is a specific mode of soft handover which was included in Release '99 specifications but was completed from the UTRAN point of view only in Release 5 specification. The main principle of the SSDT feature is that, based on the feedback signalling from the terminal, the Node Bs in the active set may transmit only the DPCCH (control) part of the transmission and use DTX for the data part.

The terminal will send in the uplink the ID of the strongest Node B, based on the measurements on the CPICH, with the intervals between 2 ms and 10 ms, depending on the length of the ID code word selected and the number of feedback bits allocated for SSDT use in the uplink direction. In the network side all Node Bs receive the data from RNC but only the one receiving the correct code word with sufficient quality shall send the data onwards on the DPDCH of the downlink DCH. The *Qth* parameter determines the minimum quality

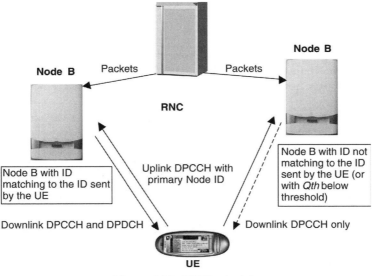

Figure 6.31. SSDT principle

(SIR) level that Node B must receive in order to consider the non-primary commands valid. The SSDT principle is illustrated in Figure 6.31, which shows the example case of two Node Bs in the active set.

Besides the original purpose of SSDT transmission in the downlink, the uplink SSDT code words may be used for other purposes as well. The Release 4 version of [4] covers the use of the received SSDT commands for the power control of DSCH and the Release 5 version of [4] covers the use of the received SSDT commands for TFCI field power control. The key enhancement is to allow optimisation of the use of DSCH or TFCI field power offset based on the short term path loss based on the physical layer feedback information.

The Release '99 and Release 4 specifications do not contain the Qth definition as that was completed in Release 5 specifications. To ensure sufficient reliability of the performance, the Release '99 and 4 UTRAN side uses only long (5 or 10 ms duration) ID words, while with Release 5 the other lengths are also supported. The potential SSDT benefits and their utilisation is covered in connection with Chapter 9. The specifications do not allow simultaneous use of SSDT with HSDPA. This configuration would not make good sense because the DCH with HSDPA is typically rather low rate, thus leaving very small potential for SSDT improvements.

6.7   Terminal Radio Access Capabilities

As explained in Chapter 2, the class mark approach of GSM is not applied in the same way with UMTS. Instead a terminal, upon connection establishment, informs a network of a large set of capability parameters instead of simply one or more class mark values. The reason for this approach has been the large variety of capabilities and data rates with UMTS terminals, which would have resulted in a very high number of different class marks. For practical guidance, reference classes were specified anyway.

The reference classes in [12] have a few common values as well, which are not covered here. For example, the support for spreading factor 512 is not expected to be covered by any of the classes by default. For the channel coding methods, the turbo coding is supported with classes above 32 kbps and with higher classes, the higher data rates above 64 kbps are supported with turbo coding only as can be seen in Tables 6.6 and 6.7. For the convolutional coding, all the classes have the value of 640 bits at an arbitrary time instant for both

Table 6.6. Terminal radio access capability parameter combinations for downlink decoding

Reference combination	32 kbps class	64 kbps class	128 kbps class	384 kbps class	768 kbps class	2048 kbps class
Transport channel parameters						
Maximum sum of number of bits of all transport blocks being received at an arbitrary time instant	640	3840	3840	6400	10 240	20 480
Maximum sum of number of bits of all turbo coded transport blocks being received at an arbitrary time instant	Not supported	3840	3840	6400	10 240	20 480
Maximum number of simultaneous Coded Composite Transport Channels (CCTrCHs), higher value with PDSCH support	1	2/1	2/1	2/1	2	2
Maximum total number of transport blocks received within TTIs that end at the same time	8	8	16	32	64	96
Maximum number of Transport Format Combinations (TFC) in the TFC Set (TFCS)	32	48	96	128	256	1024
Maximum number of Transport Formats	32	64	64	64	128	256
Physical channel parameters						
Maximum number of DPCH/PDSCH codes simultaneously received, higher value with DSCH support	1	2/1	2/1	3	3	3
Maximum number of physical channel bits received in any 10 ms interval (DPCH, PDSCH, S-CCPCH), higher value with DSCH support.	1200	3600/2400	7200/4800	19 200	28 800	57 600
Support of Physical DSCH	No	Yes/No	Yes/No	Yes/No	Yes	Yes

Table 6.7. Terminal radio access capability parameter combinations for uplink encoding

Reference combination	32 kbps class	64 kbps class	128 kbps class	384 kbps class	768 kbps class
Transport channel parameters					
Maximum sum of number of bits of all transport blocks being transmitted at an arbitrary time instant	640	3840	3840	6400	10 240
Maximum sum of number of bits of all turbo coded transport blocks being transmitted at an arbitrary time instant	Not supported	3840	3840	6400	10 240
Maximum total number of transport blocks transmitted within TTIs that start at the same time	4	8	8	16	32
Maximum number of Transport Format Combinations (TFC) in the TFC Set (TFCS)	16	32	48	64	128
Maximum number of Transport Formats	32	32	32	32	64
Physical channel parameters					
Maximum number of DPDCH bits transmitted per 10 ms	1200	2400	4800	9600	19 200

encoding and decoding. This is needed in any case for decoding of broadcast channels. All the classes, except 32 kbps uplink, support at least eight parallel transport channels.

The value given for the number of bits received at an arbitrary time instant needs to be converted to the maximum data rate supported by considering at the same time the interleaving length (or TTI length with 3GPP terminology). For example, the value 6400 bits for 384 kbps class can be converted to the maximum data rate with a particular TTI as follows. The data rate of the application is 256 kbps, thus the number of bits per 10 ms is 2560 bits. With 10 ms or 20 ms TTI lengths the number of bits per interleaving period stays below 6400 bits, but with 40 ms TTI, the 6400 limit would be exceeded and the terminal would not have enough memory to operate with such a configuration. Respectively, a 384 kbps data rate with a terminal of the same class could be maintained with 10 ms TTI, but 20 ms TTI would exceed the limit.

The values given in [12] range to beyond what the classes contain, for example, it is possible for a terminal to indicate values allowing 2 Mbps with 80 ms TTI. The minimum values have been determined by the necessary capabilities needed to access the system, e.g. to listen to the BCH or to access the RACH.

The key physical channel parameter is the maximum number of physical channel bits received/transmitted per 10 ms interval. This determines which spreading factors are supported. For example, value 1200 bits for the 32 kbps class indicated that in the downlink the spreading factors supported are 256, 128 and 64, while in the uplink the smallest value supported would be 64. The difference comes from the use of QPSK modulation in the downlink and BPSK modulation in the uplink, as explained earlier in this chapter in the section on modulation.

There are also parameters that are not dependent on a particular reference combination. Such a parameters indicate, for example, support for a particular terminal position location method. In the RF side the class independent parameters allow the indication of for example, supported frequency bands or the terminal power class.

The parameters in Table 6.6 and 6.7 cover the UTRA FDD, while for UTRA TDD there are a few additional TDD-specific parameters in the complete tables [12], such as the number of slots to be received, etc.

The first terminals on the market are typically providing, in the downlink, 384 kbps, for example the Nokia 7600 illustrated on the book cover. The Nokia 7600 provides the 64 kbps uplink data rate and 384 kbps downlink data rate [13] when in WCDMA mode and, like most of the terminals on the market, is a dual-mode GSM/WCDMA terminal. The terminals available will obviously evolve and, especially in the downlink side, data rates are going to exceed the 2 Mbps limit with HSDPA, as described in Chapter 11.

References

[1] 3GPP Technical Specification 25.211, Physical Channels and Mapping of Transport Channels onto Physical Channels (FDD).
[2] 3GPP Technical Specification 25.212, Multiplexing and Channel Coding (FDD).
[3] 3GPP Technical Specification 25.213, Spreading and Modulation (FDD).
[4] 3GPP Technical Specification 25.214, Physical Layer Procedures (FDD).
[5] 3GPP Technical Specification 25.215, Physical Layer – Measurements (FDD).
[6] 3GPP Technical Specification 25.302, Services Provided by the Physical Layer.
[7] Adachi, F., Sawahashi, M. and Okawa, K., 'Tree-structured Generation of Orthogonal Spreading Codes with Different Lengths for Forward Link of DS-CDMA Mobile', *Electronics Letters*, 1997, Vol. 33, No. 1, pp. 27–28.
[8] 3GPP Technical Specification 25.101, UE Radio Transmission and Reception (FDD).
[9] 3GPP Technical Specification 25.104, UTRA (BS) FDD; Radio Transmission and Reception.
[10] Toskala, A., Lehtinen, O. and Kinnunen, P., 'UTRA GSM Handover from Physical Layer Perspective', *Proc. ACTS Summit 1999*, Sorrento, Italy, June 1999.
[11] Gustafsson, M., Jamal, K. and Dahlman, E., 'Compressed Mode Techniques for Inter-Frequency Measurements in a Wide-band DS-CDMA System', *Proc. IEEE Int. Conf. on Personal Indoor and Mobile Radio Communications*, PIMRC'97, Helsinki, Finland, 1–4 September 1997, Vol. 1, pp. 231–235.
[12] 3GPP Technical Specification 25.306, UE Radio Access Capabilities.
[13] www.nokia.com

7

Radio Interface Protocols

Jukka Vialén and Antti Toskala

7.1 Introduction

The radio interface protocols are needed to set up, reconfigure and release the Radio Bearer services (including the UTRA FDD/TDD service), which were discussed in Chapter 2.

The protocol layers above the physical layer are called the data link layer (Layer 2) and the network layer (Layer 3). In the UTRA FDD radio interface, Layer 2 is split into sub-layers. In the control plane, Layer 2 contains two sub-layers – Medium Access Control (MAC) protocol and Radio Link Control (RLC) protocol. In the user plane, in addition to MAC and RLC, two additional service-dependent protocols exist: Packet Data Convergence Protocol (PDCP) and Broadcast/Multicast Control Protocol (BMC). Layer 3 consists of one protocol, called Radio Resource Control (RRC), which belongs to the control plane. The other network layer protocols, such as Call Control, Mobility Management, Short Message Service, and so on, are transparent to UTRAN and are not described in this book.

In this chapter the general radio interface protocol architecture is first described before going into deeper details of each protocol. For each protocol, the logical architecture and main functions are described. In the MAC section, the logical channels (services offered by MAC) and mapping between logical channels and transport channels are also explained. For MAC and RLC, an example layer model is defined to describe what happens to a data packet passing through these protocols. In the RRC section, the RRC service states are described, together with the main (RRC) functions and signalling procedures. Releases 4 and 5 have not resulted in major modifications to Layer 2/3 protocol principles, the Release 5 High Speed Downlink Packet Access (HSDPA) features have resulted in the new MAC entity in Node B, as presented in Chapter 11. With reference to Release 6, the Multimedia Broadcast Multicast Service (MBMS) principles are introduced. This chapter is then concluded by an introduction to the early UE handling principles.

7.2 Protocol Architecture

The overall radio interface protocol architecture [1] is shown in Figure 7.1. This figure contains only the protocols that are visible in UTRAN.

WCDMA for UMTS, third edition.   Edited by Harri Holma and Antti Toskala
© 2004 John Wiley & Sons, Ltd   ISBN: 0-470-87096-6

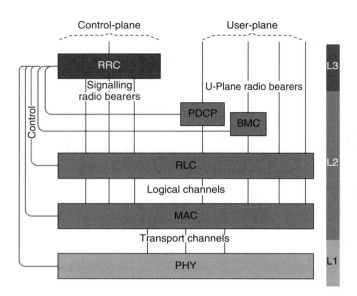

Figure 7.1. UTRA FDD Radio Interface protocol architecture

The physical layer offers services to the MAC layer via transport channels [2] that were characterised by *how and with what characteristics* data is transferred (transport channels were discussed in Chapter 6).

The MAC layer, in turn, offers services to the RLC layer by means of logical channels. The logical channels are characterised by *what type of data* is transmitted. Logical channels are described in detail in Section 7.3.3.

The RLC layer offers services to higher layers via service access points (SAPs), which describe how the RLC layer handles the data packets and if, for example, the automatic repeat request (ARQ) function is used. On the control plane, the RLC services are used by the RRC layer for signalling transport. On the user plane, the RLC services are used either by the service-specific protocol layers PDCP or BMC or by other higher-layer u-plane functions (e.g. speech codec). The RLC services are called Signalling Radio Bearers in the control plane and Radio Bearers in the user plane for services not utilising the PDCP or BMC protocols. The RLC protocol can operate in three modes – transparent, unacknowledged and acknowledged mode. These are further discussed in Section 7.4.

The Packet Data Convergence Protocol (PDCP) exists only for the PS domain services. Its main function is header compression. Services offered by PDCP are called Radio Bearers.

The Broadcast Multicast Control protocol (BMC) is used to convey over the radio interface messages originating from the Cell Broadcast Centre. In Release '99 of the 3GPP specifications, the only specified broadcasting service is the SMS Cell Broadcast service, which is derived from GSM. The service offered by BMC protocol is also called a Radio Bearer.

The RRC layer offers services to higher layers (to the Non-Access Stratum) via service access points, which are used by the higher layer protocols in the UE side and by the Iu RANAP protocol in the UTRAN side. All higher layer signalling (mobility management, call control, session management, and so on) is encapsulated into RRC messages for transmission over the radio interface.

The control interfaces between the RRC and all the lower layer protocols are used by the RRC layer to configure characteristics of the lower layer protocol entities, including parameters for the physical, transport and logical channels. The same control interfaces are used by the RRC layer, for example to command the lower layers to perform certain types of measurement and by the lower layers to report measurement results and errors to the RRC.

7.3   The Medium Access Control Protocol

In the Medium Access Control (MAC) layer [3] the logical channels are mapped to the transport channels. The MAC layer is also responsible for selecting an appropriate transport format (TF) for each transport channel depending on the instantaneous source rate(s) of the logical channels. The transport format is selected with respect to the transport format combination set (TFCS) which is defined by the admission control for each connection.

7.3.1   MAC Layer Architecture

The MAC layer logical architecture is shown in Figure 7.2.

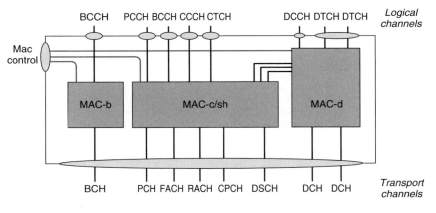

Figure 7.2. MAC layer architecture

The MAC layer consists of three *logical entities*:

1. **MAC-b** handles the broadcast channel (BCH). There is one MAC-b entity in each UE and one MAC-b in the UTRAN (located in Node B) for each cell.

2. **MAC-c/sh** handles the common channels and shared channels – paging channel (PCH), forward link access channel (FACH), random access channel (RACH), uplink Common Packet Channel (CPCH) and Downlink Shared Channel (DSCH). There is one MAC-c/sh entity in each UE that is using shared channel(s) and one MAC-c/sh in the UTRAN

(located in the controlling RNC) for each cell. Note that the BCCH logical channel can be mapped to either the BCH or FACH transport channel. Since the MAC header format for the BCCH depends on the transport channel used, two BCCH instances are shown in the figure. For PCCH, there is no MAC header, thus the only function of the MAC layer is to forward the data received from PCCH to the PCH at the time instant determined by RRC.

3. **MAC-d** is responsible for handling dedicated channels (DCH) allocated to a UE in connected mode. There is one MAC-d entity in the UE and one MAC-d entity in the UTRAN (in the serving RNC) for each UE.

7.3.2 MAC Functions

The functions of the MAC layer include:

- **Mapping between logical channels and transport channels**.

- **Selection of appropriate Transport Format (from the Transport Format Combination Set) for each Transport Channel, depending on the instantaneous source rate**.

- **Priority handling between data flows of one UE**. This is achieved by selecting 'high bit rate' and 'low bit rate' transport formats for different data flows.

- **Priority handling between UEs by means of dynamic scheduling**. A dynamic scheduling function may be applied for common and shared downlink transport channels FACH and DSCH.

- **Identification of UEs on common transport channels**. When a common transport channel (RACH, FACH or CPCH) carries data from dedicated-type logical channels (DCCH, DTCH), the identification of the UE (Cell Radio Network Temporary Identity (C-RNTI) or UTRAN Radio Network Temporary Identity (U-RNTI)) is included in the MAC header.

- **Multiplexing/demultiplexing of higher layer PDUs into/from transport blocks delivered to/from the physical layer on common transport channels**. MAC handles service multiplexing for common transport channels (RACH/FACH/CPCH). This is necessary, since it cannot be done in the physical layer.

- **Multiplexing/demultiplexing of higher layer PDUs into/from transport block sets delivered to/from the physical layer on dedicated transport channels**. MAC allows service multiplexing also for dedicated transport channels. While the physical layer multiplexing makes it possible to multiplex any type of service, including services with different quality of service parameters, MAC multiplexing is possible only for services with the same QoS parameters. Physical layer multiplexing is described in Chapter 6.

- **Traffic volume monitoring**. MAC receives RLC PDUs together with status information on the amount of data in the RLC transmission buffer. MAC compares the amount of data corresponding to a transport channel with the thresholds set by RRC. If the amount of data is too high or too low, MAC sends a measurement report on traffic volume status to RRC. The RRC can also request MAC to send these measurements periodically. The RRC uses these reports for triggering reconfiguration of Radio Bearers and/or Transport Channels.

- **Dynamic Transport Channel type switching**. Execution of the switching between common and dedicated transport channels is based on a switching decision derived by RRC.

- **Ciphering**. If a radio bearer is using transparent RLC mode, ciphering is performed in the MAC sub-layer (MAC-d entity). Ciphering is a XOR operation (as in GSM and GPRS) where data is XORed with a ciphering mask produced by a ciphering algorithm. In MAC ciphering, the time-varying input parameter (COUNT-C) for the ciphering algorithm is incremented at each transmission time interval (TTI), that is, once every 10, 20, 40 or 80 ms depending on the transport channel configuration. Each radio bearer is ciphered separately. The ciphering details are described in 3GPP specification TS 33.102 [4].

- **Access Service Class (ASC) selection for RACH transmission**. The PRACH resources (i.e. access slots and preamble signatures for FDD) may be divided between different Access Service Classes in order to provide different priorities of RACH usage. The maximum number of ASCs is eight. MAC indicates the ASC associated with a PDU to the physical layer.

7.3.3 Logical Channels

The data transfer services of the MAC layer are provided on logical channels. A set of logical channel types is defined for the different kinds of data transfer service offered by MAC. A general classification of logical channels is into two groups: Control Channels and Traffic Channels. Control Channels are used to transfer control plane information, and Traffic Channels for user plane information.

The Control Channels are:

- Broadcast Control Channel (BCCH). A downlink channel for broadcasting system control information.

- Paging Control Channel (PCCH). A downlink channel that transfers paging information.

- Dedicated Control Channel (DCCH). A point-to-point bidirectional channel that transmits dedicated control information between a UE and the RNC. This channel is established during the RRC connection establishment procedure.

- Common Control Channel (CCCH). A bidirectional channel for transmitting control information between the network and UEs. This logical channel is always mapped onto RACH/FACH transport channels. A long UTRAN UE identity is required (U-RNTI, which includes SRNC address), so that the uplink messages can be routed to the correct serving RNC even if the RNC receiving the message is not the serving RNC of this UE.

The Traffic Channels are:

- Dedicated Traffic Channel (DTCH). A Dedicated Traffic Channel (DTCH) is a point-to-point channel, dedicated to one UE, for the transfer of user information. A DTCH can exist in both uplink and downlink.

- Common Traffic Channel (CTCH). A point-to-multipoint downlink channel for transfer of dedicated user information for all, or a group of specified, UEs.

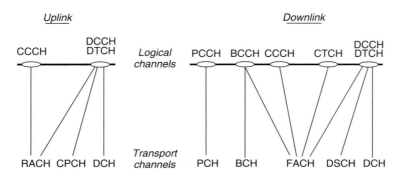

Figure 7.3. Mapping between logical channels and transport channels, uplink and downlink directions

7.3.4   Mapping Between Logical Channels and Transport Channels

The mapping between logical channels and transport channels is shown in Figure 7.3. The following connections between logical channels and transport channels exist:

- PCCH is connected to PCH.

- BCCH is connected to BCH and may also be connected to FACH.

- DCCH and DTCH can be connected to either RACH and FACH, to CPCH and FACH, to RACH and DSCH, or to a DCH and DCH.

- CCCH is connected to RACH and FACH.

- CTCH is connected to FACH.

7.3.5   Example Data Flow Through the MAC Layer

To illustrate the operation of the MAC layer, a block diagram, Figure 7.4, shows the MAC functions when data is processed through the layer. To keep the figure readable, the viewpoint is selected to be a network side transmitting entity, and uplink transport channels RACH and CPCH are omitted. The right-hand side of the figure describes the building of a MAC PDU when a packet received from DCCH or DTCH logical channel is processed by the MAC functions, which are shown in the left-hand side of the figure. In this example the MAC PDU is forwarded to the FACH transport channel.

A data packet arriving from the DCCH/DTCH logical channel triggers first the transport channel type selection in the MAC layer. In this example, the FACH transport channel is selected. In the next phase, the multiplexing unit adds a C/T field indicating the logical channel instance where the data originates. For common transport channels, such as FACH, this field is always needed. For dedicated transport channels (DCH) it is needed only if several logical channel instances are configured to use the same transport channel. The C/T field is 4 bits, allowing up to 15 simultaneous logical channels per transport channel (the value '1111' for the C/T field is reserved for future use). The priority tag (not part of the MAC PDU) for FACH and DSCH is set in MAC-d and used by MAC-c/sh when scheduling data onto transport channels. Priority for FACH can be set per UE; for DSCH it can be set

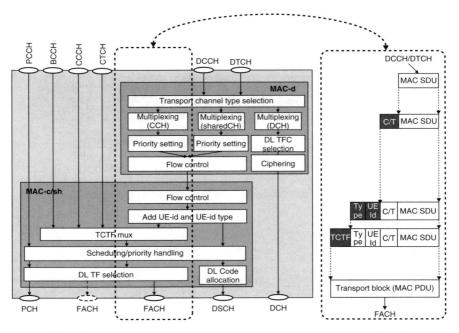

Figure 7.4. UTRAN side MAC entity (left side of figure) and building of MAC PDU when data received from DTCH or DCCH is mapped to FACH (right side of figure)

per PDU. A flow control function in Iur interface (Chapter 5) is needed to limit buffering between MAC-d and MAC-c/sh (which can be located in different RNCs). After receiving the data from MAC-d, the MAC-c/sh entity first adds the UE identification type (2 bits), the actual UE identification (C-RNTI 16 bits, or U-RNTI 32 bits), and the Target Channel Type Field (TCTF, in this example 2 bits) which is needed to separate the logical channel type using the transport channel (for FACH, the possible logical channel types could be BCCH, CCCH, CTCH or DCCH/DTCH). Now the MAC PDU is ready and the task for the scheduling/priority handling function is to decide the exact time when the PDU is passed to Layer 1 via the FACH transport channel (with an indication of the transport format to be used).

7.4   The Radio Link Control Protocol

The radio link control protocol [5] provides segmentation and retransmission services for both user and control data. Each RLC instance is configured by RRC to operate in one of three modes: transparent mode (Tr), unacknowledged mode (UM) or acknowledged mode (AM). The service the RLC layer provides in the control plane is called Signalling Radio Bearer (SRB). In the user plane, the service provided by the RLC layer is called a Radio Bearer (RB) only if the PDCP and BMC protocols are not used by that service, otherwise the RB service is provided by the PDCP or BMC.

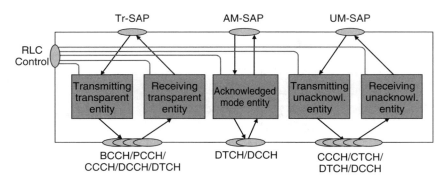

Figure 7.5. RLC layer architecture

7.4.1 RLC Layer Architecture

The RLC layer architecture is shown in Figure 7.5. All three RLC entity types and their connection to RLC-SAPs and to logical channels (MAC-SAPs) are shown. Note that the transparent and unacknowledged mode RLC entities are defined to be unidirectional, whereas the acknowledged mode entities are described as bidirectional.

For all RLC modes, the CRC error detection is performed on the physical layer and the result of the CRC check is delivered to RLC, together with the actual data.

In *transparent mode* no protocol overhead is added to higher layer data. Erroneous protocol data units (PDUs) can be discarded or marked erroneous. Transmission can be of the streaming type, in which higher layer data is not segmented, though in special cases transmission with limited segmentation/reassembly capability can be accomplished. If segmentation/reassembly is used, it has to be negotiated in the radio bearer set-up procedure. The UMTS Quality of Service classes, including the streaming class, were introduced in Chapter 2.

In *unacknowledged mode* no retransmission protocol is in use and data delivery is not guaranteed. Received erroneous data is either marked or discarded depending on the configuration. On the sender side, a timer-based discard without explicit signalling function is applied, thus RLC SDUs which are not transmitted within a specified time are simply removed from the transmission buffer. The PDU structure includes sequence numbers so that the integrity of higher layer PDUs can be observed. Segmentation and concatenation are provided by means of header fields added to the data. An RLC entity in unacknowledged mode is defined as unidirectional, because no association between uplink and downlink is needed. The unacknowledged mode is used, for example, for certain RRC signalling procedures, where the acknowledgement and retransmissions are part of the RRC procedure. Examples of user services that could utilise unacknowledged mode RLC are the cell broadcast service (see Section 7.6) and voice over IP (VoIP).

In the *acknowledged mode* an automatic repeat request (ARQ) mechanism is used for error correction. The quality vs. delay performance of the RLC can be controlled by RRC through configuration of the number of retransmissions provided by RLC. In case RLC is unable to deliver the data correctly (max number of retransmissions reached or the transmission time exceeded), the upper layer is notified and the RLC SDU is discarded. Also the peer entity is informed of an SDU discard operation by sending a Move Receiving

Window command (in a STATUS message), so that also the receiver removes all AMD PDUs belonging to the discarded RLC SDU. An acknowledged mode RLC entity is bidirectional and capable of 'piggybacking' an indication of the status of the link in the opposite direction into user data. RLC can be configured for both in-sequence and out-of-sequence delivery. With in-sequence delivery, the order of higher layer PDUs is maintained, whereas out-of-sequence delivery forwards higher layer PDUs as soon as they are completely received. In addition to data PDU delivery, *status* and *reset* control procedures can be signalled between peer RLC entities. The control procedures can even use a separate logical channel, thus one AM RLC entity can use either one or two logical channels. The acknowledged mode is the normal RLC mode for packet-type services, such as Internet browsing and email downloading, for example.

7.4.2   RLC Functions

The functions of the RLC layer are:

- **Segmentation and reassembly**. This function performs segmentation/reassembly of variable-length higher layer PDUs into/from smaller RLC Payload Units (PUs). One RLC PDU carries one PU. The RLC PDU size is set according to the smallest possible bit rate for the service using the RLC entity. Thus, for variable rate services, several RLC PDUs need to be transmitted during one transmission time interval when any bit rate higher than the lowest one is used.

- **Concatenation**. If the contents of an RLC SDU do not fill an integral number of RLC PUs, the first segment of the next RLC SDU may be put into the RLC PU in concatenation with the last segment of the previous RLC SDU.

- **Padding**. When concatenation is not applicable and the remaining data to be transmitted does not fill an entire RLC PDU of given size, the remainder of the data field is filled with padding bits.

- **Transfer of user data**. RLC supports acknowledged, unacknowledged and transparent data transfer. Transfer of user data is controlled by QoS setting.

- **Error correction**. This function provides error correction by retransmission in the acknowledged data transfer mode.

- **In-sequence delivery of higher layer PDUs**. This function preserves the order of higher layer PDUs that were submitted for transfer by RLC using the acknowledged data transfer service. If this function is not used, out-of-sequence delivery is provided.

- **Duplicate detection**. This function detects duplicated received RLC PDUs and ensures that the resultant higher layer PDU is delivered only once to the upper layer.

- **Flow control**. This function allows an RLC receiver to control the rate at which the peer RLC transmitting entity may send information.

- **Sequence number check (Unacknowledged data transfer mode)**. This function guarantees the integrity of reassembled PDUs and provides a means of detecting corrupted RLC SDUs through checking the sequence number in RLC PDUs when they are reassembled into an RLC SDU. A corrupted RLC SDU is discarded.

- **Protocol error detection and recovery**. This function detects and recovers from errors in the operation of the RLC protocol.

- **Ciphering** is performed in the RLC layer for acknowledged and unacknowledged RLC modes. The same ciphering algorithm is used as for MAC layer ciphering, the only difference being the time-varying input parameter (COUNT-C) for the algorithm, which for RLC is incremented together with the RLC PDU numbers. For retransmission, the same ciphering COUNT-C is used as for the original transmission (resulting in the same ciphering mask); this would not be so if ciphering were on the MAC layer. An identical ciphering mask for retransmissions is essential from Release 5 onwards when the HSDPA feature with physical layer retransmission combining, as described in Chapter 11, is used. The ciphering details are described in 3GPP specification TS 33.102 [4].

- **Suspend/resume function for data transfer**. Suspension is needed during the security mode control procedure so that the same ciphering keys are always used by the peer entities. Suspensions and resumptions are local operations commanded by RRC via the control interface.

7.4.3 Example Data Flow Through the RLC Layer

This section takes a closer look at how data packets pass through the RLC layer. Figure 7.6 shows a simplified block diagram of an AM-RLC entity. The figure shows only how an

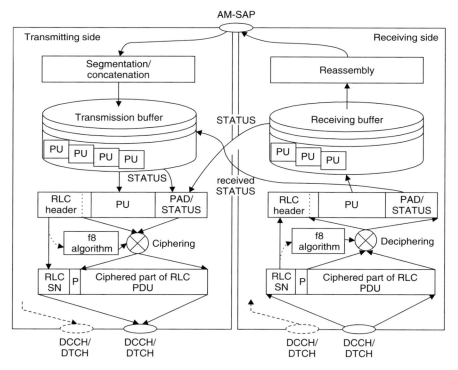

Figure 7.6. A simplified block diagram of an RLC AM entity

AMD PDU can be constructed. It does not show how separate control PDUs (*status, reset*) between RLC entities are built.

Data packets (RLC SDUs) received from higher layers via AM-SAP are segmented and/or concatenated to payload units (PU) of fixed length. The PU length is a semi-static value that is decided in the Radio Bearer set-up and can only be changed through the (RRC) Radio Bearer reconfiguration procedure. For concatenation or padding purposes, bits carrying information on the length and extension are inserted into the beginning of the last PU where data from an SDU is included. If several SDUs fit into one PU, they are concatenated and the appropriate length indicators are inserted into the beginning of the PU. The PUs are then placed in the transmission buffer, which, in this example, also takes care of retransmission management.

An RLC AMD PDU is constructed by taking one PU from the transmission buffer, adding a header for it and, if the data in the PU does not fill the whole RLC AMD PDU, a PADding field or piggybacked STATUS message is appended. The piggybacked STATUS message can originate either from the receiving side (if the peer entity has requested a status report) or from the transmitting side to indicate an RLC SDU discard. The header contains the RLC PDU sequence number SN (12 bits for AM-RLC), poll bit P (which is used to request STATUS from the peer entity) and optionally a length indicator (7 or 15 bits), which is used if concatenation of SDUs, padding or a piggybacked STATUS PDU takes place in the RLC PDU.

Next, the AM RLC PDU is ciphered, excluding the first two octets which comprise the PDU sequence number (SN) and the poll bit (P). The PDU sequence number is one input parameter to the ciphering algorithm (forming the least significant bits of a COUNT-C parameter), and it must be readable by the peer entity to be able to perform deciphering. The details of the ciphering process are described in 3GPP specification TS 33.102 [4].

After this, the PDU is ready to be forwarded to the MAC layer via a logical channel. In Figure 7.6, extra logical channels are shown by dashed lines, indicating that one RLC entity can be configured to send control PDUs and data PDUs using different logical channels. Note, however, that Figure 7.6 does not describe how the separate control PDUs are constructed.

The receiving side of the AM entity receives RLC AMD PDUs through one of the logical channels from the MAC sub-layer. Errors are checked with the (physical layer) CRC, which is calculated over the whole RLC PDU. The actual CRC check is performed in the physical layer and the RLC entity receives the result of this CRC check together with the data. After deciphering, the whole header and possible piggybacked status information can be extracted from the RLC PDU. If the received PDU was a control message, or if status information was piggybacked to an AMD PDU, the control information (STATUS message) is passed to the transmitting side, which will check its retransmission buffer against the received status information. The PDU number from the RLC header is needed for deciphering and also when storing the deciphered PU into the receiving buffer. Once all PUs belonging to a complete SDU are in the receiving buffer, the SDU is reassembled. After this (not shown in the figure), the checks for in-sequence delivery and duplicate detection are performed before the RLC SDU is delivered to the higher layer.

As described in Chapter 11, the same RLC is used with HSDPA. The packets not successfully transmitted from the MAC-hs, when the discard timer expires in Node B, will be retransmitted to the Node B from the RLC layer when acknowledged mode is used. Also, in connection with the various HSDPA mobility cases, there can be packets that are not

transmitted from the Node B, and the Node B MAC-hs buffer will be flushed. The RLC layer
is used to recover the lost data.

7.5   The Packet Data Convergence Protocol

The Packet Data Convergence Protocol (PDCP) [6] exists only in the user plane and only for
services from the PS domain. The PDCP contains compression methods, which are needed
to get better spectral efficiency for services requiring IP packets to be transmitted over the
radio. For 3GPP Release '99 standards, a header compression method is defined, for which
several header compression algorithms can be used. As an example of why header compression is valuable, the size of the combined RTP/UDP/IP headers is at least 40 bytes for IPv4
and at least 60 bytes for IPv6, while the payload, for example for IP voice service, can be
about 20 bytes or less.

7.5.1   PDCP Layer Architecture

An example of the PDCP layer architecture is shown in Figure 7.7. Multiplexing of Radio
Bearers in the PDCP layer is not part of 3GPP Release '99 but is one possible feature for

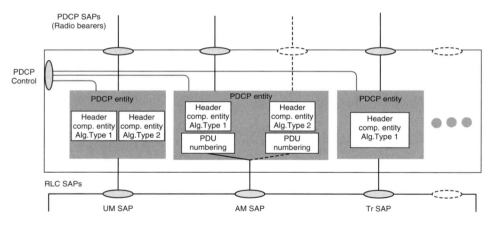

Figure 7.7. The PDCP layer architecture

future releases. The multiplexing possibility is illustrated in Figure 7.7 with the two PDCP
SAPs (one with dashed lines) provided by one PDCP entity using AM RLC. Every PDCP
entity uses zero, one or several header compression algorithm types with a set of
configurable parameters. Several PDCP entities may use the same algorithm types. The
algorithm types and their parameters are negotiated during the RRC Radio Bearer establishment or reconfiguration procedures and indicated to the PDCP through the PDCP Control
Service Access Point.

7.5.2   PDCP Functions

The main PDCP functions are:

- Compression of redundant protocol control information (e.g. TCP/IP and RTP/UDP/IP headers) at the transmitting entity, and decompression at the receiving entity. The header compression method is specific to the particular network layer, transport layer or upper layer protocol combinations, for example TCP/IP and RTP/UDP/IP. The only compression method that is mentioned in the PDCP Release '99 specification is RFC2507 [7].

- Transfer of user data. This means that the PDCP receives a PDCP SDU from the non-access stratum and forwards it to the appropriate RLC entity and vice versa.

- Support for lossless SRNS relocation. In practice this means that those PDCP entities which are configured to support lossless SRNS relocation have PDU sequence numbers, which, together with unconfirmed PDCP packets are forwarded to the new SRNC during relocation. Only applicable when PDCP is using acknowledged mode RLC with in-sequence delivery.

7.6   The Broadcast/Multicast Control Protocol

The other service-specific Layer 2 protocol – he Broadcast/Multicast Control (BMC) protocol [8] – exists also only in the user plane. This protocol is designed to adapt broadcast and multicast services, originating from the Broadcast domain, on the radio interface. In Release '99 of the standard, the only service utilising this protocol is the SMS Cell Broadcast service. This service is directly taken from GSM. It utilises UM RLC using the CTCH logical channel which is mapped into the FACH transport channel. Each SMS CB message is targeted to a geographical area, and RNC maps this area into cells.

7.6.1   BMC Layer Architecture

The BMC protocol, shown in Figure 7.8, does not have any special logical architecture.

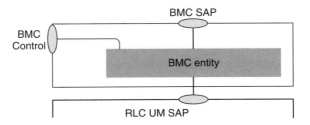

Figure 7.8. The Broadcast/Multicast Control layer architecture

7.6.2   BMC Functions

The main functions of the BMC protocol are:

- **Storage of Cell Broadcast messages**. The BMC in RNC stores the Cell Broadcast messages received over the CBC–RNC interface for scheduled transmission.

- **Traffic volume monitoring and radio resource request for CBS.** On the UTRAN side, the BMC calculates the required transmission rate for the Cell Broadcast Service based on the messages received over the CBC–RNC interface, and requests appropriate CTCH/FACH resources from RRC.

- **Scheduling of BMC messages.** The BMC receives scheduling information together with each Cell Broadcast message over the CBC–RNC interface. Based on this scheduling information, on the UTRAN side the BMC generates schedule messages and schedules BMC message sequences accordingly. On the UE side, the BMC evaluates the schedule messages and indicates scheduling parameters to RRC, which are used by RRC to configure the lower layers for CBS discontinuous reception.

- **Transmission of BMC messages to UE.** This function transmits the BMC messages (Scheduling and Cell Broadcast messages) according to the schedule.

- **Delivery of Cell Broadcast messages to the upper layer.** This UE function delivers the received non-corrupted Cell Broadcast messages to the upper layer.

When sending SMS CB messages to a cell for the first time, appropriate capacity has to be allocated in the cell. The CTCH has to be configured and the transport channel used has to be indicated to all UEs via (RRC) system information broadcast on the BCH. The capacity allocated for SMS CB is cell-specific and may vary over time to allow efficient use of the radio resources.

7.7   Multimedia Broadcast Multicast Service

In Release 6 the Multimedia Broadcast Multicast Service (MBMS) is being added to the standard [9], as first introduced in Chapter 2. The principle is similar to the CBS, in that it enables transmission of content to multiple users in a point-to-multipoint manner. The difference from CBS is that MBMS enables UTRAN also to control and monitor the users receiving the data, and thus enables charging for the content being delivered via MBMS. Typically, CBS has been used for low rate information, like sending cell location name etc., but with MBMS, the mostly quoted data rate has been 64 kbps, which enables more sophisticated content to be distributed. Depending on the number of users that have joined to receive the content via the MBMS, the network can select whether to use point-to-point or point-to-multipoint transmission. In the former case, DCH is used as the transport channel (or should be used) and in the case where several UEs want to receive the same service, FACH is used as the transport channel in a particular cell for the MBMS content. On the physical layer, the FACH is mapped on S-CCPCH and DCH respectively of the DPDCH, see Chapter 6 for the physical channel description of the DPDCH and S-CCPCH.

In the case of point-to-point connection, the logical channel can be DCCH or DTCH with all the mapping in Release '99 possible. For the point-to-multipoint case, there are two new logical channels, namely:

- MBMS point-to-multipoint Control Channel (MCCH), which carries the related control information;

- MBMS point-to-multipoint Traffic Channel (MTCH), which carries the actual user data.

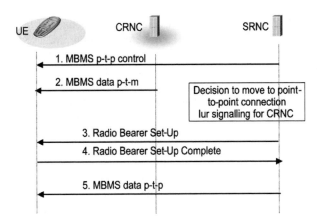

Figure 7.9. MBMS point-to-point and point-to-multipoint transmissions

The UTRAN shall decide, based on the number of UEs in a particular cell, which mode of MBMS operation to use, and if the situation changes, the network can transfer the UEs between different states of MBMS reception, as indicated in Figure 7.9. Typically, there need to be more than just a few UEs to receive the same content in order to make the use of a broadcast channel without power control efficient enough. An example scenario is shown in Figure 7.10 where one cell uses point-to-multipoint while an other cell has only one joined UE which is kept in the point-to-point state. From the MBMS operation point of view,

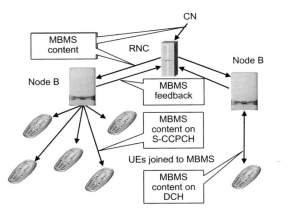

Figure 7.10 MBMS signalling example for changing to point-to-point connection

procedures are obviously simpler if the content is always provided in a point-to-multipoint manner without shifting users back and forth between different states.

As the MBMS content flow may vary, MBMS-specific paging can be used to save the terminal power. The bits not used in Release '99 on the Paging Indicator Channel (PICH), can be used for MBMS purposes to prevent UEs from continuous FACH decoding.

As such, the MBMS does not cause direct changes to the physical layer and thus it is not addressed in Chapter 6. There are, however, a few items to note. MBMS data can be detected with selective combining. The physical layer is not impacted as this is based on selection in the RLC layer and a UE may decode the MBMS data from a new cell quickly, assuming the UTRAN provides on the MCCH the MBMS neighbouring cell information to allow the UE the necessary details for decoding the data from the other cell if the same service was provided there. Otherwise, UE would need to wait until related information was broadcast in the new cell, and only then determine whether the same service could be continued there or not.

In 3GPP standardisation, the issue of whether there should be additional channel coding (outer coding) for extra protection of MBMS data was also discussed. It was decided, however, that at most there could be some application level added protection to tolerate frame losses due to mobility and other challenges related to non-power controlled point-to-multipoint transmission.

7.8   The Radio Resource Control Protocol

The major part of the control signalling between UE and UTRAN is Radio Resource Control (RRC) [10, 11] messages. RRC messages carry all parameters required to set up, modify and release Layer 2 and Layer 1 protocol entities. RRC messages carry in their payload also all higher layer signalling (MM, CM, SM, etc.). The mobility of user equipment in the connected mode is controlled by RRC signalling (measurements, handovers, cell updates, etc.).

7.8.1   RRC Layer Logical Architecture

The RRC layer logical architecture is shown in Figure 7.11.

The RRC layer can be described with four *functional entities:*

- The Dedicated Control Function Entity (DCFE) handles all functions and signalling specific to one UE. In the SRNC there is one DCFE entity for each UE having an RRC connection with this RNC. DCFE uses mostly acknowledged mode RLC (AM-SAP), but

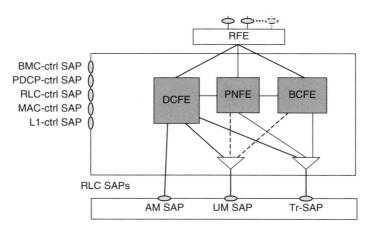

Figure 7.11. RRC layer architecture

some messages are sent using unacknowledged mode SAP (e.g. RRC Connection Release) or transparent SAP (e.g. Cell Update). DCFE can utilise services from all Signalling Radio Bearers, which are described in Section 7.8.3.4.

- The Paging and Notification control Function Entity (PNFE) handles paging of idle mode UE(s). There is at least one PNFE in the RNC for each cell controlled by this RNC. The PNFE uses the PCCH logical channel normally via transparent SAP of RLC. However, the specification mentions that PNFE could utilise also UM-SAP. In this example architecture the PNFE in RNC, when receiving a paging message from an Iu interface, needs to check with the DCFE whether or not this UE already has an RRC connection (signalling connection with another CN domain); if it does, the paging message is sent (by the DCFE) using the existing RRC connection.

- The broadcast control function entity (BCFE) handles the system information broadcasting. There is at least one BCFE for each cell in the RNC. The BCFE uses either BCCH or FACH logical channels, normally via transparent SAP. The specification mentions that BFCE could utilise also UM-SAP.

- The fourth entity is normally drawn outside of the RRC protocol, but still belonging to access stratum and 'logically' to the RRC layer, since the information required by this entity is part of RRC messages. The entity is called Routing Function Entity (RFE) and its task is the routing of higher layer (non-access stratum) messages to different MM/CM entities (UE side) or different core network domains (UTRAN side). Every higher layer message is piggybacked into the RRC *Direct Transfer* messages (three types of Direct Transfer message are specified, *Initial Direct Transfer* (uplink), *Uplink Direct Transfer* and *Downlink Direct Transfer*).

7.8.2   RRC Service States

The two basic operational modes of a UE are *idle mode* and *connected mode*. The connected mode can be further divided into service states, which define what kind of physical channels a UE is using. Figure 7.12 shows the main RRC service states in the connected mode. It also

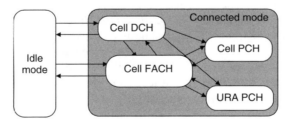

Figure 7.12. UE modes and RRC states in connected mode

shows the transitions between idle mode and connected mode and the possible transitions within the connected mode.

In the idle mode [12], after the UE is switched on, it selects (either automatically or manually) a PLMN to contact. The UE looks for a suitable cell of the chosen PLMN, chooses that cell to provide available services, and tunes to its control channel. This choosing is known as 'camping on a cell'. The cell search procedure described in Chapter 6 is part of this camping process. After camping on a cell in idle mode, the UE is able to receive system information and cell broadcast messages. The UE stays in idle mode until it transmits a request to establish an RRC connection (Section 7.8.3.4). In idle mode the UE is identified by non-access stratum identities such as IMSI, TMSI and P-TMSI. In addition, the UTRAN has no information of its own about the individual idle mode UEs and can only address, for example, all UEs in a cell or all UEs monitoring a paging occasion.

In the *Cell_DCH* state a dedicated physical channel is allocated to the UE, and the UE is known by its serving RNC on a cell or active set level. The UE performs measurements and sends measurement reports according to measurement control information received from RNC. The DSCH can also be used in this state, and UEs with certain capabilities are also able to monitor the FACH channel for system information messages.

In the *Cell_FACH* state no dedicated physical channel is allocated for the UE, but RACH and FACH channels are used instead, for transmitting both signalling messages and small amounts of user plane data. In this state the UE is also capable of listening to the broadcast channel (BCH) to acquire system information. The CPCH channel can also be used when instructed by UTRAN. In this state the UE performs cell reselections, and after a reselection always sends a Cell Update message to the RNC, so that the RNC knows the UE location on a cell level. For identification, a C-RNTI in the MAC PDU header separates UEs from each other in a cell. When the UE performs cell reselection it uses the U-RNTI when sending the Cell Update message, so that UTRAN can route the Cell Update message to the current serving RNC of the UE, even if the first RNC receiving the message is not the current SRNC. The U-RNTI is part of the RRC message, not in the MAC header. If the new cell belongs to another radio access system, such as GPRS, the UE enters idle mode and accesses the other system according to that system's access procedure.

In the *Cell_PCH* state the UE is still known on a cell level in SRNC, but it can be reached only via the paging channel (PCH). In this state the UE battery consumption is less than in the Cell_FACH state, since the monitoring of the paging channel includes a discontinuous reception (DRX) functionality. The UE also listens to system information on BCH. A UE supporting the Cell Broadcast Service (CBS) is also capable of receiving BMC messages in this state. If the UE performs a cell reselection, it moves autonomously to the Cell_FACH state to execute the Cell Update procedure, after which it re-enters the Cell_PCH state if no other activity is triggered during the Cell Update procedure. If a new cell is selected from another radio access system, the UTRAN state is changed to idle mode and access to the other system is performed according to that system's specifications.

The *URA_PCH* state is very similar to the Cell_PCH, except that the UE does not execute Cell Update after each cell reselection, but instead reads UTRAN Registration Area (URA) identities from the broadcast channel, and only if the URA changes (after cell reselection) does UE inform its location to the SRNC. This is achieved with the URA Update procedure, which is similar to the Cell Update procedure (the UE enters the Cell_FACH state to execute the procedure and then reverts to the URA_PCH state). One cell can belong to one or many URAs, and only if the UE cannot find its latest URA identification from the list of URAs in a cell does it need to execute the URA Update procedure. This 'overlapping URA' feature is needed to avoid ping-pong effects in a possible network configuration, where geographically succeeding base stations are controlled by different RNCs.

The UE leaves the connected mode and returns to idle mode when the RRC connection is released or at RRC connection failure.

7.8.2.1   Enhanced State Model for Multimode Terminals

Figure 7.13 presents an overview of the possible state transitions of a multimode terminal, in this example a UTRA FDD–GSM/GPRS terminal. With these terminal types it is possible to perform inter-system handover between UTRA FDD and GSM, and inter-system cell reselection from UTRA FDD to GPRS. The actual signalling procedures that relate to the thick arrows in Figure 7.13 are described in Section 7.8.3.

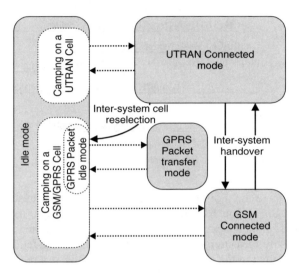

Figure 7.13. UE RRC states for a dual mode UTRA FDD–GSM/GPRS terminal

7.8.2.2   Example state transition cases with packet data

Understanding what is involved in the signalling for the RRC state changes is essential when analysing the system performance in the case of packet data operation. When sending or receiving reasonable amounts of data, UE will stay in Cell_DCH state but once the data runs out and timers have elapsed, the UE will be moved away from Cell_DCH state.

Moving back to the Cell_DCH state always requires signalling between UE and SRNC, as well as for the network to set up the necessary links to Node B. The use of Cell_DCH or Cell_FACH state is always a trade-off between terminal power consumption, service delay, signalling load and network resource utilisation. The timing impacts from state changes are analysed in Chapter 10.

The first case is based on the UE-initiated state change, where an application has created data to be transmitted to the network and the amount is such that going to Cell_FACH state and sending the data on RACH is not sufficient, a DCH needs to be set up. The signalling flow is illustrated in Figure 7.14. For changing to Cell_FACH state there is no need to send signalling to the network. Once in Cell_FACH state, the UE initiates signalling on the RACH and after the network has received the measurement report on RACH and a radio link has been set up between Node B and RNC, the reconfiguration message is sent on FACH to inform of the DCH parameters to be used.

The network-initiated RRC state change occurs when there is too much downlink data to be transmitted, and using FACH is not enough. The network first transmits the paging

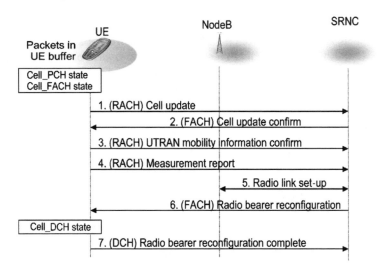

Figure 7.14. UE-initiated RRC state transition

message in the cell where the terminal is located (as the terminal location is known at cell level in Cell_PCH state). Upon reception of the paging message, the terminal moves to Cell_FACH state and initiates signalling on the RACH, as illustrated in Figure 7.15. Now there is no need for any measurement report as transition is initated by the network. The response from the terminal in the example case of Figure 7.15 is a reconfiguration complete message, assuming the DCH parameters have been altered in connection to the state transition.

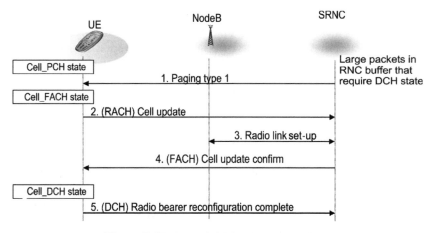

Figure 7.15 Network-initiated state transition

7.8.3 RRC Functions and Signalling Procedures

Since the RRC layer handles the main part of control signalling between the UEs and UTRAN, it has a long list of functions to perform. Most of these functions are part of the

RRM algorithms, which are discussed in Chapters 9 and 10, but since the information is carried in RRC layer messages, the specifications list the functions as part of the RRC protocol. The main RRC functions are:

- Broadcast of system information, related to access stratum and non-access stratum;

- Paging;

- Initial cell selection and reselection in idle mode;

- Establishment, maintenance and release of an RRC connection between the UE and UTRAN;

- Control of Radio Bearers, transport channels and physical channels;

- Control of security functions (ciphering and integrity protection);

- Integrity protection of signalling messages;

- UE measurement reporting and control of the reporting;

- RRC connection mobility functions;

- Support of SRNS relocation;

- Support for downlink outer loop power control in the UE;

- Open loop power control;

- Cell broadcast service related functions;

- Support for UE Positioning functions.

In the following sections, these functions and related signalling procedures are described in more detail.

7.8.3.1 Broadcast of System Information

The broadcast system information originates from the Core Network, from RNC and from Node Bs. The *System Information* messages are sent on a BCCH logical channel, which can be mapped to the BCH or FACH transport channel. A System Information message carries *system information blocks* (SIBs), which group together system information elements of the same nature. Dynamic (i.e. frequently changing) parameters are grouped into different SIBs from the more static parameters. One System Information message can carry either several SIBs or only part of one SIB, depending on the size of the SIBs to be transmitted. One System Information message will always fit into the size of a BCH or FACH transport block. If padding is required, it is inserted by the RRC layer.

The system information blocks are organised as a tree (Figure 7.16). A *master information block* (MIB) gives references and scheduling information to a number of system information blocks in a cell. It may also include reference and scheduling information to one or two *scheduling blocks*, which give references and scheduling information for all additional system information blocks. The master information block is sent regularly on the BCH and its scheduling is static. In addition to scheduling information of other SIBs and scheduling blocks, the master information block contains only the parameters 'Supported PLMN Types' and, depending on which PLMN types are supported, either 'PLMN identity' (GSM MAP)

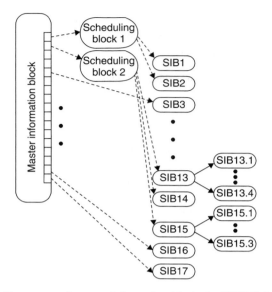

Figure 7.16. The overall structure of system information blocks in 3GPP Release '99. Dotted arrows show an example where scheduling information for each SIB could be included

or 'ANSI-41 Core Network Information'. The system information blocks contain all the other actual system information.

The scheduling information (included in the MIB or in scheduling blocks) for SIBs containing frequently changing parameters contains SIB-specific timers (value in frames), which can be used by the UE to trigger re-reading of each block.

For the other SIBs (with more 'static' parameters) the master information block, or the 'parent' SIB, contains, as part of the scheduling information, a 'value tag' that the UE compares to the latest read 'value tag' of this system information block. Only if the value tag has changed after the last reading of the SIB in question does the UE re-read it. Thus, by monitoring the master information block and the scheduling blocks, the UE can notice if any of the system information blocks (of the more 'static' nature) has changed. UTRAN can also inform of the change in system information with *Paging* messages sent on the PCH transport channel (see Section 7.8.3.2) or with a *System Information Change Indication* message on the FACH transport channel. With these two messages, all UEs needing information about a change in the system information (all UEs in the Cell_FACH, Cell_PCH and URA_PCH states) can be reached.

The number of system information blocks in 3GPP Release '99 is one master information block, two scheduling blocks and 17 SIBs. Only SIB number 10, containing information needed only in Cell_DCH state, is sent using FACH transport channel, all other SIBs (including MIB and scheduling blocks) are sent on BCH. Scheduling information for each SIB can be included only in one place, either in MIB or in one of the scheduling blocks.

7.8.3.2 Paging

The RRC layer can broadcast paging information on the PCCH from the network to selected UEs in a cell. The paging procedure can be used for three purposes:

1. At core network-originated call or session set-up. In this case the request to start paging comes from the Core Network via the Iu interface.

2. To change the UE state from Cell_PCH or URA_PCH to Cell_FACH. This can be initiated, for example, by downlink packet data activity.

3. To indicate change in the system information. In this case RNC sends a paging message with no paging records but with information describing a new 'value tag' for the master information block. This type of paging is targeted to all UEs in a cell.

7.8.3.3   Initial Cell Selection and Reselection in Idle Mode

The most suitable cell is selected, based on idle mode measurements and cell selection criteria. The cell search procedure described in Chapter 6 is part of the cell selection process.

7.8.3.4   Establishment, Maintenance and Release of RRC Connection

The establishment of an RRC connection and Signalling Radio Bearers (SRB) between UE and UTRAN (RNC) is initiated by a request from higher layers (non-access stratum) on the UE side. In a network-originated case, the establishment is preceded by an RRC *Paging* message. The request from non-access stratum is actually a request to set up a Signalling Connection between UE and CN (Signalling Connection consists of an RRC connection and an Iu connection). Only if the UE is in idle mode, thus no RRC connection exists, does the UE initiate RRC Connection Establishment procedure. There can always be only zero or one RRC connections between one UE and UTRAN. If more than one signalling connection between UE and CN nodes exist, they all 'share' the same RRC connection.

The 'maintenance' of RRC connection refers to the RRC Connection Re-establishment functionality, which can be used to re-establish a connection after radio link failure. Timers are used to control the allowed time for a UE to return to 'in-service-area' and to execute the re-establishment. The re-establishment functionality is included in the Cell Update procedure (7.8.3.9).

The RRC connection establishment procedure is shown in Figure 7.17. There is no need for a contention resolution step such as in GSM [13], since the UE identifier used in the connection request and set-up messages is a unique UE identity (for GSM-based core network P-TMSI+RAI, TMSI+LAI or IMSI). In the RRC connection establishment procedure this initial UE identifier is used only for the purpose of uniqueness and can be discarded by UTRAN after the procedure ends. Thus, when these UE identities are later needed for the higher layer (non-access stratum) signalling, they must be resent (in the higher layer messages). The RRC *Connection Set-up* message may include a dedicated physical channel assignment for the UE (move to Cell_DCH state), or it can command the UE to use common channels (move to Cell_FACH state). In the latter case, a radio network temporary identity (U-RNTI and possibly C-RNTI) to be used as UE identity on common transport channels is allocated to the UE.

The channel names in Figure 7.17 indicate either the logical channel or logical/transport channel used for each message.

The RRC connection establishment procedure creates three (optionally four) Signalling Radio Bearers (SRBs) designated by the RB identities #1 ... #4 (RB identity #0 is reserved for signalling using CCCH). The SRBs can later be created, reconfigured or even deleted

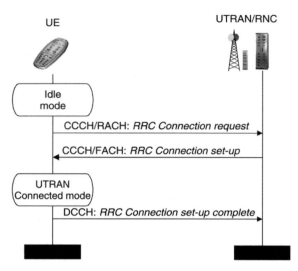

Figure 7.17. RRC connection establishment procedure

with the normal Radio Bearer control procedures. The SRBs are used for RRC signalling according to the following rules:

1. RB#1 is used for all messages sent on the DCCH and RLC-UM.

2. RB#2 is used for all messages sent on the DCCH and RLC-AM, except for the *Direct Transfer* messages.

3. RB#3 is used for the *Direct Transfer* messages (using DCCH and RLC-AM), which carries higher layer signalling. The reason for reserving a dedicated signalling radio bearer for the *Direct Transfer* is to enable prioritisation of UE–UTRAN signalling over the UE–CN signalling by using the RLC services (no need for extra RRC functionality).

4. RB#4 is optional and, if it exists, is also used for the *Direct Transfer* messages (using DCCH and RLC-AM). With two SRBs carrying higher layer signalling, UTRAN can handle prioritisation on signalling, RB#4 being used for 'low priority' and RB#3 for 'high priority' NAS signalling. The priority level is indicated to RRC with the actual NAS message to be carried over the radio. An example of low priority signalling could be the SMS.

5. For RRC messages utilising transparent mode RLC and CCCH logical channel (e.g. Cell Update, URA Update), RB identity #0 is used. A special function required in the RRC layer for these messages is padding, because RLC in transparent mode neither imposes size requirements nor performs padding, but the message size must still be equal to a Transport Block size.

7.8.3.5   Control of Radio Bearers, Transport Channels and Physical Channels

On request from higher layers, RRC performs the establishment, reconfiguration and release of Radio Bearers. At establishment and reconfiguration, UTRAN (RNC) performs admission control and selects parameters describing the Radio Bearer processing in Layer 2 and Layer 1. The SRBs are normally set up during the RRC Connection Establishment procedure (Section 7.8.3.4) but can also be controlled with the normal Radio Bearer procedures.

The transport channel and physical channel parameters are included in the Radio Bearer procedures but can also be configured separately with transport channel and physical channel dedicated procedures. These are needed, for example, if temporary congestion occurs in the network or when switching the UE between Cell_DCH and Cell_FACH states.

7.8.3.6 Control of Security Functions

The RRC *Security Mode Control* procedure is used to start ciphering and integrity protection between the UE and UTRAN and to trigger the change of the ciphering and integrity keys during the connection.

The ciphering key is CN domain specific; thus in a typical network configuration (see Chapter 5), two ciphering keys can be used simultaneously for one UE – one for the PS domain services and one for the CS domain services. For the signalling (that uses common Radio Bearer(s) for both CN domains) the newer of these two keys is used. Ciphering is executed on the RLC layer for services using unacknowledged or acknowledged RLC and on the MAC layer for services using transparent RLC.

Integrity protection (see next section) is used only for signalling. In a typical network configuration two integrity keys would be available, but since only one RRC Connection can exist per UE, all signalling is protected with one and the same integrity key, which is always the newer of the keys IK_{CS} and IK_{PS}.

7.8.3.7 Integrity Protection of Signalling

The RRC layer inserts a 32-bit integrity checksum, called a Message Authentication Code MAC-I, into most RRC PDUs. The integrity checksum is used by the receiving RRC entity to verify the origin and integrity of the messages. The receiving entity also calculates MAC-I and compares it to the one received with the signalling message. Messages received with wrong or missing message authentication codes are discarded. Since all higher layer (non-access stratum) signalling is carried in the *RRC Direct Transfer* messages, all higher layer messages are automatically also integrity protected. The only exception to this is the initial higher layer message carried in the *Initial Direct Transfer* message.

The checksum is calculated using the UMTS integrity algorithm (UIA) that uses a secret 128-bit integrity key (IK) as one input parameter. The key is generated, together with the ciphering key (CK), during the authentication procedure [14]. Figure 7.18 illustrates the calculation of MAC-I using the integrity algorithm f9 [15]. In addition to the IK, other

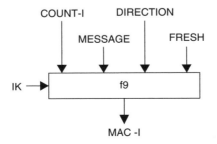

Figure 7.18. Calculation of message authentication code MAC-I

parameters used as input to the algorithm are COUNT-I, which is incremented by one for each integrity protected message, a random number FRESH generated by RNC, DIRECTION bit (uplink/downlink), and the actual signalling message. Also the signalling radio bearer identity should affect the calculation of MAC-I. Since the algorithm f9 was ready when this requirement was identified, no new input parameter could be added to the f9 algorithm. The signalling radio bearer identity is inserted into the MESSAGE before it is given to the integrity algorithm.

Only a few RRC messages cannot be integrity protected; examples are the messages exchanged during the RRC Connection Establishment procedure, since the algorithms and parameters are not yet negotiated when these messages are sent.

7.8.3.8 UE Measurement Reporting and Control

The measurements performed by the UE are controlled by the RNC using RRC protocol messages, in terms of what to measure, when to measure and how to report, including both UTRA radio interface and other systems. RRC signalling is also used in reporting of the measurements from the UE to the UTRAN (RNC).

Measurement Control

The measurement control (and reporting) procedure is designed to be very flexible. The serving RNC may start, stop or modify a number of parallel measurements in the UE and each of these measurements (including how they are reported) can be controlled independently of each other. The measurement control information is included in *System Information Block Type 12* and *System Information Block Type 11*. For UEs in Cell_DCH state also a dedicated *Measurement Control* message can be used. This is illustrated in Figure 7.19.

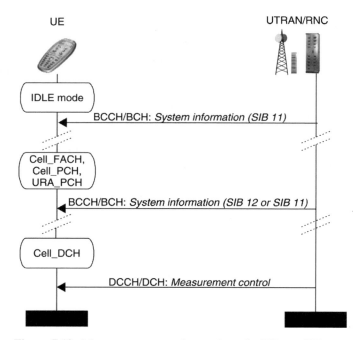

Figure 7.19. Measurement control procedures in different UE states

The measurement control information includes:

- *Measurement identity number*: A reference number that is used by the UTRAN at modification or release of the measurement and by the UE in the measurement report.

- *Measurement command*: May be either set-up, modify or release.

- *Measurement type*: One of the seven types from a predefined list, where each type describes what the UE measures. The seven types of measurement are defined as:

    - Intra-frequency measurements: measurements on downlink physical channels at the same frequency as the active set.

    - Inter-frequency measurements: measurements on downlink physical channels at frequencies that differ from the frequency of the active set.

    - Inter-system measurements: measurements on downlink physical channels belonging to a radio access system other than UTRAN, e.g. GSM.

    - Traffic volume measurements: measurements on uplink traffic volume, e.g. RLC buffer payload for each Radio Bearer.

    - Quality measurements: measurements of quality parameters, e.g. downlink transport channel block error rate.

    - Internal measurements: measurements of UE transmission power and UE received signal level.

    - Measurements for Location Services (LCS) [16]. The basic measurement provided by the UE for the network-based OTDOA-IPDL positioning method is Observed Time Difference of system frame numbers (SFN) between measured cells.

- *Measurement objects*: The objects the UE shall measure, and corresponding object information. In handover measurements this is the cell information needed by the UE to make measurements on certain intra-frequency, inter-frequency or inter-system cells. In traffic volume measurements this parameter contains transport channel identification.

- *Measurement quantity*: The quantity the UE measures.

- *Measurement reporting quantities*: The quantities the UE includes in the report.

- *Measurement reporting criteria*: The criteria that trigger the measurement report, such as periodical or event-triggered reporting.

- *Reporting mode*: This specifies whether the UE transmits the measurement report using acknowledged or unacknowledged data transfer of RLC.

Measurement Reporting

The measurement reporting procedure – shown in Figure 7.20 – is initiated from the UE side when the reporting criteria are met. The UE sends a *Measurement Report* message, including the measurement identity number and the measurement results.

The *Measurement Report* message is used in the Cell_DCH and Cell_FACH states. In the Cell_FACH state, it is used only for a traffic volume measurement report. Traffic volume measurements may be triggered also in Cell_PCH and URA_PCH states, but the UE has to first change to Cell_FACH state before being able to send a measurement report. In order to

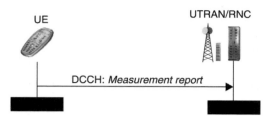

Figure 7.20. Measurement reporting procedure

receive measurement information needed for the immediate establishment of macro diversity when establishing a dedicated physical channel, the UTRAN may also request the UE to append radio link-related (intra-frequency) measurement reports to the following messages when they are sent on the RACH channel:

- *RRC Connection Request* message sent to establish an RRC connection.
- *Initial Direct Transfer* and *Uplink Direct Transfer* messages.
- *Cell Update* message.
- *Measurement Report* message sent to report uplink traffic volume in Cell_FACH state.

7.8.3.9 RRC Connection Mobility Functions

RRC 'connection mobility' means keeping track of a UE's location (on a cell or active set level) while the UE is in UTRAN Connected mode. For this, a number of RRC procedures are defined. When dedicated channels are allocated to a UE, a normal way to perform mobility control is to use *Active Set Update* and *Hard Handover* procedures. When the UE is using only common channels (RACH/FACH/PCH) while in the UTRAN Connected mode, specific procedures are used to keep track of UE location either on cell or on UTRAN Registration Area (URA) level.

The UE mobility-related RRC procedures include:

- *Active Set Update* to update the UE's active set while in the Cell_DCH state.

- *Hard Handover* to make inter-frequency or intra-frequency hard handovers while in the Cell_DCH state.

- *Inter-system handover* between UTRAN and another radio access system (e.g. GSM).

- *Inter-system cell reselection* between UTRAN and another radio access system (e.g. GPRS).

- *Inter-system cell change order* between UTRAN and another radio access system (e.g. GPRS).

- *Cell Update* to report the UE location to RNC while in the Cell_FACH or Cell_PCH state.

- *URA Update* to report the UE location to RNC while in the URA_PCH state.

These procedures are described in the following sections.

Active Set Update

The purpose of the active set update procedure is to update the active set of the connection between the UE and UTRAN while the UE is in the Cell_DCH state. The procedure – shown in Figure 7.21 – can have one of the following three functions: radio link addition; radio link removal; or combined radio link addition and removal. The maximum number of simultaneous radio links is eight and it is possible to remove even all of them with one Active Set Update command. The soft handover algorithm and its performance are discussed in Section 9.3.1.

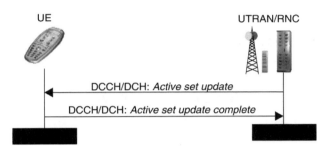

Figure 7.21. Active Set Update procedure

Hard Handover

The Hard Handover procedure can be used to change the radio frequency band of the connection between the UE and UTRAN or to change the cell on the same frequency when no network support of macro diversity exists. It can also be used to change the mode between FDD and TDD. This procedure is used only in the Cell_DCH state. No dedicated signalling messages have been defined for the Hard Handover but the functionality can be performed as part of the following RRC procedures: Physical channel reconfiguration; Radio bearer establishment; Radio bearer reconfiguration; Radio bearer release and Transport channel reconfiguration.

Inter-System Handover from UTRAN

The inter-system handover from UTRAN procedure is shown in Figure 7.22. This procedure is used for handover from UTRAN to another radio access system when the UE has at least one RAB in use for a CS domain service. For Release '99 UE, only support of handover of one RAB is expected, although the specification allows also handover of multiple RABs and even RABs from CS and PS domains simultaneously. In this example the target system is GSM, but the specifications also support handover to PCS 1900 and cdma2000 radio access systems. This procedure may be used in the Cell_DCH and Cell_FACH states. The UE receives the GSM neighbour cell parameters [13] either on *System Information* or in a *Measurement Control* message. These parameters are required to be able to measure candidate GSM cells. Based on the measurement report from UE, including GSM measurements, RNC makes a handover decision. After resources have been reserved from GSM BSS, the RNC sends a *Handover From UTRAN Command* message that carries a piggybacked *GSM Handover Command*. At this point the GSM RR protocol in UE takes control and sends a *GSM Handover Access* message to GSM BSC. After successful

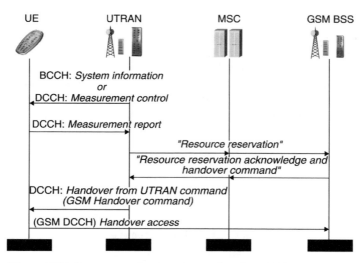

Figure 7.22. Inter-system handover procedure from UTRAN to GSM

completion of the handover procedure, GSM BSS initiates resource release from UTRAN which will release the radio connection and remove all context information for the UE concerned.

Inter-System Handover to UTRAN

The inter-system handover to UTRAN procedure is shown in Figure 7.23. This procedure is used for handover from a non-UTRAN system to UTRAN. In this example, the other system

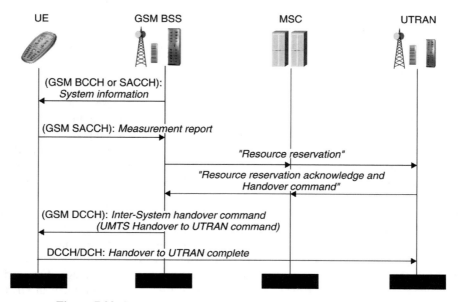

Figure 7.23. Inter-system handover procedure from GSM to UTRAN

is again GSM. The dual-mode UE receives the UTRAN neighbour cell parameters on *GSM System Information* messages. The parameters required to be able to measure UTRA FDD cells include Downlink Centre frequency or UTRA Absolute Radio Frequency Channel Number (UARFCN), Downlink Bandwidth (in 3GPP Release '99, only 5 MHz allowed, but in future other bandwidths may appear), Downlink Scrambling Code or scrambling code group for the Primary Common Pilot Channel (CPICH), and reference time difference for UTRA cell (timing between the current GSM cell and the UMTS cell that is to be measured).

After receiving a measurement report from a GSM MS, including UTRA measurements, and after making a handover decision, the GSM BSC initiates resource reservation from UTRAN RNC. In the next phase, GSM BSC sends a *GSM Inter-System Handover Command* [13], including a piggybacked UMTS *Handover To UTRAN Command* message, which contains all the information required to set up connection to a UTRA cell. The GSM handover message (*Inter-System Handover Command*) must fit into one 23-octet data link layer PDU. Since the amount of information that could be included in the *Handover To UTRAN Command* is great, a preconfiguration mechanism is included in the standards. The preconfiguration means that only a reference number to a predefined set of UTRA parameters (Radio Bearer, Transport Channel and Physical Channel parameters) is included in the message. Naturally, the preconfiguration has to be transmitted to the UE beforehand. This can be done by GSM signalling or, if the UE has earlier been in UMTS mode it has been able to read the preconfiguration information from System Information Block type 16. The UE completes the procedure with a *Handover to UTRAN Complete* message to RNC. After successful completion of the handover procedure, RNC initiates resource release from GSM BSS.

Inter-System Cell Reselection from UTRAN

The inter-system cell reselection procedure from UTRAN is used to transfer a connection between the UE and UTRAN to another radio access system, such as GSM/GPRS. This procedure may be initiated in states Cell_FACH, Cell_PCH or URA_PCH. It is controlled mainly by the UE, but to some extent also by UTRAN. When UE has initiated an establishment of a connection to the other radio access system, it shall release all UTRAN-specific resources.

Inter-System Cell Reselection to UTRAN

The inter-system cell reselection procedure to UTRAN is used to transfer a connection between the UE and another radio access system, such as GSM/GPRS, to UTRAN. This procedure is controlled mainly by the UE, but to some extent also by the other radio access system. The UE initiates an RRC connection establishment procedure to UTRAN with cause value 'Inter-system cell reselection', and releases all resources specific to the other radio access system.

Inter-System Cell Change Order from UTRAN

The inter-system cell change order procedure – illustrated in Figure 7.24 – can be used by the UTRAN to order UE to another radio access system. This procedure is used for UEs having at least one RAB for PS domain services. This procedure may be used in Cell_DCH and Cell_FACH states. As in the case of inter-system handover from UTRAN, Release '99 UE is expected to be able to perform inter-system cell change with only one PS domain RAB, but the specification does not restrict this.

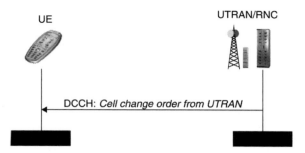

Figure 7.24. Inter-system cell change order from UTRAN

The procedure is initiated by UTRAN with the *Cell Change Order from UTRAN* message. The message contains at least required information of the target cell. After successful establishment of connection between UE and the other radio access system (e.g. GSM/GPRS), the other radio access system initiates release of the used UTRAN radio resources and UE context information.

Inter-System Cell Change Order to UTRAN

This procedure is used by the other radio access system (e.g. GSM/GPRS) to command UE to move to UTRAN cell. The 'cell change order' message in the other radio access system shall include the identity of the target UTRAN cell. On the UTRAN side, the UE shall initiate an RRC connection establishment procedure with 'establishment cause' set to 'Inter-RAT cell change order'.

Cell Update

The Cell Update procedure can be triggered by several reasons, including cell reselection, expiry of periodic cell update timer, initiation of uplink data transmission, UTRAN-originated paging and radio link failure in Cell_DCH state.

The *Cell Update Confirm* may include UTRAN mobility information elements (new U-RNTI and C-RNTI) for the UE. In this case, it responds with a *UTRAN Mobility Information Confirm* message so that the RNC knows that the new identities are taken into use.

The Cell Update Confirm may also include a radio bearer release, radio bearer reconfiguration, transport channel reconfiguration or physical channel reconfiguration. In these cases, the UE responds with a suitable 'complete' message, see Figure 7.25.

URA Update

The UTRAN Registration Area (URA) Update procedure is used in the URA_PCH state. It can be triggered either after cell reselection, if the new cell does not broadcast the URA identifier that the UE is following, or by expiry of the periodical URA Update timer. Since no uplink activity is possible in URA_PCH state, the UE has to temporarily switch to Cell_FACH state to execute the signal processing procedure, as shown in Figure 7.26.

UTRAN registration areas may be hierarchical to avoid excessive signalling. This means that several URA identifiers may be broadcast in one cell and that different UEs in one cell may reside in different URAs. A UE in the URA_PCH state always has one and only one valid URA. If a cell broadcasts several URAs, the RNC assigns one URA to a UE in the *URA Update Confirm* message.

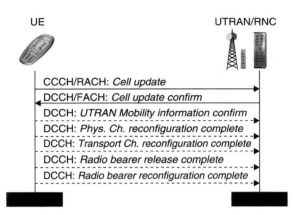

Figure 7.25. Cell Update procedure

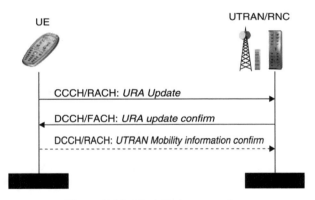

Figure 7.26. URA Update procedure

The *URA Update Confirm* may assign a new URA Identity that the UE has to follow. It may also assign new RNTIs for the UE. In these cases UE responds with a *UTRAN Mobility Information Confirm* message so that the RNC knows that the new identities are taken into use.

7.8.3.10 Support of SRNS Relocation

In the serving RNS relocation procedure (see Chapter 5), the SRNC RRC layer builds a special RRC message – *RRC Information to Target RNC*. The issue that makes this message a 'special' one is it is not targeted for UE but for the new SRNC. Thus, this message is not sent over the air, but carried from the old SRNC to the new one via the Core Network. The initialisation information contains, e.g., RRC state information and all the required protocol parameters (RRC, RLC, MAC, PDCP, PHY) that are needed to set up the UE context in the new SRNC. In addition, the expected PDCP sequence numbers (which are normally maintained locally in UE and UTRAN) need to be sent between UE and UTRAN, in any RRC messages which are sent during the SRNS relocation.

7.8.3.11   Support for Downlink Outer Loop Power Control

All RRC messages that can be used to add or reconfigure downlink transport channels (e.g. *Radio Bearer Set-up/Reconfiguration/Release, Transport Channel Reconfiguration*) include a parameter 'Quality Target' (BLER quality value) that is used to configure the quality requirement (initial downlink SIR target) for each downlink transport channel separately.

The outer loop power control algorithm and its performance are discussed in Section 9.2.2.

7.8.3.12   Open Loop Power Control

Prior to PRACH transmission (see Chapter 6), the UE calculates the power for the first preamble as:

$$\text{Preamble\_Initial\_Power} = \text{Primary CPICH DL TX power} - \text{CPICH\_RSCP}$$
$$+ \text{UL interference} + \text{constant value}$$

The value for the CPICH_RSCP is measured by the UE, all other parameters being received on *System Information*.

As long as the physical layer is configured for PRACH transmission, the UE continuously recalculates the Preamble_Initial_Power when any of the broadcast parameters used in the above formula changes. The new Preamble_Initial_Power is then resubmitted to the physical layer.

When establishing the first DPCCH the UE shall start the UL inner loop power control at a power level according to:

$$\text{DPCCH\_Initial\_power} = \text{DPCCH\_Power\_offset} - \text{CPICH\_RSCP}$$

The value for the DPCCH_Power_offset is received from UTRAN on various signalling messages. The value for the CPICH_RSCP shall be measured by the UE. The open loop power control is not used once the inner loop power control is running.

7.8.3.13   Cell Broadcast Service Related Functions

The CBS-related functions of the RRC layer are as follows:

- Initial configuration of the BMC layer.

- Allocation of radio resources for CBS, in practice allocating the schedule for mapping the CTCH logical channel into the FACH transport channel and further into the S-CCPCH physical channel.

- Configuration of Layer 1 and Layer 2 for CBS discontinuous reception in the UE.

7.8.3.14   UE Positioning Related Functions

Although the full set of Release '99 UTRAN specifications support only the Cell_ID based positioning method, RRC protocol already is capable of supporting also both UE-based and UE-assisted OTDOA and GPS methods [16]. This includes capability to transfer positioning-related UE measurements to UTRAN and delivery of assistance data for OTDOA and/or GPS from UTRAN to UE, which can be done either with *System Information* or with a dedicated message, called *Assistance Data Delivery*.

7.9   Early UE Handling Principles

The topic which created lively discussion during 2002 and the first part of 2003 in 3GPP standardisation was how to handle the potential problems with terminals that are launched on the networks before full testing coverage exists. An example of the problem case is a feature that is mandatory for the terminals but which is not implemented in any of the networks or available in test equipment before the terminal is put on the market. In such a case there can potentially be problems detected in the field because of this. In order to ensure smooth system evolution it was therefore desired to have a method that copes with such terminals without being forced to deactivate a particular feature where only a single UE is having problems.

The basic method chosen was to create a bit map in the core network side from the non-access stratum (NAS) signalling which contains the IMEISV information of a particular UE. The 3GPP standards define the bit map generated either by the SGSN or the MSC. Based on the bit map, RNC can, for example, decide not to activate a particular feature for a given UE, while allowing it to be used by other UEs. The method is illustrated in Figure 7.27.

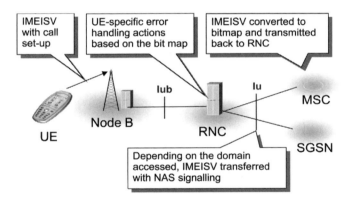

Figure 7.27. Early UE handling principle

To cope with the case where problems related to the first phase of the call set-up arise, a few bits are received in the initial access signalling. However, these bits have not been needed in the field so far. In general, the first approach, familiar from the GSM experience, is to see whether general parameterisation can avoid the problem before using the bit map solution. In 3GPP, preparation has been carried out to create technical reports which can describe either the actions recommended by the bits in the bit map or the recommended specific parameterisation. References [17] and [18] have been proposed (February 2004) for specific cases, but obviously as networks evolve and turn on features not used in available test models or in the networks active today, some problem cases are likely to surface.

References

[1] 3G TS 25.301 Radio Interface Protocol Architecture.
[2] 3G TS 25.302 Services Provided by the Physical Layer.

[3]   3G TS 25.321 MAC Protocol Specification.
[4]   3G TS 24.008 Mobile Radio Interface Layer 3 Specification, Core Network Protocols – Stage 3.
[5]   3G TS 25.322 RLC Protocol Specification.
[6]   3G TS 25.323 PDCP Protocol Specification.
[7]   IETF RFC 2507 IP Header Compression.
[8]   3G TS 25.324 Broadcast/Multicast Control Protocol (BMC) Specification.
[9]   3G TS 25.346 Introduction of the Multimedia Broadcast Multicast Service (MBMS) in the Radio
      Access Network, Stage 2; Release 6. March 2004.
[10] 3G TS 25.303 UE Functions and Interlayer Procedures in Connected Mode.
[11] 3G TS 25.331 RRC Protocol Specification.
[12] 3G TS 25.304 UE Procedures in Idle Mode.
[13] GSM 04.18 Digital Cellular Telecommunications System (Phase 2+); Mobile Radio Interface
      Layer 3 Specification, Radio Resource Control Protocol.
[14] 3G TS 33.102 3G Security: Security Architecture.
[15] 3G TS 33.105 3G Security: Cryptographic Algorithm Requirements.
[16] 3G TS 25.305 Stage 2 Functional Specification of Location Services in UTRAN.
[17] 3GPP TR 25.994: 'Measures employed by the UMTS Radio Access Network (UTRAN) to
      overcome early User Equipment (UE) implementation faults.'
[18] 3GPP TR 25.995: 'Measures employed by the UMTS Radio Access Network (UTRAN) to cater
      for legacy User Equipment (UE) which conforms to superceded versions of the RAN interface
      specification.'

8

Radio Network Planning

Harri Holma, Zhi-Chun Honkasalo, Seppo Hämäläinen, Jaana Laiho, Kari Sipilä and Achim Wacker

8.1 Introduction

This chapter presents WCDMA radio network planning, including dimensioning, detailed capacity and coverage planning, and network optimisation. The WCDMA radio network planning process is shown in Figure 8.1. In the dimensioning phase an approximate number of base station sites, base stations and their configurations and other network elements is estimated, based on the operator's requirements and the radio propagation in the area. The dimensioning must fulfil the operator's requirements for coverage, capacity and quality of service. Capacity and coverage are closely related in WCDMA networks, and therefore both must be considered simultaneously in the dimensioning of such networks. The dimensioning of WCDMA networks is introduced in Section 8.2.

In Section 8.3, detailed capacity and coverage planning is presented, together with a WCDMA planning tool. In detailed planning, real propagation maps and operator's traffic estimates in each area are needed. The base station locations and network parameters are selected by the planning tool and/or the planner. Capacity and coverage can be analysed for each cell after the detailed planning. One case study of the detailed planning is presented in Section 8.3 with capacity and coverage analysis. When the network is in operation, its performance can be observed by measurements, and the results of those measurements can be used to visualise and optimise network performance. The planning and the optimisation process can also be automated with intelligent tools and network elements. The optimisation is introduced in Section 8.3.

As most WCDMA-based networks will be deployed on top of the GSM network, the GSM co-planning issues need to be considered. Co-planning is discussed in Section 8.4.

The adjacent channel interference must be considered in designing any wideband systems where large guard bands are not possible. In Section 8.5, the effect of interference between operators is analysed and network planning solutions are presented.

Section 8.6 presents the WCDMA frequency variants and their differences. These frequency variants are needed in the first place to deploy WCDMA in the USA.

WCDMA for UMTS, third edition.   Edited by Harri Holma and Antti Toskala
© 2004 John Wiley & Sons, Ltd   ISBN: 0-470-87096-6

Input   Output

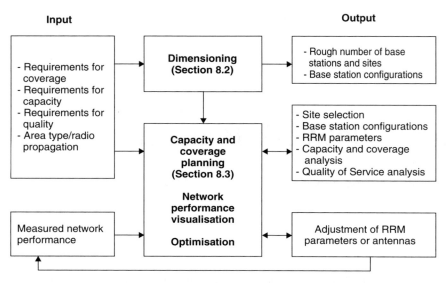

Figure 8.1. WCDMA radio network planning process

8.2 Dimensioning

WCDMA radio network dimensioning is a process through which possible configurations and the amount of network equipment are estimated, based on the operator's requirements related to the following.

Coverage:

- coverage regions;
- area type information;
- propagation conditions.

Capacity:

- spectrum available;
- subscriber growth forecast;
- traffic density information.

Quality of Service:

- area location probability (coverage probability);
- blocking probability;
- end user throughput.

Dimensioning activities include radio link budget and coverage analysis, capacity estimation and, finally, estimations on the amount of sites and base station hardware, radio network controllers (RNC), equipment at different interfaces, and core network elements (i.e. Circuit Switched Domain and Packet Switched Domain Core Networks).

8.2.1 Radio Link Budgets

The link budget of the WCDMA uplink is presented in this section. There are some WCDMA-specific parameters in the link budget that are not used in a TDMA-based radio access system such as GSM. The most important ones are as follows.

- **Interference margin**: The interference margin is needed in the link budget because the loading of the cell, the load factor, affects the coverage: see Section 8.2.2. The more loading is allowed in the system, the larger is the interference margin needed in the uplink, and the smaller is the coverage area. For coverage-limited cases a smaller interference margin is suggested, while in capacity-limited cases a larger interference margin should be used. In the coverage-limited cases, the cell size is limited by the maximum allowed path loss in the link budget, and the maximum air interface capacity of the base station site is not used. Typical values for the interference margin in the coverage-limited cases are 1.0–3.0 dB, corresponding to 20–50 % loading.

- **Fast fading margin (= power control headroom)**: Some headroom is needed in the mobile station transmission power for maintaining adequate closed loop fast power control. This applies especially to slow-moving pedestrian mobiles where fast power control is able to effectively compensate the fast fading. The power control headroom was studied in [1]. The performance of fast power control is discussed in Section 9.2.1. Typical values for the fast fading margin are 2.0–5.0 dB for slow-moving mobiles.

- **Soft handover gain**: Handovers – soft or hard – give a gain against slow fading (= log-normal fading) by reducing the required log-normal fading margin. This is because the slow fading is partly uncorrelated between the base stations, and by making a handover the mobile can select a better base station. Soft handover gives an additional macro diversity gain against fast fading by reducing the required E_b/N_0 relative to a single radio link, due to the effect of macro diversity combining. The total soft handover gain is assumed to be between 2.0 and 3.0 dB in the examples below, including the gain against slow and fast fading. The handovers are discussed in Section 9.3 and the macro diversity gain for the coverage in Section 12.2.

Other parameters in the link budget are discussed in Chapter 7 in [2]. Below, three examples of link budgets are given for typical UMTS services: 12.2 kbps voice service using AMR speech codec, 144 kbps real-time data and 384 kbps non-real-time data, in an urban macro-cellular environment at the planned uplink noise rise of 3 dB. An interference margin of 3 dB is reserved for the uplink noise rise. The assumptions that have been used in the link budgets for the receivers and transmitters are shown in Tables 8.1 and 8.2.

Table 8.1. Assumptions for the mobile station

	Speech terminal	Data terminal
Maximum transmission power	21 dBm	24 dBm
Antenna gain	0 dBi	2 dBi
Body loss	3 dB	0 dB

Table 8.2. Assumptions for the base station

Noise figure	5.0 dB
Antenna gain	18 dBi (3-sector base station)
E_b/N_0 requirement	Speech: 5.0 dB
	144 kbps real-time data: 1.5 dB
	384 kbps non-real-time data: 1.0 dB
Cable loss	2.0 dB

The link budget in Table 8.3 is calculated for 12.2 kbps speech for in-car users, including 8.0 dB in-car loss. No fast fading margin is reserved in this case, since at 120 km/h the fast power control is unable to compensate for the fading. The required E_b/N_0 is assumed to be 5.0 dB. The E_b/N_0 requirement depends on the bit rate, service, multipath profile, mobile speed, receiver algorithms and base station antenna structure. For low mobile speeds, the E_b/N_0 requirement is low but, on the other hand, a fast fading margin is required. Typically, the low mobile speeds are the limiting factor in the coverage dimensioning because of the required fast fading margin. Table 8.4 shows the link budget for a 144 kbps real-time data service when an indoor location probability of 80 % is provided by the outdoor base stations. The main differences between Tables 8.3 and 8.4 are the different processing gain, a higher mobile transmission power and a lower E_b/N_0 requirement. Additionally, a headroom of

Table 8.3. Reference link budget of AMR 12.2 kbps voice service (120 km/h, in-car users, Vehicular A type channel, with soft handover)

Transmitter (mobile)		
Max. mobile transmission power [W]	0.125	
As above in dBm	21.0	a
Mobile antenna gain [dBi]	0.0	b
Body loss [dB]	3.0	c
Equivalent Isotropic Radiated Power (EIRP) [dBm]	18.0	$d = a + b - c$
Receiver (base station)		
Thermal noise density [dBm/Hz]	−174.0	e
Base station receiver noise figure [dB]	5.0	f
Receiver noise density [dBm/Hz]	−169.0	$g = e + f$
Receiver noise power [dBm]	−103.2	$h = g + 10^* \log (3\,840\,000)$
Interference margin [dB]	3.0	i
Total effective noise + interference [dBm]	−100.2	$j = h + i$
Processing gain [dB]	25.0	$k = 10^* \log (3840/12.2)$
Required E_b/N_0 [dB]	5.0	l
Receiver sensitivity [dBm]	−120.2	$m = l - k + j$
Base station antenna gain [dBi]	18.0	n
Cable loss in the base station [dB]	2.0	o
Fast fading margin [dB]	0.0	p
Max. path loss [dB]	**154.2**	$q = d - m + n - o - p$
Log-normal fading margin [dB]	7.3	r
Soft handover gain [dB], multicell	3.0	s
In-car loss [dB]	8.0	t
Allowed propagation loss for cell range [dB]	**141.9**	$u = q - r + s - t$

Table 8.4. Reference link budget of 144 kbps real-time data service (3 km/h, indoor user covered by outdoor base station, Vehicular A type channel, with soft handover)

Transmitter (mobile)		
Max. mobile transmission power [W]	0.25	
As above in dBm	24.0	a
Mobile antenna gain [dBi]	2.0	b
Body loss [dB]	0.0	c
Equivalent Isotropic Radiated Power (EIRP) [dBm]	26.0	$d = a + b - c$
Receiver (base station)		
Thermal noise density [dBm/Hz]	−174.0	e
Base station receiver noise figure [dB]	5.0	f
Receiver noise density [dBm/Hz]	−169.0	$g = e + f$
Receiver noise power [dBm]	−103.2	$h = g + 10^* \log (3\,840\,000)$
Interference margin [dB]	3.0	i
Total effective noise + interference [dBm]	−100.2	$j = h + i$
Processing gain [dB]	14.3	$k = 10^* \log (3840/144)$
Required E_b/N_0 [dB]	1.5	l
Receiver sensitivity [dBm]	−113.0	$m = 1 - k + j$
Base station antenna gain [dBi]	18.0	n
Cable loss in the base station [dB]	2.0	o
Fast fading margin [dB]	4.0	p
Max. path loss [dB]	**151.0**	$q = d - m + n - o - p$
Log-normal fading margin [dB]	4.2	r
Soft handover gain [dB], multicell	2.0	s
Indoor loss [dB]	15.0	t
Allowed propagation loss for cell range [dB]	**133.8**	$u = q - r + s - t$

4.0 dB is reserved for the fast power control to be able to compensate for the fading at 3 km/h. An average building penetration loss of 15 dB is assumed here.

The value on row *q* gives the maximum path loss between the mobile and the base station antennas. The additional margins on rows *r* and *t* are needed to guarantee indoor coverage in the presence of shadowing. The shadowing is caused by buildings, hills etc. and is modelled as log-normal fading. The value on row *u* is used in the calculation of the cell range.

Table 8.5 presents a link budget for a 384 kbps non-real-time data service for outdoors. The processing gain is lower than in the previous tables because of the higher bit rate. Also, the E_b/N_0 requirement is lower than that of the lower bit rates. The effect of the bit rate to the coverage is discussed in Section 12.2. This link budget is calculated assuming no soft handover.

From the link budgets above, the cell range *R* can be readily calculated for a known propagation model, for example the Okumura–Hata model or the Walfish–Ikegami model. For more on propagation models see e.g. [3]. The propagation model describes the average signal propagation in that environment, and it converts the maximum allowed propagation loss in dB on the row *u* to the maximum cell range in kilometres. As an example we can take the Okumura–Hata propagation model for an urban macro cell with base station antenna height of 30 m, mobile antenna height of 1.5 m and carrier frequency of 1950 MHz [4]:

$$L = 137.4 + 35.2 \ \log_{10} (R) \tag{8.1}$$

Table 8.5. Reference link budget of non-real-time 384 kbps data service (3 km/h, outdoor user, Vehicular A type channel, no soft handover)

Transmitter (mobile)		
Max. mobile transmission power [W]	0.25	
As above in dBm	24.0	a
Mobile antenna gain [dBi]	2.0	b
Body loss [dB]	0.0	c
Equivalent Isotropic Radiated Power (EIRP) [dBm]	26.0	$d = a + b - c$
Receiver (base station)		
Thermal noise density [dBm/Hz]	−174.0	e
Base station receiver noise figure [dB]	5.0	f
Receiver noise density [dBm/Hz]	−169.0	$g = e + f$
Receiver noise power [dBm]	−103.2	$h = g + 10^* \log (3\,840\,000)$
Interference margin [dB]	3.0	i
Total effective noise + interference [dBm]	−100.2	$j = h + i$
Processing gain [dB]	10.0	$k = 10^* \log (3840/384)$
Required E_b/N_0 [dB]	1.0	l
Receiver sensitivity [dBm]	−109.2	$m = l - k + j$
Base station antenna gain [dBi]	18.0	n
Cable loss in the base station [dB]	2.0	o
Fast fading margin [dB]	4.0	p
Max. path loss [dB]	**147.2**	$q = d - m + n - o - p$
Log-normal fading margin [dB]	7.3	r
Soft handover gain [dB], multicell	0.0	s
Indoor loss [dB]	0.0	t
Allowed propagation loss for cell range [dB]	**139.9**	$u = q - r + s - t$

where L is the path loss in dB and R is the range in km. For suburban areas we assume an additional area correction factor of 8 dB and obtain the path loss as:

$$L = 129.4 + 35.2 \, \log_{10} (R) \tag{8.2}$$

According to Equation (8.2), the cell range of 12.2 kbps speech service with 141.9 dB path loss in Table 8.3 in a suburban area would be 2.3 km. The range of 144 kbps indoors would be 1.4 km. Once the cell range R is determined, the site area, which is also a function of the base station sectorisation configuration, can then be derived. For a cell of hexagonal shape covered by an omnidirectional antenna, the coverage area can be approximated as $2.6R^2$.

The process of evaluating the cell range is summarised in Figure 8.2.

8.2.2   Load Factors

The second phase of dimensioning is estimating the amount of supported traffic per base station site. When the frequency reuse of a WCDMA system is 1, the system is typically interference-limited and the amount of interference and delivered cell capacity must thus be estimated.

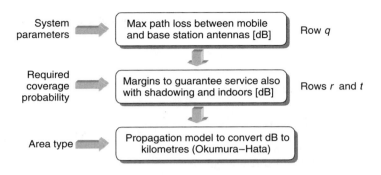

Figure 8.2. Cell range calculation

8.2.2.1 Uplink Load Factor

The theoretical spectral efficiency of a WCDMA cell can be calculated from the load equation whose derivation is shown below. We first define the E_b/N_0, energy per user bit divided by the noise spectral density:

$$(E_b/N_0)_j = \text{Processing gain of user } j \cdot \frac{\text{Signal of user } j}{\text{Total received power (excl.own signal)}} \qquad (8.3)$$

This can be written:

$$(E_b/N_0)_j = \frac{W}{v_j R_j} \cdot \frac{P_j}{I_{\text{total}} - P_j} \qquad (8.4)$$

where W is the chip rate, P_j is the received signal power from user j, v_j is the activity factor of user j, R_j is the bit rate of user j, and I_{total} is the total received wideband power including thermal noise power in the base station. Solving for P_j gives

$$P_j = \frac{1}{1 + \dfrac{W}{(E_b/N_0)_j \cdot R_j \cdot v_j}} I_{\text{total}} \qquad (8.5)$$

We define $P_j = L_j \cdot I_{\text{total}}$ and obtain the load factor L_j of one connection

$$L_j = \frac{1}{1 + \dfrac{W}{(E_b/N_0)_j \cdot R_j \cdot v_j}} \qquad (8.6)$$

The total received interference, excluding the thermal noise P_N, can be written as the sum of the received powers from all N users in the same cell

$$I_{\text{total}} - P_N = \sum_{j=1}^{N} P_j = \sum_{j=1}^{N} L_j \cdot I_{\text{total}} \qquad (8.7)$$

The noise rise is defined as the ratio of the total received wideband power to the noise power

$$\text{Noise rise} = \frac{I_{\text{total}}}{P_N} \tag{8.8}$$

and using Equation (8.7) we can obtain

$$\text{Noise rise} = \frac{I_{\text{total}}}{P_N} = \frac{1}{1 - \sum_{j=1}^{N} L_j} = \frac{1}{1 - \eta_{\text{UL}}} \tag{8.9}$$

where we have defined the load factor η_{UL} as

$$\eta_{\text{UL}} = \sum_{j=1}^{N} L_j \tag{8.10}$$

When η_{UL} becomes close to 1, the corresponding noise rise approaches infinity and the system has reached its pole capacity.

Additionally, in the load factor the interference from the other cells must be taken into account by the ratio of other cell to own cell interference, i:

$$i = \frac{\text{other cell interference}}{\text{own cell interference}} \tag{8.11}$$

The uplink load factor can then be written as

$$\eta_{\text{UL}} = (1 + i) \cdot \sum_{j=1}^{N} L_j = (1 + i) \cdot \sum_{j=1}^{N} \frac{1}{1 + \dfrac{W}{(E_b/N_0)_j \cdot R_j \cdot v_j}} \tag{8.12}$$

The load equation predicts the amount of noise rise over thermal noise due to interference. The noise rise is equal to $-10 \cdot \log_{10}(1 - \eta_{\text{UL}})$. The interference margin on row i in the link budget must be equal to the maximum planned noise rise.

The required E_b/N_0 can be derived from link level simulations, from measurements and from the 3GPP performance requirements. It includes the effect of the closed loop power control and soft handover. The effect of soft handover is measured as the macro diversity combining gain relative to the single link E_b/N_0 result. The other cell to own (serving) cell interference ratio i is a function of cell environment or cell isolation (e.g. macro/micro, urban/suburban) and antenna pattern (e.g. omni, 3-sector or 6-sector [5]). The parameters are further explained in Table 8.6.

The load equation is commonly used to make a semi-analytical prediction of the average capacity of a WCDMA cell, without going into system-level capacity simulations. This load equation can be used for the purpose of predicting cell capacity and planning noise rise in the dimensioning process.

For a classical all-voice-service network, where all N users in the cell have a low bit rate of R, we can note that

$$\frac{W}{E_b/N_0 \cdot R \cdot v} \gg 1 \tag{8.13}$$

Table 8.6. Parameters used in uplink load factor calculation

	Definitions	Recommended values
N	Number of users per cell	
v_j	Activity factor of user j at physical layer	0.67 for speech, assumed 50 % voice activity and DPCCH overhead during DTX 1.0 for data
E_b/N_0	Signal energy per bit divided by noise spectral density that is required to meet a predefined Block error rate, *BLER*. Noise includes both thermal noise and interference	Dependent on service, bit rate, multipath fading channel, receive antenna diversity, mobile speed, etc. See Section 12.5.
W	WCDMA chip rate	3.84 Mcps
R_j	Bit rate of user j	Dependent on service
i	Other cell to own cell interference ratio seen by the base station receiver	Macro cell with omnidirectional antennas: 55 %. Macro cell with 3 sectors: 65 %.

and the above uplink load equation can be approximated and simplified to

$$\eta_{\mathrm{UL}} = \frac{E_b/N_0}{W/R} \cdot N \cdot v \cdot (1+i). \tag{8.14}$$

An example uplink noise rise is shown in Figure 8.3 for data service, assuming an E_b/N_0 requirement of 1.5 dB and $i = 0.65$. The noise rise of 3.0 dB corresponds to a 50 % load factor, and the noise rise of 6.0 dB to a 75 % load factor. Instead of showing the number of users N, we show the total data throughput per cell of all simultaneous users. In this example, a throughput of 860 kbps can be supported with 3.0 dB noise rise, and 1300 kbps with 6.0 dB noise rise.

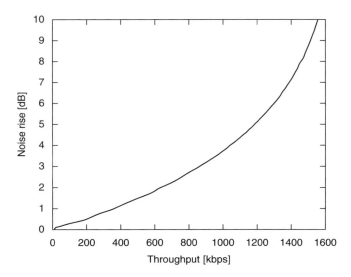

Figure 8.3. Uplink noise rise as a function of uplink data throughput

8.2.2.2 Downlink Load Factor

The downlink load factor, η_{DL}, can be defined based on a similar principle as for the uplink, although the parameters are slightly different [6]:

$$\eta_{DL} = \sum_{j=1}^{N} v_j \cdot \frac{(E_b/N_0)j}{W/R_j} \cdot [(1 - \alpha_j) + i_j] \qquad (8.15)$$

where $-10 \cdot \log_{10}(1 - \eta_{DL})$ is equal to the noise rise over thermal noise due to multiple access interference. The parameters are further explained in Table 8.7. Compared to the uplink load equation, the most important new parameter is α_j, which represents the orthogonality factor in the downlink. WCDMA employs orthogonal codes in the downlink to separate users, and without any multipath propagation the orthogonality remains when the base station signal is received by the mobile. However, if there is sufficient delay spread in the radio channel, the mobile will see part of the base station signal as multiple access interference. The orthogonality of 1 corresponds to perfectly orthogonal users. Typically, the orthogonality is between 0.4 and 0.9 in multipath channels.

Table 8.7. Parameters used in downlink load factor calculation

	Definitions	Recommended values for dimensioning
N	Number of users per cell	
v_j	Activity factor of user j at physical layer	0.58 for speech, assumed 50 % voice activity and DPCCH overhead during DTX 1.0 for data
E_b/N_0	Signal energy per bit divided by noise spectral density, required to meet a predefined Block error rate, *BLER*. Noise includes both thermal noise and interference	Dependent on service, bit rate, multipath fading channel, transmit antenna diversity, mobile speed, etc. See Section 12.5.
W	WCDMA chip rate	3.84 Mcps
R_j	Bit rate of user j	Dependent on service
α_j	Orthogonality of channel of user j	Dependent on the multipath propagation 1: fully orthogonal 1-path channel 0: no orthogonality
i_j	Ratio of other cell to own cell base station power, received by user j	Each user sees a different i_j, depending on its location in the cell and log-normal shadowing
$\bar{\alpha}$	Average orthogonality factor in the cell	ITU Vehicular A channel: ~50 % ITU Pedestrian A channel: ~90 %
$\bar{i}$	Average ratio of other cell to own cell base station power received by user. Own cell interference is here wideband	Macro cell with omnidirectional antennas: 55 %. Macro cell with 3 sectors: 65 %.

Note: The own cell is defined as the best serving cell. If a user is in soft handover, all the other base stations in the active set are counted as part of the 'other cell'.

In the downlink, the ratio of other cell to own cell interference, i, depends on the user location and is therefore different for each user j. The load factor can be approximated by its average value across the cell, that is

$$\overline{\eta_{DL}} = \sum_{j=1}^{N} v_j \cdot \frac{(E_b/N_0)_j}{W/R_j} \cdot [(1 - \bar{\alpha}) + \bar{i}] \qquad (8.16)$$

In downlink interference modelling, the effect of soft handover transmission can be modelled in two different ways:

1. Increase the number of connections by soft handover overhead, and reduce the E_b/N_0 requirement per link with soft handover gain.

2. Keep the number of connections fixed, i.e. equal to the number of users, and use the combined E_b/N_0 requirement.

If the soft handover gain per link is assumed to be 3 dB, the combined E_b/N_0 is the same both with and without soft handover. In that case we do not need to include the effect of soft handover in the air interface dimensioning. This simplified approach is used in the examples later in this chapter. Figure 8.4 illustrates the soft handover modelling in dimensioning with two cells.

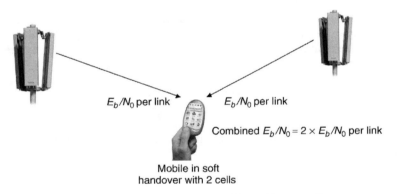

Figure 8.4. Soft handover modelling with two cells

The downlink load factor η_{DL} exhibits very similar behaviour to the uplink load factor η_{UL}, in the sense that when approaching unity, the system reaches its pole capacity and the noise rise over thermal goes to infinity.

For downlink dimensioning, it is important to estimate the total amount of base station transmission power required. This should be based on the *average* transmission power for the user, not the *maximum* transmission power for the cell edge shown by the link budget. The reason is that the wideband technology gives trunking gain in the power amplifier dimensioning: while some users at the cell edge are requiring high power, other users close to the base station need much less power at the same time. The difference between the maximum and the average path loss is typically 6 dB in macro cells and it is illustrated in

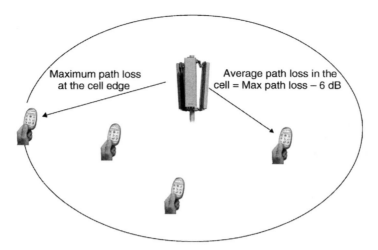

Figure 8.5. Maximum and average path loss in macro cells

Figure 8.5. This effect can be considered as power trunking gain of the wideband technology allowing use of a smaller base station power amplifier than in narrowband technologies.

The minimum required transmission power for each user is determined by the average attenuation between base station transmitter and mobile receiver, that is $\bar{L}$, and the mobile receiver sensitivity, in the absence of multiple access interference (intra- or inter-cell). Then the effect of noise rise due to interference is added to this minimum power and the total represents the transmission power required for a user at an 'average' location in the cell. Mathematically, the total base station transmission power can be expressed by the following equation:

$$ BS\_TxP = \frac{N_{\text{rf}} \cdot W \cdot \bar{L} \cdot \sum_{j=1}^{N} v_j \frac{(E_b/N_0)_j}{W/R_j}}{1 - \overline{\eta_{\text{DL}}}} \tag{8.17} $$

where N_{rf} is the noise spectral density of the mobile receiver front-end. The value of N_{rf} can be obtained from

$$
\begin{aligned}
N_{\text{rf}} &= \text{k} \cdot T + NF \\
&= -174.0\,\text{dBm} + NF \ (\text{assuming } T = 290\,K)
\end{aligned}
\tag{8.18}
$$

where k is the Boltzmann constant of $1.381 \cdot 10^{-23}$ J/K, T is temperature in Kelvin and NF is the mobile station receiver noise figure with typical values of 5–9 dB.

Downlink Common Channels

Part of the downlink power has to be allocated for the common channels that are transmitted independently of the traffic channels. The common channels were introduced in Section 6.5. The amount of power of the common channels affects synchronisation time, channel estimation accuracy, and the reception quality of the broadcast channel. On the other hand, the common channels eat up the capacity of the cell that could otherwise be allocated for the traffic channels. Typical power allocations for the common channels are shown in Table 8.8.

Table 8.8. Typical powers for the downlink common channels [7]

Downlink common channel	Relative to CPICH	Activity	Average power allocation with 20 W maximum power
Common pilot channel CPICH	0 dB	100 %	2.0 W
Primary synchronisation channel SCH	−3 dB	10 %	0.1 W
Secondary synchronisation channel SCH	−3 dB	10 %	0.1 W
Primary common control physical channel P-CCPCH	−5 dB	90 %	0.6 W
Paging indicator channel PICH	−8 dB	100 %[1]	0.3 W
Acquisition indicator channel AICH	−8 dB	100 %[1]	0.3 W
Secondary common control physical channel S-CCPCH	0 dB[2]	10 %[3]	0.2 W
Total common channel powers			3.6 W

[1]Worst case
[2]Depends on the FACH bit rate, 32 kbps assumed here
[3]Depends on the amount of PCH and FACH traffic

8.2.2.3 Example Load Factor Calculation

Example downlink load factor calculations are demonstrated in this section. The assumptions are shown in Table 8.9.

The results are obtained as follows:

1. Assume the required aggregate cell throughput in kbps.

2. Calculate load factor $\overline{\eta_{DL}}$ from Equation (8.16). Throughput is equal to the number of users $N \times$ bit rate $R \times (1 - \text{BLER})$.

Table 8.9. Assumptions in example calculation

Parameter	Data	Voice
Activity factor v_j	1.0	Downlink: 0.58
		Uplink: 0.67
E_b/N_0	5.0 dB	7.0 dB
Block error rate BLER	10 %	1 %
Bit rate of user R_j	64 kbps	12.2 kbps
Mobile antenna gain	2 dBi	0 dBi
WCDMA chip rate W	3.84 Mcps	
Orthogonality $\bar{\alpha}$	0.5	
Other cell to own cell interference ratio i	0.65	
Base station output power	20 W	
Common channel power allocation	15 % of base station max power = 3 W	
Base station cable loss in downlink	3 dB	
Base station cable loss in uplink	0 dB, cable loss compensated with mast head amplifier	
Average mobile noise figure N_{rf}	7 dB	
Maximum vs. average path loss	6 dB	

3. Calculate average path loss from Equation (8.17).

4. Calculate maximum path loss by adding 6 dB.

The results are shown for data in Table 8.10. It is assumed that the soft handover gain is 3 dB per link, i.e. the total transmission power is the same with and without soft handover. In this case we need not consider the soft handover in the air interface dimensioning.

Table 8.10. Maximum path loss calculations for data

Throughput $N \cdot R \cdot (1 - \text{BLER})$	Load factor $\overline{\eta_{DL}}$	Average path loss $\bar{L}$	Max path loss
100 kbps	12 %	170.7 dB	176.7 dB
200 kbps	25 %	167.1 dB	173.1 dB
300 kbps	37 %	164.5 dB	170.5 dB
400 kbps	50 %	162.3 dB	168.3 dB
500 kbps	62 %	160.1 dB	161.1 dB
600 kbps	74 %	157.7 dB	163.6 dB
700 kbps	87 %	154.1 dB	160.1 dB
800 kbps	99 %	142.1 dB	148.1 dB
808 kbps	100 % = pole capacity	—	—

The results in Table 8.10 are plotted in Figure 8.6 together with the corresponding uplink calculations. The uplink is calculated for 64 kbps data and the link budget is shown in Table 8.19. In both uplink and downlink the air interface load affects the coverage but the effect is not exactly the same. In the downlink, the coverage depends more on the load than in the uplink, according to Figure 8.6. The reason is that in the downlink the power of 20 W is

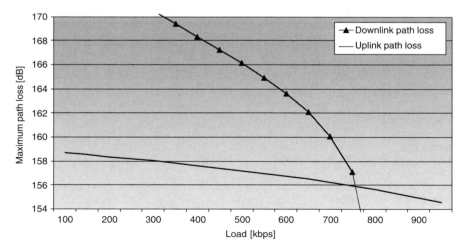

Figure 8.6. Example coverage vs. capacity relationship in downlink and uplink in macro cells

shared between the downlink users: the more users, the less power per user. Therefore, even with low load in the downlink, the coverage decreases as a function of the number of users.

We note that with the above assumptions the coverage is clearly limited by the uplink for loads below 760 kbps, while the capacity is downlink limited. Therefore, in Chapter 12 the coverage discussion concentrates on the uplink, while the capacity discussion concentrates on downlink.

Figure 8.7 presents the same curves as Figure 8.6 but with number of simultaneous 64-kbps users. Figure 8.8 presents the load curves for voice users.

We need to further remember that in third generation networks the data traffic can be asymmetric between uplink and downlink, and the load can be different in uplink and in downlink.

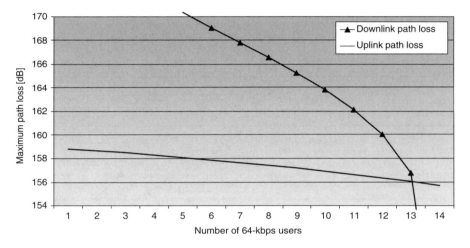

Figure 8.7. The same as Figure 8.6 for 64-kbps users with BLER of 10 %

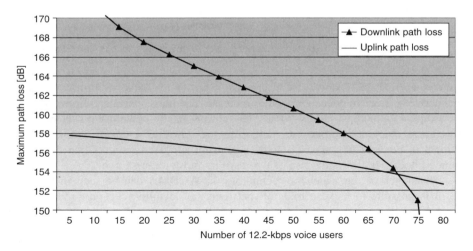

Figure 8.8. Example coverage vs. capacity relationship in macro cells for voice users

The WCDMA load equations assume that all users are allocated the same bit rate which corresponds to real time service with a guaranteed bit rate. If we allocate the same power, instead of the same bit rate, to all users, the cell throughput would be increased by 30–40 %.

Power Splitting Between Frequencies

In Figure 8.6 a base station maximum power of 20 W is assumed. What would happen to the downlink performance if the maximum power was lowered to 10 W? The difference in downlink coverage and capacity between 10 W and 20 W base station output powers is shown in Figure 8.9. If we lower the downlink power by 3.0 dB, the maximum allowed path loss is 3 dB lower. The effect on the capacity is smaller than the effect on the coverage because of the load curve. If we now keep the downlink path loss fixed at 156 dB, which is the maximum uplink path loss with 3 dB interference margin, the downlink capacity is decreased by only 5 % (0.2 dB) from 760 kbps to 720 kbps. Increasing downlink transmission power is an inefficient approach to increasing the interference limited downlink capacity, since the available power does not affect the pole capacity.

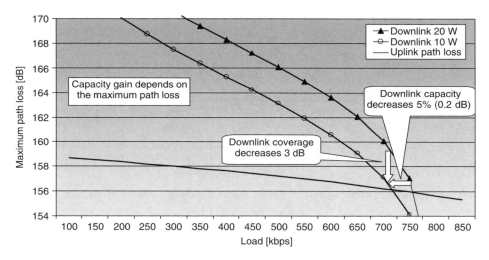

Figure 8.9. Effect of base station output power to downlink capacity and coverage

Assume we had 20 W downlink transmission power available. Splitting the downlink power between two frequencies would increase downlink capacity from 760 kbps to 2×720 kbps $= 1440$ kbps, i.e. by 90 %. The splitting of the downlink power between two carriers is an efficient approach to increasing the downlink capacity without any extra investment in power amplifiers. The power splitting approach requires that the operator's frequency allocation allows the use of two carriers in the base station.

The advantages of the wideband technology in WCDMA can be seen in the example above. It is possible to make a trade-off between the downlink capacity and coverage: if there are fewer users, more power can be allocated for one user allowing a higher path loss. The wideband power amplifier also allows the addition of a second carrier without adding a power amplifier. On the other hand, WCDMA requires very linear base station power amplifiers, which is challenging for the implementation.

Power Splitting Between Sectors

In the initial deployment phase the sectorised uplink is needed to improve the coverage, but the sectorisation may bring more capacity than is required by the initial traffic density. In WCDMA it is possible to use sectorised uplink reception while having only one common wideband power amplifier for all the sectors. This solution is illustrated in Figure 8.10.

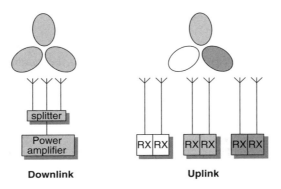

Figure 8.10. 3-sector uplink with receive diversity, single power amplifier in downlink

This solution is low-cost compared to the real sectorisation, where one wideband power amplifier is needed for each sector. The performance of this low-cost solution can be estimated from Figure 8.9. Let's take an example where the uplink uses three sectors while only one 20 W power amplifier is used in downlink: the available power per sector is $20/3 = 6.7$ W. That power gives 680 kbps capacity in Figure 8.9 with 156 dB maximum path loss. The curve for 6.7 W can be obtained by moving the 20 W curve down by $10^* \log 10$ $(6.7/20) = -4.7$ dB. Since the same power amplifier is shared between all sectors, the capacity of 680 kbps is the total capacity of the site. This low-cost solution provides 680/ $(3^* 760) = 30\%$ of the capacity of the real sectorised solution, and from RNC point of view it is equal to a single sector solution. The performance of the solution is summarised in Table 8.11. RNC is not involved in softer handovers between the sectors in this solution, but it is the base station baseband that decides which uplink sectors are used in receiving the user signal.

Table 8.11. 3-sector uplink, omni downlink

Uplink coverage	Equal to normal 3-sector configuration
Downlink capacity	30% of the 3-sector downlink
Number of logical sectors (RNC)	One

From the downlink coverage point of view, the downlink power splitting has less effect, see the downlink coverage discussion in Section 12.2.2. The downlink coverage is only slightly reduced as the power is reduced from 20 W to 6.7 W. We can conclude that power splitting is a feasible option to start the network operation: the uplink coverage is equal to the real sectorised solution and the downlink coverage is only slightly reduced.

8.2.3   Capacity Upgrade Paths

When the amount of traffic increases, the downlink capacity can be upgraded in a number of different ways. The most typical upgrade options are:

- more power amplifiers if initially the power amplifier is split between sectors;
- two or more carriers if the operator's frequency allocation permits;
- transmit diversity with a 2nd power amplifier per sector.

The availability of these capacity upgrade solutions depends on the base station manufacturer. All these capacity upgrade options may not be available in all base station types. Chapters 6 and 7 in [3] provide an extensive overview of coverage and capacity enhancement methods.

An example capacity upgrade path is shown in Figure 8.11. The initial solution with one power amplifier and one carrier gives 680 kbps capacity while the two-carrier three-sector transmit diversity solution gives $2*3*760*1.2 = 5.5$ Mbps assuming 20 % capacity increase from the transmit diversity. The transmit diversity procedure is described in Section 6.6.7 and its performance is discussed in Section 12.3.2.

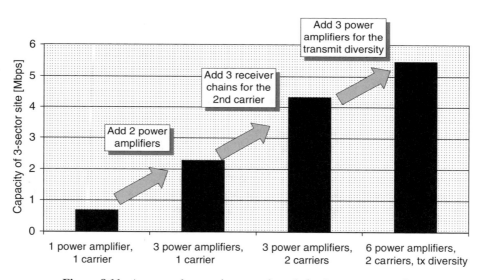

Figure 8.11. An example capacity upgrade path for 3-sector macro site

These capacity upgrade solutions do not require any changes to the antenna configurations, only upgrades within the base station cabinet are needed on the site. The uplink coverage is not affected by these upgrades.

The capacity can be improved also by increasing the number of antenna sectors, for example, starting with omni-directional antennas and upgrading to 3-sector and finally to 6-sector antennas. The drawback of increasing the number of sectors is that the antennas must be replaced. The sectorisation is analysed in more detail in [5]. The increased number of sectors also brings improved coverage through a higher antenna gain.

Table 8.12. Capacities per km^2 with macro and micro layers in an urban area

	Macro cell layer	Micro cell layer
Capacity per site per carrier with one power amplifier	630 kbps	—
Maximum capacity per site per carrier	3 Mbps with three sectors	2 Mbps
Capacity per site with three UMTS frequencies	9 Mbps	6 Mbps
Initial sparse site density	0.5 sites/km^2	—
Maximum dense site density	5 sites/km^2	30 sites/km^2
Maximum capacity	45 Mbps/km^2	180 Mbps/km^2

8.2.4 Capacity per km^2

Providing high capacity will be challenging in urban areas where the offered amount of traffic per km^2 can be very high. In this section we evaluate the maximal capacity that can be provided per km^2 using macro and micro sites. The results are shown in Table 8.12 and illustrated in Figure 8.12.

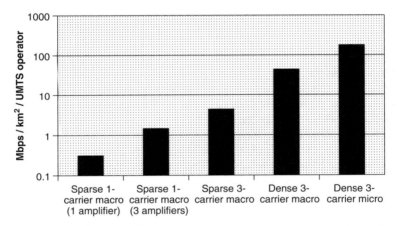

Figure 8.12. Capacity per km^2 for one UMTS operator with macro and micro layers

Let us assume that the maximum capacity per sector per carrier is 1 Mbps in macro cells and 2 Mbps in micro cells with transmit diversity. If the UMTS operation is started using 1-carrier macro cells with one power amplifier per site with 2 km^2 site area, the available capacity will be 630 kbps/carrier/2 km^2 = 315 kbps/km^2/carrier. If more capacity is needed, the operator can install more power amplifiers and deploy several frequencies per site, his frequency license permitting. With three frequencies the capacity will be 3 carriers * 3.0 Mbps/carrier/2 km^2 = 4.5 Mbps/km^2.

When more capacity is needed, the operator can add more macro sites. If we assume a maximum macro site density of 5 sites per km^2, the capacity of the macro cell layer is 3 Mbps/carrier * 3 carriers/site * 5 site/km^2 = 45 Mbps/km^2. If clearly more capacity is needed, the micro cell layer needs to be deployed.

For the micro cell layer we assume a maximum site density of 30 sites per km². Having an even higher site density is challenging because the other-to-own cell interference tends to increase and the capacity per site decreases. Also, the site acquisition may be difficult if more sites are needed.

Using three carriers, the micro cell layer would offer 2 Mbps/carrier * 3 carriers/site * 30 site/km² = 180 Mbps/km². The total capacity of the UMTS band of 12 carriers could be up to 700 Mbps/km² if all operators deploy dense micro cell networks. Using indoor pico cells with FDD or with TDD can further enhance the capacity. The new frequency bands shown in Chapter 1 will allow further capacity enhancements in the future. Chapter 2 discusses the achievable capacity per subscriber in terms of MB/sub/month.

8.2.5 Soft Capacity

8.2.5.1 Erlang Capacity

In the dimensioning in Section 8.2 the number of channels per cell was calculated. Based on those figures, we can calculate the maximum traffic density that can be supported with a given blocking probability. The traffic density can be measured in Erlang and is defined ([8], p. 270) as:

$$\text{Traffic density[Erlang]} = \frac{\text{Call arrival rate[calls/s]}}{\text{Call departure rate[calls/s]}} \tag{8.19}$$

If the capacity is hard blocked, i.e. limited by the amount of hardware, the Erlang capacity can be obtained from the Erlang B model [8]. If the maximum capacity is limited by the amount of interference in the air interface, it is by definition a soft capacity, since there is no single fixed value for the maximum capacity. For a soft capacity limited system, the Erlang capacity cannot be calculated from the Erlang B formula, since it would give too pessimistic results. The total channel pool is larger than just the average number of channels per cell, since the adjacent cells share part of the same interference, and therefore more traffic can be served with the same blocking probability. The soft capacity can be explained as follows. The less interference is coming from the neighbouring cells, the more channels are available in the middle cell, as shown in Figure 8.13. With a low number of channels per cell, i.e. for high bit rate real time data users, the average loading must be quite low to guarantee low blocking probability. Since the average loading is low, there is typically extra capacity available in the neighbouring cells. This capacity can be borrowed from the adjacent cells,

Equally loaded cells   Less interference in the neighbouring cells
                                          → higher capacity in the middle cell

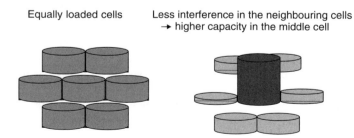

Figure 8.13. Interference sharing between cells in WCDMA

therefore the interference sharing gives soft capacity. Soft capacity is important for high bit rate real time data users, e.g. for video connections. It can also be obtained in GSM if the air interface capacity is limited by the amount of interference instead of the number of time slots; this assumes low frequency reuse factors in GSM with fractional loading.

In the soft capacity calculations below it is assumed that the number of subscribers is the same in all cells but the connections start and end independently. In addition, the call arrival interval follows a Poisson distribution. This approach can be used in dimensioning when calculating Erlang capacities. There is an additional soft capacity in WCDMA if the number of users in the neighbouring cells is smaller.

The difference between hard blocking and soft blocking is shown with a few examples in the uplink below. WCDMA soft capacity is defined in this section as the increase of Erlang capacity with soft blocking compared to that with hard blocking with the same maximum number of channels per cell on average with both soft and hard blocking:

$$\text{Soft capacity} = \frac{\text{Erlang capacity with soft blocking}}{\text{Erlang capacity with hard blocking}} - 1 \qquad (8.20)$$

Uplink soft capacity can be approximated based on the total interference at the base station. This total interference includes both own cell and other cell interference. Therefore, the total channel pool can be obtained by multiplying the number of channels per cell in the equally loaded case by $1 + i$, which gives the single isolated cell capacity, since

$$\begin{aligned} i + 1 &= \frac{\text{other cell interference}}{\text{own cell interference}} + 1 \\ &= \frac{\text{other cell interference} + \text{own cell interference}}{\text{own cell interference}} = \frac{\text{isolated cell capacity}}{\text{multicell capacity}} \end{aligned} \qquad (8.21)$$

The basic Erlang B formula is then applied to this larger channel pool ($=$ interference pool). The Erlang capacity obtained is then shared equally between the cells. The procedure for estimating the soft capacity is summarised below.

1. Calculate the number of channels per cell, N, in the equally loaded case, based on the uplink load factor, Equation (8.12).

2. Multiply that number of channels by $1 + i$ to obtain the total channel pool in the soft blocking case.

3. Calculate the maximum offered traffic from the Erlang B formula.

4. Divide the Erlang capacity by $1 + i$.

8.2.5.2 Uplink Soft Capacity Examples

A few numerical examples of soft capacity calculations are given, with the assumptions shown in Table 8.13.

The capacities obtained, in terms of both channels based on Equation (8.12) and Erlang per cell, are shown in Table 8.14. The trunking efficiency shown in Table 8.14 is defined as the hard blocked capacity divided by the number of channels. The lower the trunking efficiency, the lower is the average loading, the more capacity can be borrowed from the neighbouring cells, and the more soft capacity is available.

Table 8.13. Assumptions in soft capacity calculations

Bit rates	Speech: 7.95 and 12.2 kbps
	Real time data: 16–144 kbps
Voice activity	Speech 67 %
	Data 100 %
E_b/N_0	Speech: 4 dB
	Data 16–32 kbps: 3 dB
	Data 64 kbps: 2 dB
	Data 144 kbps: 1.5 dB
i	0.55
Noise rise	3 dB ($= 50$ % load factor)
Blocking probability	2 %

Table 8.14. Soft capacity calculations in the uplink

Bit rate (kbps)	Channels per cell	Hard blocked capacity	Trunking efficiency	Soft blocked capacity	Soft capacity
7.95	92.7	80.9 Erl	87 %	84.2 Erl	4 %
12.2	60.5	50.1 Erl	83 %	52.8 Erl	5 %
16	39.0	30.1 Erl	77 %	32.3 Erl	7 %
32	19.7	12.9 Erl	65 %	14.4 Erl	12 %
64	12.5	7.0 Erl	56 %	8.2 Erl	17 %
144	6.4	2.5 Erl	39 %	3.2 Erl	28 %

We note that there is more soft capacity for higher bit rates than for lower bit rates. This relationship is shown in Figure 8.14.

It should be noted that the amount of soft capacity depends also on the propagation environment and on the network planning which affect the value of i. The soft capacity can be obtained only if the radio resource management algorithms can utilise a higher capacity in one cell if the adjacent cells have lower loading. This can be achieved if the radio resource

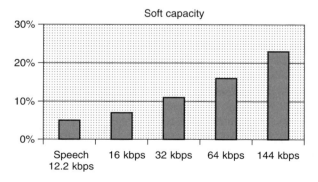

Figure 8.14. Soft capacity as a function of bit rate for real time connections

management algorithms are based on the wideband interference, not on the throughput or the number of connections. Interference-based admission control is presented in Section 9.5.

Similar soft capacity is also available in the WCDMA downlink as well as in GSM if interference-based radio resource management algorithms are applied.

8.2.6 Network Sharing

The cost of the network deployment can be reduced by network sharing. An example of a network sharing approach is illustrated in Figure 8.15 where both operators have their own core networks and share a common radio access network, RAN. This solution offers cost savings in site acquisition, civil works, transmission, RAN equipment costs and operation expenses. Both operators can still keep their full independence in core network, services and have dedicated radio carrier frequencies. When the amount of traffic increases in the future, the operators can exit the shared RAN and continue with separate RANs.

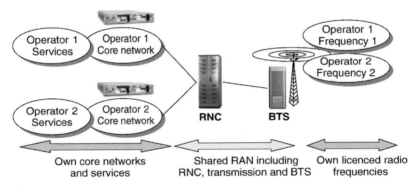

Figure 8.15. Sharing of a WCDMA radio access network

In the case of shared RAN, the RNC is connected with multiple Iu-CS and Iu-PS interfaces to each operator's core network. The subscribers of different operators are separated to their own licensed frequencies. This feature is supported in 3GPP Release '99.

The base station configuration of a shared 3-sector site is typically 2+2+2, both operators using 1+1+1 layers. The radio coverage will be the same for both operators and they share the radio capacity and the hardware resources including base station, transmission and RNC resources. Consequently, a combined radio network dimensioning and planning is needed, based on the total capacity figures from both operators.

The shared RAN allows some flexibility in setting the parameters for the radio resource management algorithms. For example, the maximum allowed load threshold for the packet scheduling can be different, and handover parameters and neighbourlists can be optimised separately.

Other levels of network sharing are also possible. One geographical area could be covered by only one operator that allows roaming for the other operators' users. It is also possible to share the sites and the related costs while having independent RANs.

8.3   Capacity and Coverage Planning and Optimisation

8.3.1   Iterative Capacity and Coverage Prediction

In this section, detailed capacity and coverage planning are presented. In the detailed planning phase real propagation data from the planned area is needed, together with the estimated user density and user traffic. Also, information about the existing base station sites is needed in order to utilise the existing site investments. The output of the detailed capacity and coverage planning are the base station locations, configurations and parameters. A comprehensive treatment of the topic can be found in Chapter 3 in [3].

Since, in WCDMA, all users are sharing the same interference resources in the air interface, they cannot be analysed independently. Each user is influencing the others and causing their transmission powers to change. These changes themselves again cause changes, and so on. Therefore, the whole prediction process has to be done iteratively until the transmission powers stabilise. This iterative process is illustrated in Figure 8.16.

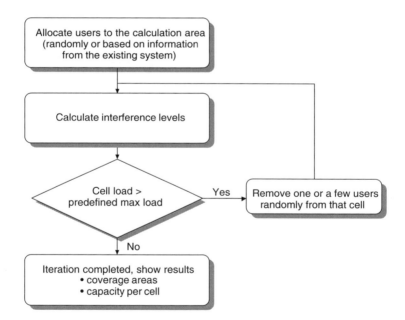

Figure 8.16. Iteration capacity and coverage calculations

Also, the mobile speeds, multipath channel profiles, and bit rates and type of services used play a more important role than in second generation TDMA/FDMA systems. Furthermore, in WCDMA fast power control in both uplink and downlink, soft/softer handover and orthogonal downlink channels are included, which also impact on system performance. The main difference between WCDMA and TDMA/FDMA coverage prediction is that the interference estimation is already crucial in the coverage prediction phase in WCDMA. In the current GSM coverage planning processes the base station sensitivity is typically assumed to be constant and the coverage threshold is the same for each base station. In the case of WCDMA the base station sensitivity depends on the number of users and used bit rates in all cells, thus it is cell- and service-specific. Note also that in third generation networks, the downlink can be loaded higher than the uplink or vice versa.

8.3.2   Planning Tool

In second generation systems, detailed planning concentrated strongly on coverage planning. In third generation systems, a more detailed interference planning and capacity analysis than simple coverage optimisation is needed. The tool should aid the planner to optimise the base station configurations, the antenna selections and antenna directions and even the site locations, in order to meet the quality of service and the capacity and service requirements at minimum cost. To achieve the optimum result the tool must have knowledge of the radio resource algorithms. Uplink and downlink coverage probability is determined for a specific service by testing the service availability in each location of the plan. A detailed description of the planning tool can be found in [9]. An example of a commercial WCDMA planning tool is shown in Figure 8.17.

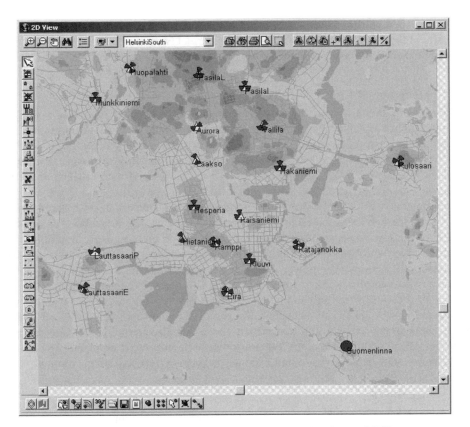

Figure 8.17. Commercial WCDMA network planning tool [10]

The actual detailed planning phase does not differ very much from second generation planning. The sites and sectors are placed in the tool. The main difference is the importance of the traffic layer. The proposed detailed analysis methods (see the following sections) use discrete mobile stations in the WCDMA analysis. The mobile station density in different cells should be based on actual traffic information. The hotspots should be identified as an input for accurate analysis. One source of information concerning user density would be the data from the operator's second generation network or later from the third generation.

8.3.2.1 Uplink and Downlink Iterations

The target in the uplink iteration is to allocate the simulated mobile stations' transmission powers so that the interference levels and the base station sensitivity values converge. The base station sensitivity level is corrected by the estimated uplink interference level (noise rise) and therefore is cell specific. The impact of the uplink loading on the sensitivity is taken into account with a term $-10 \cdot \log_{10}(1 - \eta_{UL})$, where η_{UL} is given by Equation (8.12). In the uplink iteration the transmission powers of the mobile stations are estimated based on the sensitivity level of the best server, the service, the speed and the link losses. Transmission powers are then compared to the maximum allowed transmission power of the mobile stations, and mobile stations exceeding this limit are put to outage. The interference can then be re-estimated and new loading values and sensitivities for each base station assigned. If the uplink load factor is higher than the set limit, the mobile stations are randomly moved from the highly loaded cell to another carrier (if the spectrum allows) or to outage.

The aim of the downlink iteration is to allocate correct base station transmission powers to each mobile station until the received signal at the mobile station meets the required E_b/N_0 target.

8.3.2.2 Modelling of Link Level Performance

In radio network dimensioning and planning it is necessary to make simplifying assumptions concerning the multipath propagation channel, transmitter and receiver. A traditional model is to use the average received E_b/N_0 ensuring the required quality of service as the basic number, which includes the effect of the power delay profile. In systems using fast power control the average received E_b/N_0 is not enough to characterise the influence of the radio channel on network performance. Also, the transmission power distribution must be taken into account when modelling link level performance in network level calculations. An appropriate approach is presented in [1] for the WCDMA uplink. It has been demonstrated that, due to the fast power control in the multipath fading environment, in addition to the average received E_b/N_0 requirement, an average transmission power rise is needed in interference calculations. The power rise is presented in detail in Section 9.2.1.2. Further-more, a power control headroom must be included in the link budget estimation to allow power control to follow the fast fading at the cell edge.

Multiple links are taken into account in the simulator when estimating the soft handover gains in the average received and transmitted power and also in the required power control headroom. During the simulations the transmission powers are corrected by the voice activity factor, soft handover gain and average power rise for each mobile station.

8.3.3 Case Study

In this case study an area in Espoo, Finland, was planned, comprising roughly $12 \times 12 \text{ km}^2$, as shown in Figure 8.18. The network planning tool described in Section 8.3.2 was utilised in this case study.

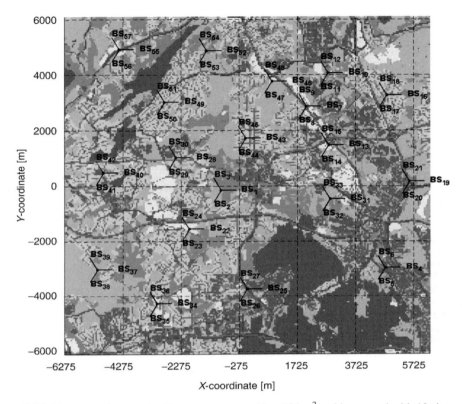

Figure 8.18. The network scenario. The area measures $12 \times 12 \text{ km}^2$ and is covered with 19 sites, each with three sectors

The operator's coverage probability requirement for the 8 kbps, 64 kbps and 384 kbps services was set, respectively, to 95 %, 80 % and 50 %, or better. The planning phase started with radio link budget estimation and site location selections. In the next planning step the dominance areas for each cell were optimised. In this context the dominance is related only to the propagation conditions. Antenna tilting, bearing and site locations can be tuned to achieve clear dominance areas for the cells. Dominance area optimisation is crucial for interference and soft handover area and soft handover probability control. The improved soft/softer handover and interference performance is automatically seen in the improved network capacity. The plan consists of 19 three-sectored macro sites, and the average site area is 7.6 km². In the city area, the uplink loading limitation was set to 75 %, corresponding to a 6 dB noise rise. In case the loading was exceeded, the necessary number of mobile stations was randomly set to outage (or moved to another carrier) from the highly loaded cells. Table 8.15 shows the user distribution in the simulations and the other simulation parameters are listed in Table 8.16.

In all three simulation cases the cell throughput in kbps and the coverage probability for each service were of interest. Furthermore, the soft handover probability and loading results were collected. Tables 8.17 and 8.18 show the simulation results for cell throughput

Table 8.15. The user distribution

Service in kbps	Users per service
8	1735
64	250
384	15

Table 8.16. Parameters used in the simulator

Uplink loading limit	75 %
Base station maximum transmission power	20 W (43 dBm)
Mobile station maximum transmission power	300 mW ($=$ 25 dBm)
Mobile station power control dynamic range	70 dB
Slow (log-normal) fading correlation between base stations	50 %
Standard deviation for the slow fading	6 dB
Multipath channel profile	ITU Vehicular A
Mobile station speeds	3 km/h and 50 km/h
Mobile/base station noise figures	7 dB/5 dB
Soft handover addition window	−6 dB
Pilot channel power	30 dBm
Combined power for other common channels	30 dBm
Downlink orthogonality	0.5
Activity factor speech/data	50 %/100 %
Base station antennas	65°/17 dBi
Mobile antennas speech/data	Omni/1.5 dBi

and coverage probabilities. The maximum uplink loading was set to 75 % according to Table 8.16. Note that in Table 8.17 in some cells the loading is lower than 75 %, and, correspondingly, the throughput is also lower than the achievable maximum value. The reason is that there was not enough offered traffic in the area to fully load the cells. The loading in cell 5 was 75 %. Cell 5 is located in the lower right corner in Figure 8.18, and there is no other cell close to cell 5. Therefore, that cell can collect more traffic than the other cells. For example, cells 2 and 3 are in the middle of the area and there is not enough traffic to fully load the cells.

Table 8.18 shows that mobile station speed has an impact on both throughput and coverage probability. When mobile stations are moving at 50 km/h, fewer can be served, the throughput is lower and the resulting loading is higher than when mobile stations are moving at 3 km/h. If the throughput values are normalised to correspond to the same loading value, the difference between the 3 km/h and 50 km/h cases is more than 20 %. The better capacity with the slower-moving mobile stations can be explained by the better E_b/N_0 performance. The fast power control is able to follow the fading signal and the required E_b/N_0 target is reduced. The lower target value reduces the overall interference level and more users can be served in the network.

Table 8.17. The cell throughput, loading and soft handover (SHO) overhead. UL = uplink, DL = downlink

Basic loading: mobile speed 3 km/h, served users: 1805

Cell ID	Throughput UL (kbps)	Throughput DL (kbps)	UL loading	SHO overhead
cell 1	728.00	720.00	0.50	0.34
cell 2	208.70	216.00	0.26	0.50
cell 3	231.20	192.00	0.24	0.35
cell 4	721.60	760.00	0.43	0.17
cell 5	1508.80	1132.52	0.75	0.22
cell 6	762.67	800.00	0.53	0.30
MEAN (all cells)	519.20	508.85	0.37	0.39

Basic loading: mobile speed 50 km/h, served users: 1777

Cell ID	Throughput UL (kbps)	Throughput DL (kbps)	UL loading	SHO overhead
cell 1	672.00	710.67	0.58	0.29
cell 2	208.70	216.00	0.33	0.50
cell 3	226.67	192.00	0.29	0.35
cell 4	721.60	760.00	0.50	0.12
cell 5	1101.60	629.14	0.74	0.29
cell 6	772.68	800.00	0.60	0.27
MEAN	531.04	506.62	0.45	0.39

Basic loading: mobile speed 50 km/h and 3 km/h, served users: 1802

Cell ID	Throughput UL (kbps)	Throughput DL (kbps)	UL loading	SHO overhead
cell 1	728.00	720.00	0.51	0.34
cell 2	208.70	216.00	0.29	0.50
cell 3	240.00	200.00	0.25	0.33
cell 4	730.55	760.00	0.44	0.20
cell 5	1162.52	780.92	0.67	0.33
cell 6	772.68	800.00	0.55	0.32
MEAN	525.04	513.63	0.40	0.39

Comparing coverage probability, the faster-moving mobile stations experience better quality than the slow-moving ones, because for the latter a headroom is needed in the mobile transmission power to be able to maintain the fast power control – see Section 8.2.1. The impact of the speed can be seen, especially if the bit rates used are high, because for low bit rates the coverage is better due to a larger processing gain. The coverage is tested in this planning tool by using a test mobile after the uplink iterations have converged. It is assumed that this test mobile does not affect the loading in the network.

This example case demonstrates the impact of the user profile, i.e. the service used and the mobile station speed, on network performance. It is shown that the lower mobile station speed provides better capacity: the number of mobile stations served and the cell throughput are higher in the 3 km/h case than in the 50 km/h case. Comparing coverage probability, the impact of the mobile station speed is different. The higher speed reduces the required fast

Table 8.18. The coverage probability results

Basic loading: mobile speed 3 km/h	Test mobile speed:	
	3 km/h	50 km/h
8 kbps	96.6 %	97.7 %
64 kbps	84.6 %	88.9 %
384 kbps	66.9 %	71.4 %

Basic loading: mobile speed 50 km/h	Test mobile speed:	
	3 km/h	50 km/h
8 kbps	95.5 %	97.1 %
64 kbps	82.4 %	87.2 %
384 kbps	63.0 %	67.2 %

Basic loading: mobile 3 and 50 km/h	Test mobile speed:	
	3 km/h	50 km/h
8 kbps	96.0 %	97.5 %
64 kbps	83.9 %	88.3 %
384 kbps	65.7 %	70.2 %

fading margin and thus the coverage probability is improved when the mobile station speed is increased.

8.3.4 Network Optimisation

Network optimisation is a process to improve the overall network quality as experienced by the mobile subscribers and to ensure that network resources are used efficiently. Optimisation includes:

1. Performance measurements.
2. Analysis of the measurement results.
3. Updates in the network configuration and parameters.

The optimisation process is shown in Figure 8.19.

A clear picture of the current network performance is needed for the performance optimisation. Typical measurement tools are shown in Figure 8.20. The measurements can be obtained from the test mobile and from the radio network elements. The WCDMA mobile can provide relevant measurement data, e.g. uplink transmission power, soft handover rate and probabilities, CPICH E_c/N_0 and downlink BLER. Also, scanners can be used to provide some of the downlink measurements, like CPICH measurements for the neighbourlist optimisation.

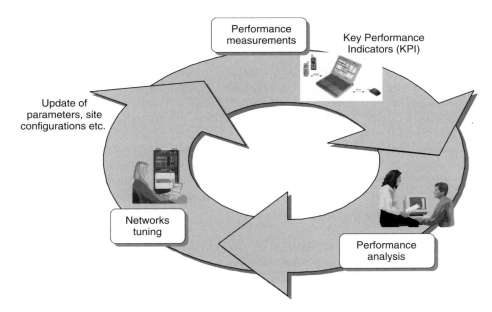

Figure 8.19. Network optimisation process

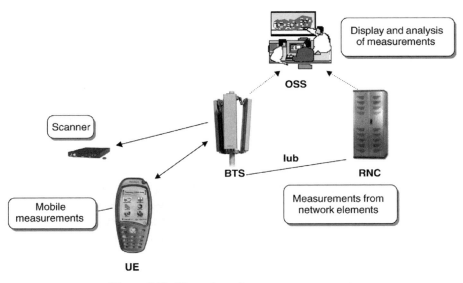

Figure 8.20. Network performance measurements

The radio network can typically provide connection level and cell level measurements. Examples of the connection measurements include uplink BLER and downlink transmission power. The connection level measurements both from the mobile and from the network are important to get the network running and provide the required quality for the end users. The cell level measurements become more important in the capacity optimisation phase. The cell level measurements may include total received power and total transmitted power, the same parameters that are used by the radio resource management algorithms.

The measurement tools can provide lots of results. In order to speed up the measurement analysis it is beneficial to define those measurement results that are considered the most important ones, Key Performance Indicators, KPIs. Examples of KPIs are total base station transmission power, soft handover overhead, drop call rate and packet data delay. The comparison of KPIs and desired target values indicates the problem areas in the network where the network tuning can be focused.

The network tuning can include updates of RRM parameters, e.g. handover parameters, common channel powers or packet data parameters. The tuning can also include changes of antenna directions. It may be possible to adjust the antenna tilts remotely without any site visits. An example case is illustrated in Figure 8.21. If there is too much overlapping of the

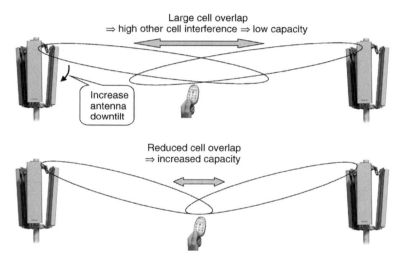

Figure 8.21. Network tuning with antenna tilts

adjacent cells, the other cell interference is high and the system capacity is low. The effect of other cell interference is represented with the parameter other cell to own cell interference ratio, i, in the load equations of Section 8.2, see Equation (8.16). The importance of the other cell interference is illustrated in Figure 8.22: if the other cell interference can be decreased

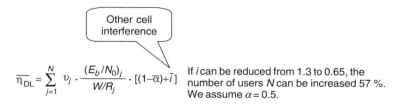

$$\overline{\eta_{DL}} = \sum_{j=1}^{N} v_j \cdot \frac{(E_b/N_0)_j}{W/R_j} \cdot [(1-\overline{\alpha})+\overline{i}]$$

If i can be reduced from 1.3 to 0.65, the number of users N can be increased 57 %. We assume $\alpha = 0.5$.

Figure 8.22. Importance of other cell interference for WCDMA downlink capacity

by 50 %, the capacity can be increased by 57 %. The large overlapping can be seen from the high number of users in soft handover between these cells.

With advanced Operations Support System (OSS) the network performance monitoring and optimisation can be automated. OSS can point out the performance problems, propose corrective actions and even make some tuning actions automatically.

The network performance can be best observed when the network load is high. With low load some of the problems may not be visible. Therefore, we need to consider artificial load generation to emulate high loading in the network. A high uplink load can be generated by increasing the E_b/N_0 target of the outer loop power control. In the normal operation the outer loop power control provides the required quality with minimum E_b/N_0. If we increase manually the E_b/N_0 target, e.g. 10 dB higher than the normal operation point, that uplink connection will cause 10 times more interference and converts 32 kbps connection into 320 kbps high bit rate connection from the interference point of view. The effect of higher E_b/N_0 can be seen in the uplink load equation of Equation (8.12). The same approach can be applied in the downlink as well in Equation (8.16). Another load generation approach in downlink is to transmit dummy data in downlink with a few code channels, even if there are no mobiles receiving that data. That approach is called Orthogonal Channel Noise Source, OCNS.

For more information on the radio network optimisation process please refer to [3], Chapter 8, and for advanced monitoring and network tuning see [3], Chapter 10.

8.4   GSM Co-planning

Utilisation of existing base station sites is important in speeding up WCDMA deployment and in sharing sites and transmission costs with the existing second generation system. The feasibility of sharing sites depends on the relative coverage of the existing network compared to WCDMA. In this section we compare the relative uplink coverage of existing GSM900 and GSM1800 full rate speech services and WCDMA speech and 64 kbps and 144 kbps data services. Table 8.19 shows the assumptions made and the results of the comparison of coverage. The maximum path loss of the WCDMA 144 kbps here is 3 dB greater than in Table 8.4. The difference comes because of a smaller interference margin, a lower base station receiver noise figure, and no cable loss. Note also that the soft handover gain is included in the fast fading margin in Table 8.19 and the mobile station power class is here assumed to be 21 dBm.

Table 8.19 shows that the maximum path loss of the 144 kbps data service is the same as for speech service of GSM1800. Therefore, a 144 kbps WCDMA data service can be provided when using GSM1800 sites, with the same coverage probability as GSM1800 speech. If GSM900 sites are used for WCDMA and 64 kbps full coverage is needed, a 3 dB coverage improvement is needed in WCDMA. Section 12.2.1 analyses the uplink coverage of WCDMA and presents a number of solutions for improving WCDMA coverage to match GSM site density. The comparison in Table 8.19 assumes that GSM900 sites are planned as coverage-limited. In densely populated areas, however, GSM900 cells are typically smaller to provide enough capacity, and WCDMA co-siting is feasible.

The downlink coverage of WCDMA is discussed in Section 12.2.2 and is shown to be better than the uplink coverage. Therefore, it is possible to provide full downlink coverage for bit rates 144 to 384 kbps using GSM1800 sites.

Table 8.19. Typical maximum path losses with existing GSM and with WCDMA

	GSM900/ speech	GSM1800/ speech	WCDMA/ speech	WCDMA/ 64 kbps	WCDMA/ 144 kbps
Mobile transmission power	33 dBm	30 dBm	21 dBm	21 dBm	21 dBm
Receiver sensitivity[1]	−110 dBm	−110 dBm	−125 dBm	−120 dBm	−117 dBm
Interference margin[2]	1.0 dB	0.0 dB	2.0 dB	2.0 dB	2.0 dB
Fast fading margin[3]	2.0 dB	2.0 dB	2.0 dB	2.0 dB	2.0 dB
Base station antenna gain[4]	16.0 dBi	18.0 dBi	18.0 dBi	18.0 dBi	18.0 dBi
Body loss[5]	3.0 dB	3.0 dB	3.0 dB	—	—
Mobile antenna gain[6]	0.0 dBi	0.0 dBi	0.0 dBi	2.0 dBi	2.0 dBi
Relative gain from lower frequency compared to UMTS frequency[7]	7.0 dB	1.0 dB	—	—	—
Maximum path loss	160.0 dB	154.0 dB	157.0 dB	157.0 dB	154.0 dB

[1]WCDMA sensitivity assumes 4.0 dB base station noise figure and E_b/N_0 of 4.0 dB for 12.2 kbps speech, 2.0 dB for 64 kbps and 1.5 dB for 144 kbps data. For the E_b/N_0 values see Section 12.5. GSM sensitivity is assumed to be −110 dBm with receive antenna diversity.
[2]The WCDMA interference margin corresponds to 37 % loading of the pole capacity: see Figure 8.3. An interference margin of 1.0 dB is reserved for GSM900 because the small amount of spectrum in 900 MHz does not allow large reuse factors.
[3]The fast fading margin for WCDMA includes the macro diversity gain against fast fading.
[4]The antenna gain assumes three-sector configuration in both GSM and WCDMA.
[5]The body loss accounts for the loss when the terminal is close to the user's head.
[6]A 2.0 dBi antenna gain is assumed for the data terminal.
[7]The attenuation in 900 MHz is assumed to be 7.0 dB lower than in UMTS band and in GSM1800 band 1.0 dB lower than in UMTS band.

Any comparison of the coverage of WCDMA and GSM depends on the exact receiver sensitivity values and on system parameters such as handover parameters and frequency hopping. The aim of this exercise is to compare the coverage of the GSM base station systems that have been deployed up to the present with WCDMA coverage in the initial deployment phase during 2002. The sensitivity of the latest GSM base stations is better than the one assumed in Table 8.19.

Since the coverage of WCDMA typically is satisfactory when reusing GSM sites, GSM site reuse is the preferred solution in practice. Let us consider next the practical co-siting of the system. Co-sited WCDMA and GSM systems can share the antenna when a dual band or wideband antenna is used. The antenna needs to cover both the GSM band and UMTS band. GSM and WCDMA signals are combined with a diplexer to the common antenna feeder. The shared antenna solution is attractive from the site solution point of view but it limits the flexibility in optimising the antenna directions of GSM and WCDMA independently. Another co-siting solution is to use separate antennas for the two networks. That solution gives full flexibility in optimising the networks separately. These two solutions are shown in Figure 8.23. The co-siting of GSM and WCDMA is taken into account in 3GPP performance requirements and the interference between the systems can be avoided.

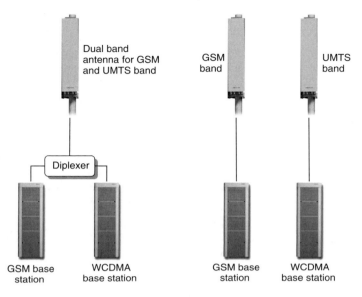

Figure 8.23. Co-siting of GSM and WCDMA

8.5   Inter-operator Interference

8.5.1   Introduction

In this section, the effect of adjacent channel interference between two operators on adjacent frequencies is studied. Adjacent channel interference needs to be considered, because it will affect all wideband systems where large guard bands are not possible, and WCDMA is no exception. If the adjacent frequencies are isolated in the frequency domain by large guard bands, spectrum is wasted due to the large system bandwidth. Tight spectrum mask requirements for a transmitter and high selectivity requirements for a receiver, in the mobile station and in the base station, would guarantee low adjacent channel interference. However, these requirements have a large impact, especially on the implementation of a small WCDMA mobile station.

Adjacent Channel Interference power Ratio (ACIR) is defined as the ratio of the transmission power to the power measured after a receiver filter in the adjacent channel(s). Both the transmitted and the received power are measured with a filter that has a Root-Raised Cosine filter response with roll-off of 0.22 and a bandwidth equal to the chip rate [11]. The adjacent channel interference is caused by transmitter non-idealities and imperfect receiver filtering. In both uplink and downlink, the adjacent channel performance is limited by the performance of the mobile. In the uplink the main source of adjacent channel interference is the non-linear power amplifier in the mobile station, which introduces adjacent channel leakage power. In the downlink the limiting factor for adjacent channel interference is the receiver selectivity of the WCDMA terminal. The requirements for adjacent channel performance are shown in Table 8.20.

Table 8.20. Requirements for adjacent channel performance [11]

Frequency separation	Required attenuation
Adjacent carrier (5 MHz separation)	33 dB both uplink and downlink
Second adjacent carrier (10 MHz separation)	43 dB in uplink, 40 dB in downlink (estimated from in-band blocking)

Such an interference scenario, where the adjacent channel interference could affect network performance, is illustrated in Figure 8.24. Operator 1's mobile is connected to a far-away base station and is at the same time located close to Operator 2's base station on the adjacent frequency. The mobile will receive interference from Operator 2's base station which may – in the worst case – block the reception of its own weak signal.

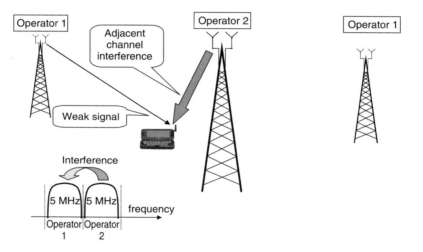

Figure 8.24. Adjacent channel interference in downlink

In the following sections the effect of the adjacent channel interference in this interference scenario is analysed by worst-case calculations and by system simulations. It will be shown that the worst-case calculations give very bad results but also that the worst-case scenario is extremely unlikely to happen in real networks. Therefore, simulations are also used to study this interference scenario. Finally, conclusions are drawn regarding adjacent channel interference and implications for network planning are discussed.

8.5.2   Uplink vs. Downlink Effects

While the mobile in Figure 8.24 receives interference, it will also cause interference in uplink to Operator 2's base station. In this section we analyse the differences between uplink and downlink in the worst-case scenario. The worst-case adjacent channel interference occurs when a mobile in uplink and a base station in downlink are transmitting on full power, and the mobile is located very close to a base station that is receiving on the adjacent carrier.

Table 8.21. Worst-case adjacent channel interference level

	Downlink	Uplink
Interferer power	43 dBm (base station)	21 dBm (mobile)
Minimum coupling loss between mobile and interfering base station in Figure 8.24	70 dB	70 dB
Adjacent channel attenuation	33 dB	33 dB
Adjacent channel interference	43 dBm − 70 dB − 33 dB =−60 dBm	21 dBm − 70 dB − 33 dB =−82 dBm

A minimum coupling loss of 70 dB is assumed here. The minimum coupling loss is defined as the minimum path loss between mobile and base station antenna connectors. The level of the adjacent channel interference is calculated in Table 8.21 and it is compared to the receiver thermal noise level of Table 8.22, both in uplink and in downlink. The worst-case increase in the receiver interference level is calculated in Table 8.23.

Table 8.22. Receiver thermal noise level

	Downlink	Uplink
Thermal noise level kTB	−108 dBm	−108 dBm
Receiver noise figure	7 dB	4 dB
Receiver noise level	−108 dBm + 7 dB =−101 dBm	−108 dBm + 4 dB =−104 dBm

Table 8.23. Worst-case desensitisation

Downlink	Uplink
−60 dBm − (−101 dBm) = 41 dB	−82 dBm − (−104 dBm) = 22 dB

The maximum desensitisation in downlink is 41 dB and in uplink 22 dB, which indicates that the downlink direction will be affected before the mobile is able to cause high interference levels in uplink. This is mainly because of higher base station power compared to the mobile power. It is also preferable to cause interference to one connection in downlink than to allow that mobile to interfere with all uplink connections of one cell. In the following sections we concentrate on the downlink analysis.

8.5.3 Local Downlink Interference

The adjacent channel interference in downlink may cause dead zones around interfering base stations. In this section we evaluate the sizes of these dead zones as a function of the

Table 8.24. Assumptions for dead zone calculation for 12.2 kbps voice

Parameter	Value
Transmission power of Operator 2's base station	33–43 dBm
Pilot power from Operator 1's base station	33 dBm
Maximum allocated power per voice connection from	33 dBm
Operator 1's base station	
Required E_b/N_0 for voice connection	7 dB
Required E_c/I_0 for voice connection	7 dB $- 10^*$ log 10(3.84 e
	6/12.2 e 3) $= -18$ dB
Path loss calculation to the interfering Operator 2's base	37 dB $+ 20^*$ log 10 (d)
station with distance d [metres] in line-of-sight	

coverage of the own signal. The coverage is defined as the received pilot power level. The assumptions in the calculations are shown in Table 8.24.

The dead zones are evaluated as follows.

1. Assume received pilot power level from Operator 1's base station.

2. Calculate maximum received signal power level for the voice connection. In this case it is equal to the pilot power level since the maximum transmission power for voice is assumed to be equal to the pilot power of 33 dBm.

3. Calculate maximum tolerated interference level I_0 on the same carrier based on the required E_c/I_0.

4. Calculate maximum tolerated interference level on the adjacent carrier based on the adjacent channel attenuation.

5. Calculate minimum required path loss to the interfering base station.

6. Calculate minimum required distance to the interfering base station.

An example calculation is shown below assuming pilot power coverage of -90 dBm.

1. Assume pilot power level of -90 dBm.

2. Maximum received power for voice connection -90 dBm.

3. Maximum tolerated interference level $I_0 = -90$ dBm $+ 18$ dB $= -72$ dBm.

4. Maximum tolerated interference level on the adjacent carrier -72 dBm $+ 33$ dB $= -39$ dBm.

5. Minimum required path loss 43 dBm $- (-39$ dBm$) = 82$ dB when Operator 2's base station transmits with 43 dBm. The required path loss is reduced to 33 dBm $- (-39$ dBm$) = 72$ dB when operator 2's base station transmits only common channels with 33 dBm.

6. Minimum required distance $d = 10^\wedge$ $((82 - 37)/20) = 178$ m or $d = 10^\wedge$ $((72 - 37)/20) = 56$ m.

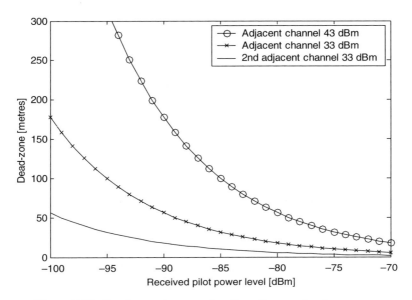

Figure 8.25. Dead zone sizes as a function of own network coverage

The results of the calculations are plotted in Figure 8.25. The results show that the dead zones can occur only if the following conditions take place at the same time: own network coverage is weak, the mobile is located close to the interfering base station that is operating on the adjacent frequency with maximum power, and UE performance is just meeting 3GPP selectivity requirements.

8.5.4 Average Downlink Interference

Since the probability of the adjacent channel interference is low, we need to resort to system simulations to evaluate the effect on the average performance. More transmission power is needed because of adjacent channel interference which leads to a lower capacity. The simulations show the reduction in average capacity when the same outage probability is maintained, with and without adjacent channel interference. The simulation results and assumptions are presented in [12]. The worst-case scenario is shown in Figure 8.26 where the site distance is 1 km and the interfering sites are just between our own sites. The best case is when the operators' sites are co-located.

The simulation results are shown in Table 8.25. The worst-case capacity loss is 2.0–3.5 %. These capacity loss figures can be reduced with the solutions shown in the following section.

8.5.5 Path Loss Measurements

The adjacent channel interference is basically about power competition between operators. The interference problems hit the connection if the interfering signal is strong at the same time as the own signal is weak. We can calculate the maximum tolerable power difference

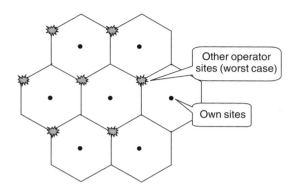

Figure 8.26. Worst-case simulation scenario

Table 8.25. Capacity loss because of adjacent channel interference

	Worst-case	Intermediate case	Co-siting
Capacity loss	3.5 %	2.5 %	No loss

between own signal and the interfering signal in Figure 8.24. When the maximum power difference is known, we can go and measure the power differences between two operators' networks and find the locations where the interference could cause problems. We show an example for WCDMA voice service in downlink with the following assumptions:

- The required E_c/I_0 for WCDMA voice $= E_b/N_0$ − processing gain $= -18$ dB from Table 8.24.

- The maximum transmission power per WCDMA connection is assumed to be 33 dBm.

- WCDMA mobile selectivity is 33 dB.

- The base stations' transmit power is 43 dBm.

The maximum allowed signal power difference between two operators can be estimated as follows:

$$= -E_c/I_0 + \text{mobile selectivity} - \text{downlink power allocation}$$
$$= 18\,\text{dB} + 33\,\text{dB} - 10\,\text{dB (the power for a connection is 10 dB below the base station}$$
$$\quad \text{max power)}$$
$$= 41\,\text{dB}$$

When the frequency separation is 10 MHz, the allowed signal power difference increases to 51 dB. Relative signal power measurements from today's network show that the probability of a larger power difference than 41 dB is typically <1–2 % and larger than

51 dB is practically non-existent. This is the probability that counter-measures are needed against interference. The measurement results are in line with the simulation results.

8.5.6 Solutions to Avoid Adjacent Channel Interference

This section presents a few network planning and radio resource management solutions that make sure that adjacent channel interference does not affect WCDMA network performance.

If the operators using adjacent frequency bands co-locate their base stations, either in the same sites or using the same masts, adjacent channel interference problems can be avoided, since the received power levels from both operators' transmissions are then very similar. Since there are no large power differences, the adjacent channel attenuation of 33 dB is enough to prevent any adjacent channel interference problems.

The nominal WCDMA carrier spacing is 5.0 MHz but can be adjusted with a 200 kHz raster according to the requirements of the adjacent channel interference. By using a larger carrier spacing, the adjacent channel interference can be reduced. If the operator has two carriers in the same base station, the carrier spacing between them could be as small as 4.0 MHz, because the adjacent channel interference problems are completely avoided if the two carriers use the same base station antennas. In that case a larger carrier spacing can be reserved between operators, as shown in Figure 8.27.

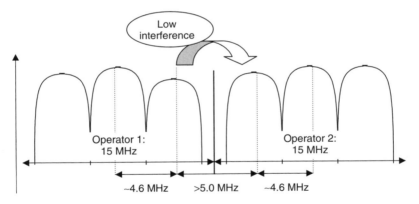

Figure 8.27. Selection of carrier spacings within operator's band and between operators

In addition to the network planning solutions, the radio resource management can also be effectively utilised to avoid the problems from inter-operator interference. The calculations in the sections above suggest the following radio resource management solutions to avoid adjacent channel interference in addition to the network planning solutions:

- make inter-frequency handover to another frequency to provide higher selectivity and more protection against interference;

- allocate more power per connection in downlink to overcome the effect of the interference;

- reduce the downlink instantaneous packet data bit rate to provide more processing gain to tolerate more interference;

- reduce the downlink AMR voice bit rate to provide more processing gain.

8.6 WCDMA Frequency Variants

8.6.1 Introduction

The 3GPP WCDMA standard covers a number of other frequency variants in addition to the UMTS core band. The frequency variants are listed in Table 8.26.

Table 8.26. WCDMA frequency variants

Frequency variant	Uplink [MHz]	Downlink [MHz]	Countries
Band I / UMTS core band	1920–1980	2110–2170	Europe, Asia, some Latin American countries like Brazil
Band II / WCDMA1900	1850–1910	1930–1990	Americas
Band III / WCDMA1800	1710–1785	1805–1880	Europe, Asia, some Latin American countries like Brazil
Band IV / WCDMA1700	1710–1755	2110–2155	Americas
Band V / WCDMA850	824–849	869–894	Americas, some Asian countries
Band VI / WCDMA800	830–840	875–885	Japan

These frequency variants use exactly the same 3GPP standard, except for the RF parameters that have been adapted for each band. The differences between 3GPP frequency variants are shown in Section 8.6.2. The frequency variants are especially relevant for the Americas market, where the uplink part of the UMTS core band is already used by the existing PCS system – like GSM, TDMA and IS-95 – see Figure 1.2. New spectrum for third generation services in the USA will be available from 1.7/2.1 GHz, where the downlink would be using the same spectrum as in Europe and in Asia, while the uplink would be in the 1.7 GHz band which is used for GSM1800 uplink in Europe and in Asia. This new band is not yet available and the third generation services need to be implemented in the existing bands – 850 and 1900 MHz – in the first place. The US spectrum allocations are illustrated in Figure 8.28.

The practical performance of WCDMA1900 using existing second generation sites in the US is evaluated with a simulation case study in Section 8.6.3.

8.6.2 Differences Between Frequency Variants

The main differences in the RF requirements between the frequency variants are summarised below. WCDMA2100 refers to WCDMA in the UMTS core band.

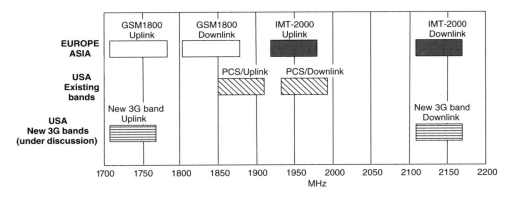

Figure 8.28. Spectrum for third generation services in the USA

- New channel numbers are defined. Also, additional channels with 100 kHz raster are defined for Bands IV, V and VI to allow WCDMA to be located exactly in the centre of the 5 MHz deployment in Figure 8.29. UMTS core band uses 200 kHz channel raster.

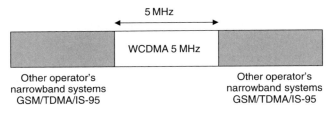

Figure 8.29. Isolated 5 MHz allocation for WCDMA

- Narrowband blocking and intermodulation requirements are specified for mobile and base stations to cope with the interference from the narrowband systems. The required interference rejection is 30 dB from a GSM carrier 2.7 MHz from the WCDMA centre frequency in Figure 8.30. The narrowband blocking requirements are defined for the Bands II, III, IV, V, where other technologies exist on the same band.

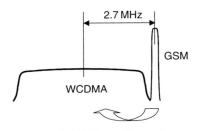

Figure 8.30. Attenuation from GSM signal 2.7 MHz from WCDMA derived from narrowband blocking requirements

- The mobile reference sensitivity requirement is relaxed by 2–3 dB from −117 dBm to −115/−114 dBm to allow high enough Duplex attenuation between uplink and downlink in Bands II, III and V. The separation between uplink and downlink is only 20 MHz in those bands.

These new requirements make the WCDMA deployment possible in an isolated 5 MHz block shown in Figure 8.29. The inter-system interference in the 1.9 GHz band is very similar to the multi-operator interference that was discussed in Section 8.5, and the same solutions can be applied. If the operator has a 10 MHz continuous block, the inter-operator interference can be completely avoided by allocating WCDMA in the middle of the 10 MHz block and narrowband 200 kHz GSM/EDGE carriers on both sides of the WCDMA. The narrowband GSM/EDGE carriers protect the WCDMA carrier from the inter-operator interference. This approach is referred to as a sandwich approach and is shown in Figure 8.31.

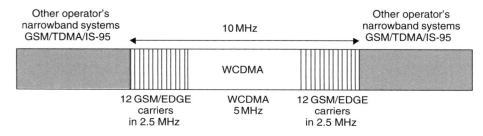

Figure 8.31. 10 MHz sandwich for WCDMA and GSM/EDGE

8.6.3   WCDMA1900 in an Isolated 5 MHz Block

The performance of WCDMA in an isolated 5 MHz block is evaluated in this section. The evaluation is based on a simulation case study using existing cell sites in a US network. The study area is a suburban area with 16 sites, each with three sectors, totalling 48 sectors. The average site covered 7 km². The other operator's site locations are randomly selected typical site locations between our own sites. The results are presented in more detail in [13]. The effect of the inter-operator interference to the capacity is studied and compared to the results in [14].

The main interference mechanism is the downlink adjacent channel interference from the interfering base station transmission to the WCDMA mobile reception. We assume here that the adjacent operator uses GSM technology and the GSM sites are transmitting at 43 dBm continuously with an average antenna height of 25 m. The simulation results are shown in Table 8.27.

Table 8.27. WCDMA1900 simulation in an isolated 5 MHz block

	Results from [13], Realistic scenario	Results from [14], Worst-case scenario
Capacity loss	<0.5%	1–2 %

The capacity loss shown in [13] is negligible. The results in 3GPP report [14] show higher capacity loss than the results in [13]. The target in 3GPP simulations has been to study the worst-case interference scenario where all the interfering sites are located at the edge of the WCDMA cells. In that case the capacity loss is 1–2 %.

Finally, note that the inter-system interference problems can be completely avoided when co-siting with other operators or when using a sandwich approach.

References

[1] Sipilä, K., Laiho-Steffens, J., Jäsberg, M. and Wacker, A., 'Modelling the Impact of the Fast Power Control on the WCDMA Uplink', *Proceedings of VTC'99*, Houston, Texas, May 1999, pp. 1266–1270.

[2] Ojanperä, T. and Prasad, R., *Wideband CDMA for Third Generation Mobile Communications*, Artech House, 1998.

[3] Laiho, J., Wacker, A. and Novosad, T., *Radio Network Planning and Optimisation for UMTS*, John Wiley & Sons, 2001.

[4] Saunders, S., *Antennas and Propagation for Wireless Communication Systems*, John Wiley & Sons, 1999.

[5] Wacker, A., Laiho-Steffens, J., Sipilä, K. and Heiska, K., 'The Impact of the Base Station Sectorisation on WCDMA Radio Network Performance', *Proceedings of VTC'99*, Amsterdam, The Netherlands, September 1999, pp. 2611–2615.

[6] Sipilä, K., Honkasalo, Z., Laiho-Steffens, J. and Wacker, A., 'Estimation of Capacity and Required Transmission Power of WCDMA Downlink Based on a Downlink Pole Equation', *Proceedings of VTC2000*, Spring 2000.

[7] Wang, Y.-P. and Ottosson, T., 'Cell Search in W-CDMA', *IEEE J. Select. Areas Commun.*, Vol. 18, No. 8, 2000, pp. 1470–1482.

[8] Lee, J. and Miller, L., *CDMA Systems Engineering Handbook*, Artech House, 1998.

[9] Wacker, A., Laiho-Steffens, J., Sipilä, K. and Jäsberg, M., 'Static Simulator for Studying WCDMA Radio Network Planning Issues', *Proceedings of VTC'99*, Houston, Texas, May 1999, pp. 2436–2440.

[10] Nokia NetAct[TM] Planner, *http://www.nokia.com/networks/services/netact/netact_planner/*

[11] 3GPP Technical Specification 25.101, UE Radio Transmission and Reception (FDD).

[12] 3GPP Technical Report 25.942, RF System Scenarios.

[13] Holma, H. and Velez, F. 'Performance of WCDMA1900 with 5-MHz Spectrum Reusing 2G Sites', presented at VTC'02 Fall, Vancouver, Canada, 24–29 September 2002.

[14] 3GPP Technical Report 25.885 'UMTS1800/1900 Work Item Technical Report'.

9

Radio Resource Management

Harri Holma, Klaus Pedersen, Jussi Reunanen, Janne Laakso
and Oscar Salonaho

9.1 Interference-Based Radio Resource Management

Radio Resource Management (RRM) algorithms are responsible for efficient utilisation of the air interface resources. RRM is needed to guarantee Quality of Service (QoS), to maintain the planned coverage area, and to offer high capacity. The family of RRM algorithms can be divided into handover control, power control, admission control, load control, and packet scheduling functionalities. Power control is needed to keep the interference levels at minimum in the air interface and to provide the required quality of service. WCDMA power control is described in Section 9.2. Handovers are needed in cellular systems to handle the mobility of the UEs across cell boundaries. Handovers are presented in Section 9.3. In third generation networks other RRM algorithms – like admission control, load control and packet scheduling – are required to guarantee the quality of service and to maximise the system throughput with a mix of different bit rates, services and quality requirements. Admission control is presented in Section 9.5 and load control in Section 9.6. WCDMA packet scheduling is described in Chapter 10.

The RRM algorithms can be based on the amount of hardware in the network or on the interference levels in the air interface. Hard blocking is defined as the case where the hardware limits the capacity before the air interface gets overloaded. Soft blocking is defined as the case where the air interface load is estimated to be above the planned limit. The difference between hard blocking and soft blocking is analysed in Section 8.2.5. It is shown that soft blocking based RRM gives higher capacity than hard blocking based RRM. If soft blocking based RRM is applied, the air interface load needs to be measured. The measurement of the air interface load is presented in Section 9.4. In IS-95 networks RRM is typically based on the available channel elements (hard blocking), but that approach is not applicable in the third generation WCDMA air interface, where various bit rates have to be supported simultaneously.

Typical locations of the RRM algorithms in a WCDMA network are shown in Figure 9.1.

WCDMA for UMTS, third edition.   Edited by Harri Holma and Antti Toskala
© 2004 John Wiley & Sons, Ltd   ISBN: 0-470-87096-6

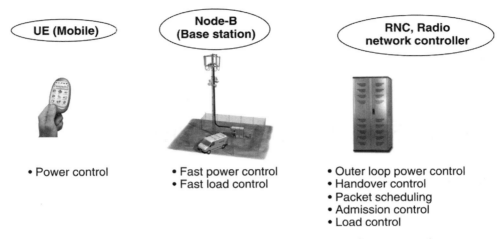

Figure 9.1. Typical locations of RRM algorithms in a WCDMA network

9.2   Power Control

Power control was introduced briefly in Section 3.5. In this chapter a few important aspects of WCDMA power control are covered. Some of these issues are not present in existing second generation systems, such as GSM and IS-95, but are new in third generation systems and therefore require special attention. In Section 9.2.1 fast power control is presented and in Section 9.2.2 outer loop power control is analysed. Outer loop power control sets the target for fast power control so that the required quality is provided.

In the following sections the need for fast power control and outer loop power control is shown using simulation results. Two special aspects of fast power control are presented in detail in Section 9.2.1: the relationship between fast power control and diversity, and fast power control in soft handover.

9.2.1   Fast Power Control

In WCDMA, fast power control with 1.5 kHz frequency is supported in both uplink and downlink. In GSM, only slow (frequency approximately 2 Hz) power control is employed. In IS-95, fast power control with 800 Hz frequency is supported only in the uplink.

9.2.1.1   Gain of Fast Power Control

In this section, examples of the benefits of fast power control are presented. The simulated service is 8 kbps with BLER = 1% and 10 ms interleaving. Simulations are made with and without fast power control with a step size of 1 dB. Slow power control assumes that the average power is kept at the desired level and that the slow power control would be able to ideally compensate for the effect of path loss and shadowing, whereas fast power control can compensate also for fast fading. Two-branch receive diversity is assumed in the Node B. ITU Vehicular A is a five-tap channel with WCDMA resolution, and ITU Pedestrian A is a two-path channel where the second tap is very weak. The required E_b/N_0 with and without fast power control are shown in Table 9.1 and the required average transmission powers in Table 9.2.

Table 9.1. Required E_b/N_0 values with and without fast power control

	Slow power control (dB)	Fast 1.5 kHz power control (dB)	Gain from fast power control (dB)
ITU Pedestrian A 3 km/h	11.3	5.5	5.8
ITU Vehicular A 3 km/h	8.5	6.7	1.8
ITU Vehicular A 50 km/h	6.8	7.3	−0.5

Table 9.2. Required relative transmission powers with and without fast power control

	Slow power control (dB)	Fast 1.5 kHz power control (dB)	Gain from fast power control (dB)
ITU Pedestrian A 3 km/h	11.3	7.7	3.6
ITU Vehicular A 3 km/h	8.5	7.5	1.0
ITU Vehicular A 50 km/h	6.8	7.6	−0.8

Fast power control gives clear gain, which can be seen from Tables 9.1 and 9.2. The gain from the fast power control is larger:

- for low UE speeds than for high UE speeds;
- in required E_b/N_0 than in transmission powers;
- for those cases where only a little multipath diversity is available, as in the ITU Pedestrian A channel. The relationship between fast power control and diversity is discussed in Section 9.2.1.2.

In Tables 9.1 and 9.2 the negative gains at 50 km/h indicate that an ideal slow power control would give better performance than the realistic fast power control. The negative gains are due to inaccuracies in the SIR estimation, power control signalling errors, and the delay in the power control loop.

The gain from fast power control in Table 9.1 can be used to estimate the required fast fading margin in the link budget in Section 8.2.1. The fast fading margin is needed in the UE transmission power for maintaining adequate closed loop fast power control. The maximum cell range is obtained when the UE is transmitting with full constant power, i.e. without the gain of fast power control. Typical values for the fast fading margin for low mobile speeds are 2–5 dB.

9.2.1.2 Power Control and Diversity

In this section the importance of diversity is analysed together with fast power control. At low UE speed the fast power control can compensate for the fading of the channel and keep the received power level fairly constant. The main sources of errors in the received powers arise from inaccurate SIR estimation, signalling errors and delays in the power control loop. The compensation of the fading causes peaks in the transmission power. The received power and the transmitted power are shown as a function of time in Figures 9.2 and 9.3 with a UE speed of 3 km/h. These simulation results include realistic SIR estimation and power control signalling. A power control step size of 1.0 dB is used. In Figure 9.2 very little diversity is assumed, while in Figure 9.3 more diversity is assumed in the simulation. Variations in the transmitted power are higher in Figure 9.2 than in Figure 9.3. This is due to the difference in the amount of diversity. The diversity can be obtained with, for example, multipath diversity, receive antenna diversity, transmit antenna diversity or macro diversity.

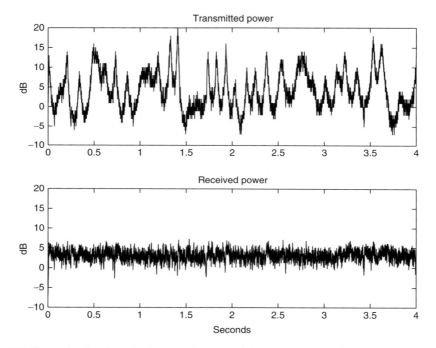

Figure 9.2. Transmitted and received powers in two-path (average tap powers 0 dB, −10 dB) Rayleigh fading channel at 3 km/h

With less diversity there are more variations in the transmitted power, but also the average transmitted power is higher. Here we define power rise to be the ratio of the average transmission power in a fading channel to that in a non-fading channel when the received power level is the same in both fading and non-fading channels with fast power control. The power rise is depicted in Figure 9.4.

The link level results for uplink power rise are presented in Table 9.3. The simulations are performed at different UE speeds in a two-path ITU Pedestrian A channel with average

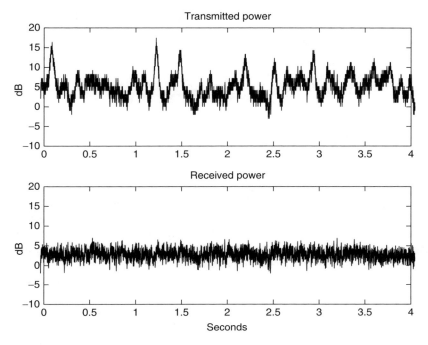

Figure 9.3. Transmitted and received powers in three-path (equal tap powers) Rayleigh fading channel at 3 km/h

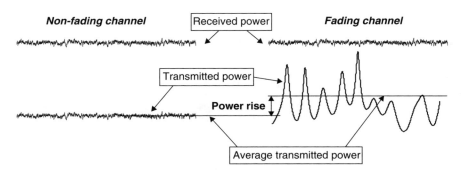

Figure 9.4. Power rise in fading channel with fast power control

multipath component powers of 0.0 dB and −12.5 dB. In the simulations the received and transmitted powers are collected slot by slot. With ideal power control the power rise would be 2.3 dB. At low UE speeds the simulated power rise values are close to the theoretical value of 2.3 dB, indicating that fast power control works efficiently in compensating the fading. At high UE speeds (>100 km/h) there is only very little power rise since the fast power control cannot compensate for the fading.

Table 9.3. Simulated power rises. Multipath channel ITU
Pedestrian A, antenna diversity assumed

UE speed (km/h)	Average power rise (dB)
3	2.1
10	2.0
20	1.6
50	0.8
140	0.2

More information about the uplink power control modelling can be found in [1].

Why is the power rise important for WCDMA system performance? In the downlink, the air interface capacity is directly determined by the required transmission power, since that determines the transmitted interference. Thus, to maximise the downlink capacity the transmission power needed by one link should be minimised. In the downlink, the received power level in the UE does not affect the capacity.

In the uplink, the transmission powers determine the amount of interference to the adjacent cells, and the received powers determine the amount of interference to other UEs in the same cell. If, for example, there were only one WCDMA cell in one area, the uplink capacity of this cell would be maximised by minimising the required received powers, and the power rise would not affect the uplink capacity. We are, however, interested in cellular networks where the design of the uplink diversity schemes has to take into account both the transmitted and received powers. The effect of received and transmission powers on network interference levels is illustrated in Figure 9.5.

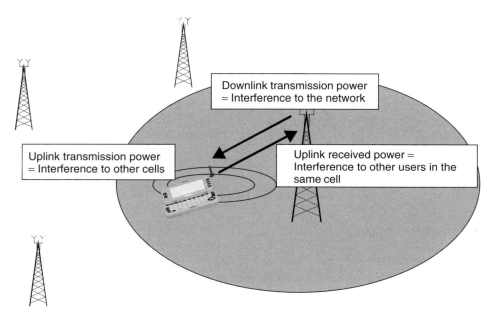

Figure 9.5. Effect of received and transmission powers on interference levels

9.2.1.3 Power Control in Soft Handover

Fast power control in soft handover has two major issues that are different from the single-link case: power drifting in the Node B powers in the downlink, and reliable detection of the uplink power control commands in the UE. These aspects are illustrated in Figure 9.6 and in Figure 9.7 and described in more detail in this section. A solution for improving the power control signalling quality is also presented in this section.

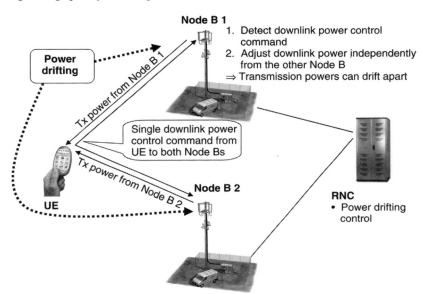

Figure 9.6. Downlink power drifting in soft handover

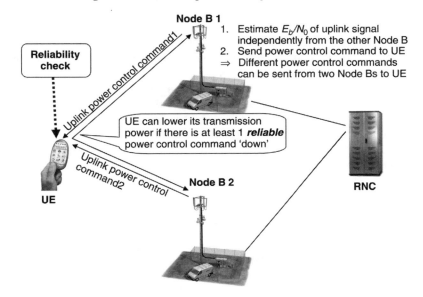

Figure 9.7. Reliability check of the uplink power control commands in UE in soft handover

Downlink Power Drifting

The UE sends a single command to control the downlink transmission powers; this is received by all Node Bs in the active set. The Node Bs detect the command independently, since the power control commands cannot be combined in RNC because it would cause too much delay and signalling in the network. Due to signalling errors in the air interface, the Node Bs may detect this power control command in a different way. It is possible that one of the Node Bs lowers its transmission power to that UE while the other Node B increases its transmission power. This behaviour leads to a situation where the downlink powers start drifting apart; this is referred to here as power drifting.

Power drifting is not desirable, since it mostly degrades the downlink soft handover performance. It can be controlled via RNC. The simplest method is to set relatively strict limits for the downlink power control dynamics. These power limits apply to the UEs specific transmission powers. Naturally, the smaller the allowed power control dynamics, the smaller the maximum power drifting. On the other hand, large power control dynamics typically improve power control performance, as shown in Table 9.2.

Another way to reduce power drifting is as follows. RNC can receive information from the Node Bs concerning the transmission power levels of the soft handover connections. These levels are averaged over a number of power control commands, e.g. over 500 ms or equivalently over 750 power control commands. Based on those measurements, RNC can send a reference value for the downlink transmission powers to the Node Bs. The soft handover Node Bs use that reference value in their downlink power control for that connection to reduce the power drifting. The idea is that a small correction is periodically performed towards the reference power. The size of this correction is proportional to the difference between the actual transmitted power and the reference power. This method will reduce the amount of power drifting. Power drifting can happen only if there is fast power control in the downlink. In IS-95 only slow power control is used in the downlink, and no method of controlling downlink power drifting is needed.

Reliability of Uplink Power Control Commands

All the Node Bs in the active set send an independent power control command to the UEs to control the uplink transmission power. It is enough if one of the Node Bs in the active set receives the uplink signal correctly. Therefore, the UE can lower its transmission power if one of the Node Bs sends a power-down command. Maximal ratio combining can be applied to the data bits in soft handover in the UE, because the same data is sent from all soft handover Node Bs, but not to the power control bits because they contain different information from each of the Node Bs. Therefore, the reliability of the power control bits is not as good as for the data bits, and a threshold in the UE is used to check the reliability of the power control commands. Very unreliable power control commands should be discarded because they are corrupted by interference. The 3GPP specifications include this requirement for the UE in [2].

Improved Power Control Signalling Quality

The power control signalling quality can be improved by setting a higher power for the dedicated physical control channel (DPCCH) than for the dedicated physical data channel (DPDCH) in the downlink if the UE is in soft handover. This power offset between DPCCH and DPDCH can be different for different DPCCH fields: power control bits, pilot bits and TFCI. The power offset is illustrated in Figure 9.8.

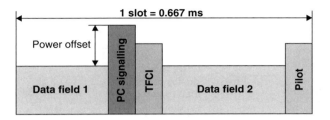

Figure 9.8. Power offset for improving downlink signalling quality

The reduction of the UE transmission power is typically up to 0.5 dB with power offsets. This reduction is obtained because of the improved quality of the power control signalling. [3]

9.2.2 Outer Loop Power Control

The outer loop power control is needed to keep the quality of communication at the required level by setting the target for the fast power control. The outer loop aims at providing the required quality: no worse, no better. Too high quality would waste capacity. The outer loop is needed in both uplink and downlink because there is fast power control in both uplink and downlink. In the following sections a few aspects of this control loop are described; these apply to both uplink and downlink. The uplink outer loop is located in RNC and the downlink outer loop in the UE. In IS-95, outer loop power control is used only in the uplink because there is no fast power control in the downlink.

An overview of uplink outer loop power control is shown in Figure 9.9. The uplink quality is observed after macro diversity combining in RNC and the SIR target is sent to the Node Bs. The frequency of the fast power control is 1.5 kHz and frequency of the outer loop power control typically 10–100 Hz. A general outer loop power control algorithm is presented in Figure 9.10.

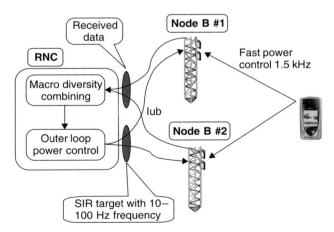

Figure 9.9. Uplink outer loop power control in RNC

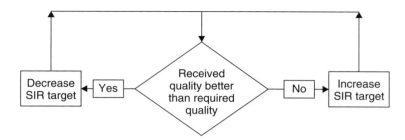

Figure 9.10. General outer loop power control algorithm

9.2.2.1 Gain of Outer Loop Power Control

In this section we analyse how much the SIR target needs to be adjusted when the UE speed or the multipath propagation environment changes. The terms SIR target and E_b/N_0 target are used interchangeably in this chapter. Simulation results with AMR speech service and BLER $= 1\%$ are shown in Table 9.4 with outer loop power control. Three different multipath profiles are used: non-fading channel corresponding to strong line-of-sight component, fading ITU Pedestrian A channel, and fading three-path channel with equal average powers of the multipath components. No antenna diversity is assumed here.

Table 9.4. Average E_b/N_0 targets in different environments

Multipath	UE speed (km/h)	Average E_b/N_0 target (dB)
Non-fading	—	5.3
ITU Pedestrian A	3	5.9
ITU Pedestrian A	20	6.8
ITU Pedestrian A	50	6.8
ITU Pedestrian A	120	7.1
3-path equal powers	3	6.0
3-path equal powers	20	6.4
3-path equal powers	50	6.4
3-path equal powers	120	6.9

The lowest average E_b/N_0 target is needed in the non-fading channel and the highest target in the ITU Pedestrian A channel with high UE speed. This result indicates that the higher the variation in the received power, the higher the E_b/N_0 target needs to be to provide the same quality. If we were to select a fixed E_b/N_0 target of 5.3 dB according to the static channel, the frame error rate of the connection would be too high in fading channels and speech quality would be degraded. If we were to select a fixed E_b/N_0 target of 7.1 dB, the quality would be good enough but unnecessary high powers would be used in most situations. We can conclude that there is clearly a need to adjust the target of the fast closed loop power control by outer loop power control.

9.2.2.2 Estimation of Received Quality

A few different approaches to measuring the received quality are introduced in this section. A simple and reliable approach is to use the result of the error detection – cyclic redundancy check (CRC) – to detect whether there is an error or not. The advantages of using the CRC check are that it is a very reliable detector of frame errors, and it is simple. The CRC-based approach is well suited for those services where errors are allowed to occur fairly frequently, at least once every few seconds, such as the non-real time packet data service, where the block error rate (BLER) can be up to 10–20 % before retransmissions, and the speech service, where typically BLER = 1 % provides the required quality. With Adaptive Multirate (AMR) speech codec the interleaving depth is 20 ms and BLER = 1 % corresponds to one error on average every 2 seconds.

The received quality can also be estimated based on soft frame reliability information. Such information could be, for example:

- Estimated bit error rate (BER) before channel decoder, called raw BER or physical channel BER.

- Soft information from Viterbi decoder with convolutional codes.

- Soft information from Turbo decoder, for example BER or BLER after an intermediate decoding iteration.

- Received E_b/N_0.

Soft information is needed for high quality services, see Section 9.2.2.4. Raw BER is used as soft information over the Iub interface. The estimation of the quality is illustrated in Figure 9.11.

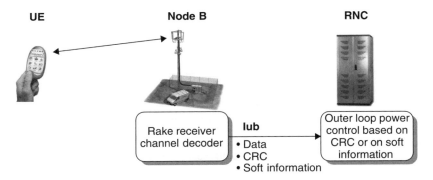

Figure 9.11. Estimation of quality in outer loop in RNC

9.2.2.3 Outer Loop Power Control Algorithm

One possible outer loop power control algorithm is presented in [4]. The algorithm is based on the result of a CRC check of the data and can be characterised in pseudocode as shown in Figure 9.12.

If the BLER of the connection is a monotonically decreasing function of the E_b/N_0 target, this algorithm will result in a BLER equal to the BLER target if the call is long enough. The step size parameter determines the convergence speed of the algorithm to the desired target and also defines the overhead caused by the algorithm. The principle is that the higher the

IF CRC check OK

Step_down = BLER_target*Step_size ;
Eb/N0_target (n+1) = Eb/N0_target (n) – Step_down ;

ELSE

Step_up = Step_size – BLER_target*Step_size ;
Eb/N0_target (n+1) = Eb/N0_target (n) + Step_up ;

END

where

Eb/N0_target (n) is the E_b/N_0 target in frame *n*,
BLER_target is the BLER target for the call and
Step_size is a parameter, typically 0.3–0.5 dB

Figure 9.12. Pseudocode of one outer loop power control algorithm

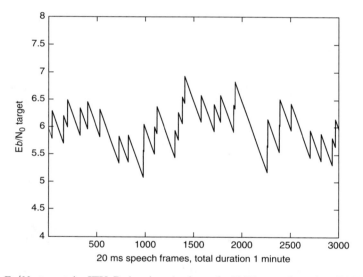

Figure 9.13. E_b/N_0 target in ITU Pedestrian A channel, AMR speech codec, BLER target 1 %, step size 0.5 dB, speed 3 km/h

step size the faster the convergence and the higher the overhead. Figure 9.13 gives an example of the behaviour of the algorithm with BLER target of 1 % and step size of 0.5 dB.

9.2.2.4   High Quality Services

High quality services with very low BLER ($<10^{-3}$) are required to be supported by third generation networks. In such services errors are very rare events. If the required BLER $= 10^{-3}$ and the interleaving depth is 40 ms, an error occurs on average every 40/ 10^{-3} ms $= 40$ seconds. If the received quality is estimated based on the errors detected by CRC bits, the adjustments of the E_b/N_0 target are very slow and the convergence of the E_b/N_0 target to the optimal value takes a long time. Therefore, for high quality services the soft frame reliability information has advantages. Soft information can be obtained from every frame even if there are no errors.

9.2.2.5 Limited Power Control Dynamics

At the edge of the coverage area the UE may hit its maximum transmission power. In that case the received BLER can be higher than desired. If we apply directly the outer loop algorithm of Figure 9.10, the uplink SIR target would be increased. The increase of the SIR target does not improve the uplink quality if the Node B is already sending only power-up commands to the UE. In that case the E_b/N_0 target might become unnecessarily high. When the UE returns closer to the Node B, the quality of the uplink connection is unnecessarily high before the outer loop lowers the E_b/N_0 target back to the optimal value. The situation in which the UE hits its maximum transmission power is shown in Figure 9.14. In this example,

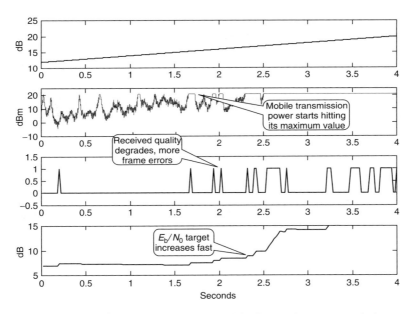

Figure 9.14. Increase of E_b/N_0 target when the UE hits its maximum transmission power. Top: attenuation between UE and Node B; second figure: UE transmission power (dBm); third figure: block errors (1 = error, 0 = no errors); bottom: uplink E_b/N_0 target

the AMR speech service with 20 ms interleaving is simulated with the outer loop power control algorithm from Figure 9.12. A BLER target of 1 % and an outer loop step size of 0.5 dB are used here. With full power control dynamics an error should occur once every 2 seconds to provide BLER of 1 % with 20 ms interleaving. The maximum transmission power of the UE is 125 mW, i.e. 21 dBm.

The same problem could also occur if the UE hits its minimum transmission power. In that case, the E_b/N_0 target would become unnecessarily low. The same problems can be observed also in the downlink if the power of the downlink connection is using its maximum or minimum value.

The outer loop problems from limited power control dynamics can be avoided by setting tight limits for the E_b/N_0 target or by an intelligent outer loop power control algorithm. Such an algorithm would not increase the E_b/N_0 target if the increase did not improve the quality. The 3GPP specifications include this requirement for the UE in [2].

9.2.2.6   Multiservice

One of the basic requirements of UMTS is to be able to multiplex several services on a single physical connection. Since all the services have a common fast power control, there can be only one common target for fast power control. This must be selected according to the service requiring the highest target. Fortunately, there should be no large differences between the required targets if unequal rate matching has been applied on Layer 1 to provide the different qualities. The multiservice scenario is shown in Figure 9.15.

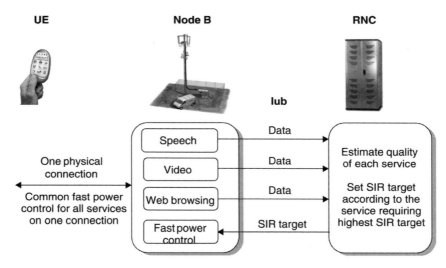

Figure 9.15. Uplink outer loop power control for multiple services on one physical connection

9.2.2.7   Downlink Outer Loop Power Control

The downlink outer loop power control runs in the UE. The network can effectively control the downlink connections even if it cannot control the downlink outer loop algorithm. First, the network sets the quality target for each downlink connection; that target can be modified during the connection. Second, the Node B does not need to increase the downlink power of that connection even if the UE sends a power-up command. The network can control the quality of the different downlink connections very quickly by not obeying the power control commands from the UE. This approach could be used, for example, during downlink overload to reduce the downlink power of those connections that have a lower priority, such as background-type services (see load control in Section 9.6). This reduction of downlink powers can take place at the frequency of fast power control, i.e. 1.5 kHz.

9.3 Handovers

9.3.1 Intra-frequency Handovers

9.3.1.1 Handover Algorithms

The WCDMA soft handover algorithm is introduced in this section, for specifications see [5]. The soft handover uses typically CPICH E_c/I_0 (= pilot E_c/I_0) as the handover measurement quantity, which is signalled to RNC by using Layer 3 signalling (see Section 7.8). The following terminology is used in the handover description:

- **Active set**: The cells in the active set form a soft handover connection to the UE.

- **Neighbour set/monitored set**: The neighbour set or monitored set is the list of cells that the UE continuously measures, but whose pilot E_c/I_0 are not strong enough to be added to the active set.

The soft handover algorithm as described in Figure 9.16 is as follows:

- If *Pilot_E_c/I_0 > Best_ Pilot_E_c/I_0 − Reporting_range + Hysteresis_event1A* for a period of ΔT and the active set is not full, the cell is added to the active set. This event is called Event 1A or Radio Link Addition.

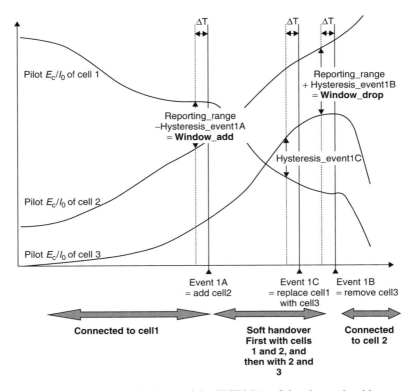

Figure 9.16. General scheme of the WCDMA soft handover algorithm

- If $Pilot\_E_c/I_0 < Best\_Pilot\_E_c/I_0 - Reporting\_range - Hysteresis\_event1B$ for a period of ΔT, then the cell is removed from the active set. This event is called Event 1B or Radio Link Removal.

- If the active set is full and $Best\_candidate\_Pilot\_E_c/I_0 > Worst\_Old\_Pilot\_E_c/I_0 + Hysteresis\_event1C$ for a period of ΔT, then the weakest cell in the active set is replaced by the strongest candidate cell (i.e. strongest cell in the monitored set). This event is called Event 1C or Combined Radio Link Addition and Removal. The maximum size of the active set in Figure 9.16 is assumed to be two.

where

- *Reporting_range* is the threshold for soft handover
- *Hysteresis_event1A* is the addition hysteresis
- *Hysteresis_event1B* is the removal hysteresis
- *Hysteresis_event1C* is the replacement hysteresis
- *Reporting_range* $-$ *Hysteresis_event1A* is also called Window_add
- *Reporting_range* $+$ *Hysteresis_event1B* is also called Window_drop
- ΔT is the time to trigger
- $Best\_Pilot\_E_c/I_0$ is the strongest measured cell in the active set
- $Worst\_Old\_Pilot\_E_c/I_0$ is the weakest measured cell in the active set
- $Best\_candidate\_Pilot\_E_c/I_0$ is the strongest measured cell in the monitored set
- $Pilot\_E_c/I_0$ is the measured and filtered quantity.

9.3.1.2 Handover Measurements

The WCDMA UE scans continuously the other cells on the same frequency when in cell_DCH state. The UE typically uses a matched filter to find the primary synchronisation channel, P-SCH, of the neighbouring cells. All cells transmit the same synchronisation code that the UE seeks. The UE further identifies the cells with secondary synchronisation channel, S-SCH and pilot channel, CPICH. The synchronisation procedure is described in detail in Chapter 6. After that procedure, the UE is able to make pilot E_c/I_0 measurements and identify to which cell that measurement result belongs.

Since the WCDMA Node Bs can be asynchronous, the UE also decodes the system frame number (SFN) from the neighbouring cells. SFN indicates the Node B timing with 10 ms frame resolution. SFN is transmitted on the Broadcast channel, BCH, carried on the Primary Common Control Physical Channel, P-CCPCH.

The intra-frequency handover measurement procedure is shown in Figure 9.17.

We evaluate the mobile handover measurement requirements in 3GPP and compare the performance requirements to the typical CPICH E_c/I_0 values with the common channel power allocations suggested in Chapter 8.

Phase (1): Cell Identification

The cell identification time in phase (1) of Figure 9.17 depends mainly on the number of cells and multipath components that the UE can receive. The UE needs to check every peak in its matched filter; the fewer peaks there are, the faster is the cell identification. The length of the neighbourlist has only a marginal effect on the handover measurement performance.

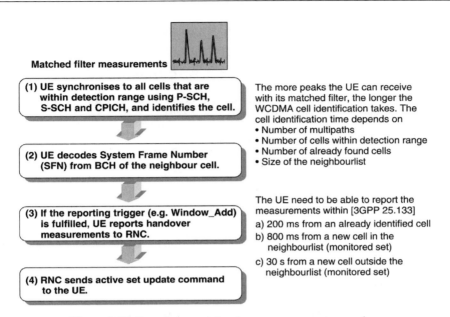

Figure 9.17. Intra-frequency handover measurement procedure

The 3GPP handover measurement performance requirements for the UE are given in [6]: with CPICH $E_c/I_0 > -20$ dB and SCH $E_c/I_0 > -20$ dB the UE shall be able to report measurements within 200 ms from an already identified cell and within 800 ms from a new cell belonging to the monitored set. We will consider these performance requirements, together with typical common channel power allocations from Chapter 8 and typical handover parameter values. The scenario is illustrated in Figure 9.18 where the UE is

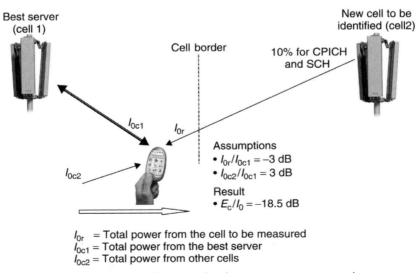

Figure 9.18. Intra-frequency handover measurement scenario

connected to cell1 and it needs to identify cell2 that is approaching Window_add value. The resulting E_c/I_0 is obtained as follows:

1. We allocate 10 % for CPICH and for SCH giving $E_c/I_{0r} = -10\,\text{dB}$.

2. We assume Window_add $= 3\,\text{dB}$ where the UE needs to identify the cell when it is 3 dB below the strongest cell. This gives $I_{0r}/I_{0c1} = -3\,\text{dB}$.

3. We further assume that the interference from other cells is 3 dB higher than the signal power from the best server. This gives $I_{0c2}/I_{0c1} = 3\,\text{dB}$.

$$\frac{E_c}{I_0} = \frac{E_c}{I_{0r} + I_{0c1} + I_{0c2}} = \frac{E_c}{I_{0r}(1 + 10^{0.3} + 10^{0.6})} = \frac{E_c}{I_{0r}} = -8.5\,\text{dB} = -18.5\,\text{dB} \qquad (9.1)$$

The E_c/I_0 in this scenario is $-18.5\,\text{dB}$ which is better than $-20\,\text{dB}$ given in the performance requirements.

Phase (2): System Frame Number (SFN) Decoding

In phase (2) of Figure 9.17 the UE decodes the system frame number from BCH that is transmitted on P-CCPCH. If we allocate $-5\,\text{dB}$ power for P-CCPCH compared to CPICH, the resulting E_c/I_0 is $-23.5\,\text{dB}$. The performance requirement for BCH decoding with BLER $= 1\,\%$ is $-22.0\,\text{dB}$ [2]. As higher BLER levels can be allowed, the planned P-CCPCH power allocation should be adequate.

These calculations of the cell identification and the system frame number decoding show that the assumed common channel powers in Chapter 8 are high enough to guarantee accurate handover measurements. It may even be possible to optimise the common channel powers to be lower than shown in Chapter 8.

Before the pilot E_c/I_0 is used by the active set update algorithm in the UE, some filtering is applied to make the results more reliable. The measurement is filtered both on Layer 1 and on Layer 3. The Layer 3 filtering can be controlled by the network. The WCDMA handover measurement filtering is described in Figure 9.19.

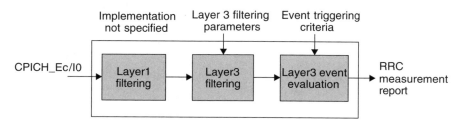

Figure 9.19. Handover measurement filtering and reporting

The handover measurement reporting from the UE to RNC can be configured to be periodic, as in GSM, or event-triggered. According to [7] the event-triggered reporting provides the same performance as periodic reporting but with less signalling load.

Before we leave the area of handover measurements we note the size of the neighbourlist. The maximum number of intra-frequency cells in the neighbourlist is 32. Additionally, there can be 32 inter-frequency and 32 GSM cells on the neighbourlist. The inter-frequency and inter-system measurements are covered later in this chapter. The maximum number of cells in the neighbourlist is shown in Table 9.5.

Table 9.5. Maximum number of monitored cells

Intra-frequency	Inter-frequency	Inter-system GSM
32	32[1]	32

[1] The 32 inter-frequency neighbours can be on two frequencies.

The neighbourlist can be defined by the network planning tool or it can also be tuned automatically by network optimisation algorithms based on the UE measurements. If there is a neighbour cell missing from the list, it can be noticed based on the UE measurements, since UE is required to identify a new detectable cell that does not belong to the monitored set within 30 s [6].

9.3.1.3 Soft Handover Link Gains

The primary purpose of soft handover is to provide seamless handover and added robustness to the system. This is mainly achieved via three types of gain provided by the soft handover mechanism, i.e.,

- Macro diversity gain: A diversity gain over slow fading and sudden drops in signal strength due to, e.g., UE movement around a corner.

- Micro diversity gain: A diversity gain over fast fading.

- Downlink load sharing: A UE in soft handover receives power from multiple Node Bs, which implies that the maximum transmit power to a UE in X-way soft handover is multiplied by factor X (i.e. improved coverage).

These three soft handover gains can be mapped to improved coverage and capacity of the WCDMA network. This section presents results of micro diversity soft handover gains that have been obtained by means of link level simulations. The micro diversity gains are presented relative to the ideal hard handover case, where the UE would be connected to the Node B with the highest pilot E_c/I_0. The results were presented and discussed in more detail in [3].

Figures 9.20 and 9.21 show the simulation results of 8 kbps speech in an ITU Pedestrian A channel, at 3 km/h, assuming that the UE is in soft handover with two Node Bs. The relative path loss from the UE to Node B #1 compared to Node B #2 is 0, −3, −6 or −10 dB. The highest gains are obtained when the path loss is the same to both Node Bs, i.e. the relative path loss difference is 0 dB. Figure 9.20 shows the soft handover gain in uplink transmission power with two branch Node B receive antenna diversity. Figure 9.21 shows the corresponding gains in downlink transmission power without transmit or receive antenna diversity. The gains are relative to the single link case where the UE only is connected to the best Node B. It should be noted that ITU Pedestrian A channel has only little multipath diversity, and thus the micro diversity soft handover gains are relatively high. With more multipath diversity, the gains tend to decrease.

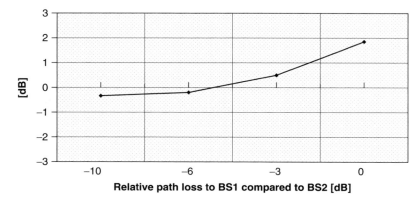

Figure 9.20. Soft handover gain in uplink transmission power (positive value = gain, negative value = loss)

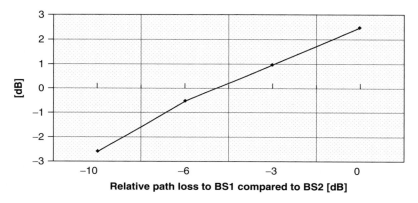

Figure 9.21. Soft handover gain in downlink transmission power (positive = gain, negative = loss)

In Figure 9.20 the maximum reduction of the UE transmission power due to soft handover is observed to be 1.8 dB if the path loss is the same to both Node Bs. If the path loss difference is very large, the uplink soft handover should – in theory – never increase the UE transmission power since there is no extra transmission but only more Node Bs trying to detect the signal. In practice, if the path loss difference is very large, the soft handover can cause an increase in the UE transmission power. This increase is caused by the signalling errors of the uplink power control commands, which are transmitted in the downlink. But, typically, the Node B would not be in the active set of the UE if the path loss were 3–6 dB larger than the path loss to the strongest Node B in the UE's active set.

In the downlink the maximum soft handover gain is 2.3 dB (Figure 9.21), which is more than in the uplink (Figure 9.20). The reason is that no antenna diversity is assumed in the downlink, and thus in the downlink there is more need for micro diversity soft handover gain.

In the downlink, soft handover causes an increase in the required downlink transmission power if the path loss difference is more than 4–5 dB for the current example. In that case the UE does not experience a gain from the signal transmitted from the Node B with the largest path loss. Hence, the power transmitted from that Node B to the UE will only contribute to the total interference in the network These simulation results also suggest values for Window_add and for Window_drop. Typical values for those parameters are shown in Table 9.6.

Table 9.6. Typical handover parameters

Window_add	Window_drop
1–3 dB	2–5 dB

9.3.1.4   Network Capacity Gains

The potential capacity gain of soft handover mainly depends on the soft handover overhead (i.e. on the relative proportion of UEs in soft handover), the soft handover link gain, and the applied power control algorithm. Note that there are two different downlink power control algorithms for UEs in soft handover;

1. Conventional power control as described in Section 9.2.1.3; and
2. Site selection diversity transmission (SSDT) scheme presented in Chapter 6.

Recall that SSDT power control relies on feedback information from the UE, so only one of the Node Bs in the active set transmits data, while the other Node Bs only transmit physical layer control information. Thus, SSDT is equivalent to selection transmit diversity, while conventional power control of UEs in soft handover may be characterised as equal gain transmit diversity. The potential gain of SSDT comes from the reduced interference in the downlink, which should compensate for the loss of diversity gain in the downlink for the user data. From the conceptual point of view, it is obvious that the gain of SSDT is larger at high data rates where the overhead from the control information is marginal.

The capacity gain of soft handover in combination with SSDT, is on the same order of magnitude as the gain of soft handover and conventional power control. No significant gain of SSDT is observed, and in some cases the gain even turns into a loss. The reason for these observations can be explained as follows. A UE in soft handover periodically sends feedback commands to the Node Bs in the active set, which dictates which of the Node Bs should be transmitting the data. This results in larger power fluctuations at the different Node Bs, because the transmission to UEs is switched on/off on a relatively fast basis, as dictated by the UEs in soft handover. The alternation of Node B transmission towards UEs in soft handover is not within the control of the network, but purely UE controlled. Hence, although the SSDT scheme results in a reduction of the average total transmit power from Node Bs, the variance of the total transmit power also increases. The increased variance of the total transmit power maps to larger required power control headroom, which tends to reduce the potential gain of SSDT. Other aspects to note from a performance point of view include the impact of UE velocity, as with higher velocities the UE feedback is not well synchronised with the actual channel state. At some velocities, resonance problems do occur so the UE constantly asks for the 'wrong' Node B to transmit, via the feedback signalling to the network. This effect does, e.g., become dominant when the fading rate is roughly equal to the feedback rate.

9.3.1.5 Soft Handover Overhead

The soft handover overhead is a common metric, which often is used to quantify the soft handover activity in a network. The soft handover overhead (β) is defined as

$$\beta = \sum_{n=1}^{N} nP_n - 1 \tag{9.2}$$

where N is the active set size and P_n is the average probability of a UE being in n-way soft handover. In this context one-way soft handover refers to a situation where the UE is connected to one Node B, while two-way soft handover means that the UE is connected to two Node Bs, etc as shown in Figure 9.22. As each connection between a UE and Node B requires logical baseband resources, reservation of transmission capacity over the Iub, and RNC resources, the soft handover overhead may also be regarded as a measure of the additional hardware/transmission resources required for implementation of soft handover. Radio network planning is responsible for proper handover parameter setting and site planning so that the soft handover overhead is planned to be on the order of 20–40 % for a standard hexagonal cell grid with three sector sites. Note that an excessive soft handover overhead could decrease the downlink capacity as shown in Figure 9.21. In the downlink each soft handover connection increases the transmitted interference to the network. When the increased interference exceeds the diversity gain, the soft handover does not provide any gain for system performance.

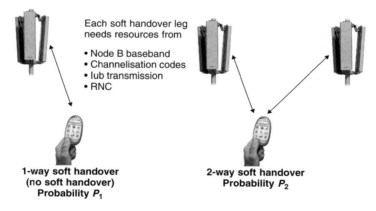

Figure 9.22. Soft handover overhead

The soft handover overhead can, to a large extent, be controlled by proper selection of the parameters Window_add, Window_drop, and the active set size. However, there are also factors which influence the soft handover overhead, which are not controllable via the soft handover parameter settings. Some of these are:

- The network topology: How are the sites located relative to each other, how many sectors per site, etc.?

- The Node B antenna radiation patterns.

- The path loss and shadow fading characteristics.

- The average number of Node Bs that a UE can synchronise to.

As an example, the soft handover overhead is plotted in Figure 9.23 for a standard hexagonal cell grid with three sector sites. These results are obtained from a dynamic network level simulator. Results are presented for a cell radii of 666 metres and 2000 metres. A standard 65 degree antenna is assumed for each sector. The deterministic path loss is modelled according to the Okumura–Hata model, while the shadow fading component is assumed to be log-normal distributed with 8 dB standard deviation. The transmit power of the CPICH is fixed at 10 % and 20 % of the maximum Node B transmit power for the small and large cell radii, respectively. The power of the SCH is −3.0 dB relative to the P-CPICH. The active set size is limited to three.

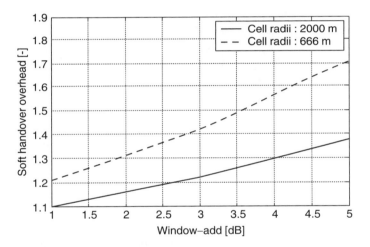

Figure 9.23. The soft handover overhead versus the soft handover parameter Window_add for a hexagonal cell grid with three sector sites, and two different cell radii. Window_drop = Window_add + 2.0 dB

It is observed that the soft handover overhead increases approximately linearly when Window_add and Window_drop are increased. For the same soft handover parameter settings, the soft handover overhead is typically larger for the scenario with small cells, compared to large cells. This behaviour is observed because UEs in the large cell grid can only synchronise to a few Node Bs, while UEs in the small cell grid typically can synchronise to many Node Bs. Assuming that the design goal is to have a soft handover overhead of 20–40 %, then the results in Figure 9.23 indicate that appropriate parameter settings are Window_add = 1–3 dB in small cells and slightly larger values in large cells. This conclusion is, however, only valid for a network topology with three sector sites. For the same soft handover parameter settings, the soft handover overhead increases when migrating from three sector sites to six sector sites. As a rule of thumb, the soft handover overhead increases by approximately 30 % when comparing results for three and six sector site configurations. This calls for selection of lower Window_add/Window_drop when increasing the number of sectors.

An example of soft handover overhead in a live WCDMA network in a dense urban area is shown in Figure 9.24. The average overhead, including softer handovers, is 38 %. From the transmission point of view the overhead is less than 38 % because softer handover combining takes place in Node B and does not require extra transmission resources. The Window_add has been 2–4 dB, the Window_drop 4–6 dB, addition timer 320 ms and drop timer 640 ms.

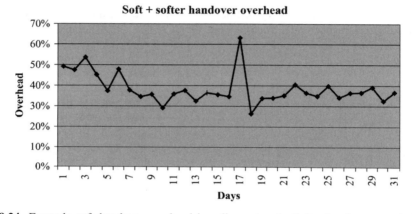

Figure 9.24. Example soft handover overhead in a live network. Softer handovers are included in this figure

9.3.1.6   Active Set Update Rate

Active set update rate is counted as the time between consecutive active set update commands and includes all the events: addition, removal and replacement. This measure is relevant for RNC dimensioning since active set update signalling causes load to the RNC. Active set update rates are shown in Figure 9.25. The average value in this example is 12 seconds, i.e. 5 active set updates per minute. The value depends on the average mobility of the users, on the cell size, on the network planning and on the handover parameters. Typically, soft handover overhead and active set update rates are related: smaller overhead can be obtained at the expense of a higher active set update rate.

9.3.2   *Inter-system Handovers Between WCDMA and GSM*

WCDMA and GSM standards support handovers both ways between WCDMA and GSM. These handovers can be used for coverage or load balancing reasons. At the start of WCDMA deployment, handovers to GSM are needed to provide continuous coverage, and handovers from GSM to WCDMA can be used to lower the loading in GSM cells. This scenario is shown in Figure 9.26. When the traffic in WCDMA networks increases, it is important to have load reason handovers in both directions. The inter-system handovers are

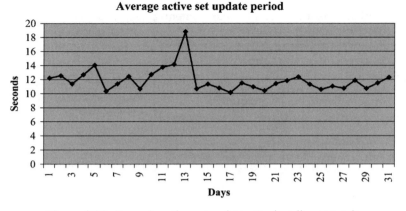

Figure 9.25. Example active set update rates in a live network

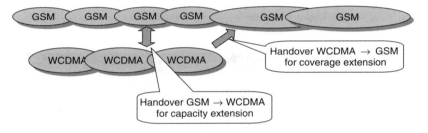

Figure 9.26. Inter-system handovers between GSM and WCDMA

triggered in the source RNC/BSC, and from the receiving system point of view, the inter-system handover is similar to inter-RNC or inter-BSC handover. The handover algorithms and triggers are not standardised.

A typical inter-system handover procedure is shown in Figure 9.27. The inter-system measurements are not active all the time but are triggered when there is a need to make inter-system handover. The measurement trigger is an RNC vendor-specific algorithm and could be based, for example, on the quality (block error rate) or on the required transmission power. When the measurements are triggered, the UE measures first the signal powers of the GSM frequencies on the neighbourlist. Once those measurements are received by RNC, it commands the UE to decode the BSIC (base station identity code) of the best GSM candidate. When the BSIC is received by RNC, a handover command can be sent to the UE. The measurements can be completed in approximately 2 seconds.

9.3.2.1 Compressed Mode

WCDMA uses continuous transmission and reception and cannot make inter-system measurements with a single receiver if there are no gaps generated in the WCDMA signals. Therefore, compressed mode is needed both for inter-frequency and for inter-system

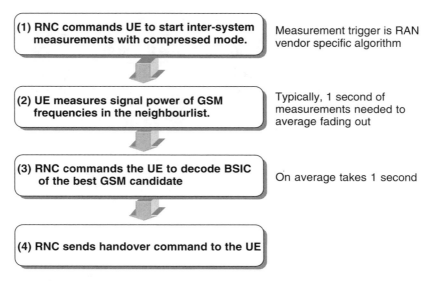

Figure 9.27. Inter-system handover procedure

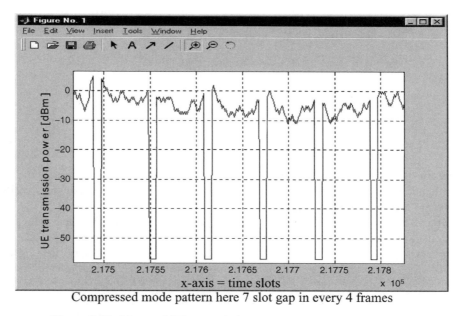

Figure 9.28. Measured UE transmission power during compressed mode

measurements. The compressed mode procedure is described in Chapter 6. UE transmission power during compressed mode is illustrated in Figure 9.28.

During the gaps of the compressed mode, the fast power control cannot be applied and part of the interleaving gain is lost. Therefore, a higher E_b/N_0 is needed during compressed frames, leading to a capacity degradation. An example calculation for the capacity effect is

Table 9.7. Effect of compressed mode to the capacity

	100 % UEs in compressed mode	10 % UEs in compressed mode
Every frame compressed	−58 %	−5.8 %
Every 3rd frame compressed	−19 %	−1.9 %

shown in Table 9.7. In here it is assumed that the E_b/N_0 needs to be 2.0 dB higher during the compressed frames. If all UEs used compressed mode in every frame, the interference levels would increase 58 % (=10^0.2), and the capacity would reduce correspondingly.

The measurements capability is typically fast enough if every 3rd frame is compressed. With every 3rd frame compressed, the UE can measure six samples in every gap, i.e. 480 ms/ 30 ms*6 = 96 samples per 480 ms [6]. That measurement capability is similar to that in GSM only mobile for intra-GSM measurements.

If we have every 3rd frame compressed, the capacity degradation is still 19 %. The results clearly show that compressed mode should only be activated when there is a need to make inter-system and inter-frequency handover. If we assume that 10 % of the UEs are simultaneously using the compressed mode, the capacity effect is reduced down to 1.9 %. These results show that the effect of the compressed mode to the capacity is negligible if the compressed mode is used only when needed. The RNC algorithms for activating the compressed mode are important to guarantee reliable handovers while maintaining low compressed mode usage.

Compressed mode also affects the uplink coverage area of the real time services where the bit rate cannot be lowered during the compressed mode. Therefore, the coverage reason inter-system handover procedure has to be initiated early enough at the cell edge to avoid any quality degradation during the compressed mode. This situation is shown in Figure 9.29. The compressed mode affects coverage in two ways:

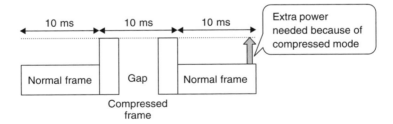

Figure 9.29. Effect of compressed mode on the coverage

1. The same amount of data is transmitted in a shorter time. This effect is 10*log10(15/ (15 − 7)) = 2.7 dB, with a 7-slot gap in a 15-slot frame.

2. The E_b/N_0 performance degrades during the compressed mode. The degradation is assumed to be 2.0 dB here.

An example effect to the coverage area of speech is shown in Table 9.8, where the coverage is reduced by 2.4 dB with 20 ms interleaving. AMR voice connection uses 20 ms

Table 9.8. Effect of compressed mode on the coverage

Interleaving	Reduction in coverage
10 ms	2.7 dB + 2.0 dB = 4.7 dB
20 ms	(2.7 dB + 2.0 dB)/2 = 2.4 dB

interleaving. This 2.4 dB coverage degradation can be compensated by lowering instanta-
neously the AMR bit rate during the compressed mode if the UE hits its maximum power,
see Section 12.2.1.2.

Inter-system handovers from GSM to WCDMA are initiated in GSM BSC. No compressed mode is needed for making WCDMA measurements from GSM because GSM uses
discontinuous transmission and reception.

The service interruption time in the inter-system handovers is 40 ms maximum. The
interruption time is the time between the last received transport block on the old frequency
and the time the UE starts transmission of the new uplink channel. The total service gap is
slightly more than the interruption time because the UE needs to get the dedicated channel
running in GSM. The service gap is typically below 80 ms which is similar to that in intra-
GSM handovers. Such a service gap does not degrade voice quality. The service gap is
illustrated in Figure 9.30.

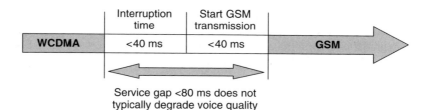

Figure 9.30. Service gap in inter-system handover

9.3.3 Inter-frequency Handovers within WCDMA

Most UMTS operators have two or three FDD frequencies available. The operation can be
started using one frequency and the second and the third frequency are needed later to
enhance the capacity. Several frequencies can be used in two different ways, as shown in
Figure 9.31: several frequencies on the same sites for high capacity sites or macro and micro
layers using different frequencies. Inter-frequency handovers between WCDMA carriers are
needed to support these scenarios.

Compressed mode measurements are used also in the inter-frequency handover in the
same way as in the inter-system handovers.

The inter-frequency handover procedure is shown in Figure 9.32. The UE uses the same
WCDMA synchronisation procedure as for intra-frequency handovers to identify the cells on
the target frequency. The synchronisation procedure is presented in Chapter 6. The cell

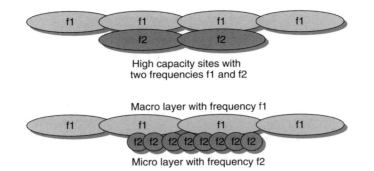

High capacity sites with
two frequencies f1 and f2

Macro layer with frequency f1

Micro layer with frequency f2

Figure 9.31. Need for inter-frequency handovers between WCDMA carriers

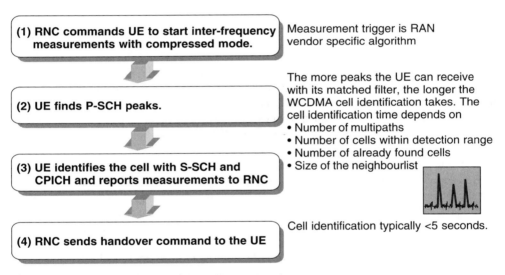

(1) RNC commands UE to start inter-frequency measurements with compressed mode.

Measurement trigger is RAN vendor specific algorithm

(2) UE finds P-SCH peaks.

The more peaks the UE can receive with its matched filter, the longer the WCDMA cell identification takes. The cell identification time depends on
• Number of multipaths
• Number of cells within detection range
• Number of already found cells
• Size of the neighbourlist

(3) UE identifies the cell with S-SCH and CPICH and reports measurements to RNC

(4) RNC sends handover command to the UE

Cell identification typically <5 seconds.

Figure 9.32. Inter-frequency handover procedure

identification time depends mainly on the number of cells and multipath components that the UE can receive, in the same way as with intra-frequency handovers. The requirement for the cell identification in 3GPP is 5 seconds with CPICH $E_c/I_0 > -20$ dB [6].

9.3.4 Summary of Handovers

The WCDMA handover types are summarised in Table 9.9. The most typical WCDMA handover is intra-frequency handover that is needed due to the mobility of the UEs. The intra-frequency handover is controlled by those parameters shown in Figure 9.16. The intra-frequency handover reporting is typically event-triggered, and RNC commands the hand-overs according to the measurement reports. In the case of intra-frequency handover, the UE should be connected to the best Node B(s) to avoid the near–far problem, and RNC does not have any freedom in selecting the target cells.

Table 9.9. WCDMA handover types

Handover type	Handover measurements	Typical handover measurement reporting from UE to RNC	Typical handover reason
WCDMA intra-frequency	Measurements all the time with matched filter	Event-triggered reporting	– Normal mobility
WCDMA → GSM inter-system	Measurements started only when needed, compressed mode used	Periodic during compressed mode	– Coverage
			– Load
			– Service
WCDMA inter-frequency	Measurements started only when needed, compressed mode used	Periodic during compressed mode	– Coverage
			– Load

Inter-frequency and inter-system measurements are typically initiated only when there is a need to make inter-system and inter-frequency handovers. Inter-frequency handovers are needed to balance loading between WCDMA carriers and cell layers, and to extend the coverage area if the other frequency does not have continuous coverage. Inter-system handovers to GSM are needed to extend the WCDMA coverage area, to balance load between systems and to direct services to the most suitable systems.

An example handover scenario is shown in Figure 9.33. The UE is first connected to cell1 on frequency f1. When it moves, intra-frequency handover to cell2 is made. The cell2, however, happens to have a high load, and RNC commands load reason inter-frequency handover to cell5 on frequency f2. The UE remains on frequency f2 and continues to cell6. When it runs out of the coverage area of frequency f2, coverage reason inter-frequency handover is made to cell4 on frequency f1.

Handovers are used in Cell_DCH state. In Cell_FACH, Cell_PCH and URA_PCH state the UE makes the cell reselections between frequencies and systems itself according to the idle mode parameters. The states are described in Section 7.8.2.

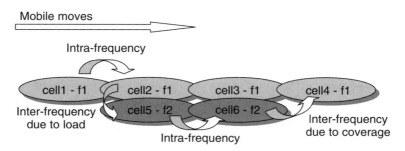

Figure 9.33. Example handover scenario

9.4 Measurement of Air Interface Load

If the radio resource management is based on the interference levels in the air interface, the air interface load needs to be measured. The estimation of the uplink load is presented in Section 9.4.1 and the estimation of the downlink load is in Section 9.4.2.

9.4.1 Uplink Load

In this section two uplink load measures are presented: load estimation based on wideband received power, and load estimation based on throughput. These are example approaches that could be used in WCDMA networks.

9.4.1.1 Load Estimation Based on Wideband Received Power

The wideband received power level can be used in estimating the uplink load. The received power levels can be measured in the Node B. Based on these measurements, the uplink load factor can be obtained. The calculations are shown below.

The received wideband interference power, I_{total}, can be divided into the powers of own-cell (= intra-cell) users, I_{own}, other-cell (= inter-cell) users, I_{oth}, and background and receiver noise, P_N:

$$I_{total} = I_{own} + I_{oth} + P_N \tag{9.3}$$

The uplink noise rise is defined as the ratio of the total received power to the noise power:

$$\text{Noise rise} = \frac{I_{total}}{P_N} = \frac{1}{1 - \eta_{UL}} \tag{9.4}$$

This equation can be rearranged to give the uplink load factor η_{UL}:

$$\eta_{UL} = 1 - \frac{P_N}{I_{total}} = \frac{\text{Noise rise} - 1}{\text{Noise rise}} \tag{9.5}$$

where I_{total} can be measured by the Node B and P_N is known beforehand.

The uplink load factor η_{UL} is normally used as the uplink load indicator. For example, if the uplink load is said to be 60 % of the WCDMA pole capacity, this means that the load factor $\eta_{UL} = 0.60$.

Load estimation based on the received power level is also presented in [8] and [9].

9.4.1.2 Load Estimation Based on Throughput

The uplink load factor η_{UL} can be calculated as the sum of the load factors of the UEs that are connected to this Node B:

$$\eta_{UL} = (1 + i) \cdot \sum_{j=1}^{N} L_j = (1 + i) \cdot \sum_{j=1}^{N} \frac{1}{1 + \dfrac{W}{(E_b/N_0)_j \cdot R_j \cdot v_j}} \tag{9.6}$$

where N is the number of UEs in the own cell, W is the chip rate, L_j is the load factor of the jth UE, R_j is the bit rate of the jth UE, $(E_b/N_0)_j$ is E_b/N_0 of the jth UE, v_j is the voice activity factor of the jth UE, and i is the other-to-own cell interference ratio.

Note that Equation (9.6) is the same as the load factor calculation in radio network dimensioning in Section 8.2.2. In dimensioning, the average number of UEs, N, of a cell needs to be estimated, and average values for E_b/N_0, i and v are used as input parameters. These values are typical for that environment and can be based on the measurements and simulations. In load estimation the instantaneous measured values for E_b/N_0, i, v and the number of UEs N are used to estimate the instantaneous air interface load.

In throughput-based load estimation, interference from other cells is not directly included in the load but needs to be taken into account with the parameter i. Also, the part of own-cell interference that is not captured by the Rake receiver can be taken into account with the parameter i. If it is assumed that $i = 0$, then only own-cell interference is taken into account.

9.4.1.3   Comparison of Uplink Load Estimation Methods

Table 9.10 compares the above two load estimation methods. In the wideband power-based approach, interference from the adjacent cells is directly included in the load estimation because the measured wideband power includes all interference that is received in that carrier frequency by the Node B. If the loading in the adjacent cells is low, this can be seen in the wideband power-based load measurement, and a higher load can be allowed in this cell, i.e. soft capacity can be obtained. The importance of soft capacity was explained in radio network dimensioning in Section 8.2.3.

Table 9.10. Comparison of uplink load estimation methods

	Wideband received power	Throughput	Number of connections
What to measure	Wideband received power I_{total} per cell	Uplink E_b/N_0 and bit rates R for each connection	Number of connections
What needs to be assumed or measured separately	Thermal noise level (=unloaded interference power) P_N	Other-to-own cell interference ratio, i	Load caused by one connection
Other-cell interference	Included in measurement of wideband received power	Assumed explicitly in i	Assumed explicitly when choosing the maximum number of connections
Soft capacity	Yes, automatically	Not directly, possible via RNC	No
Other interference sources (=adjacent channel)	Reduced capacity	Reduced coverage	Reduced coverage

The wideband power-based and throughput based load estimations are shown in Figure 9.34. The different curves represent a different loading in the adjacent cells. The larger the value of i, the more interference from adjacent cells. The wideband power-based load estimation keeps the coverage within the planned limits and the delivered capacity depends on the loading in the adjacent cells (soft capacity). This approach effectively prevents cell breathing which would exceed the planned values.

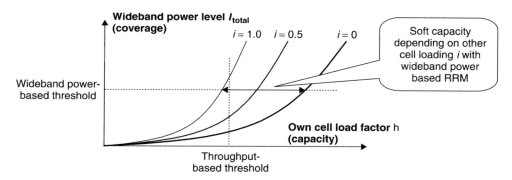

Figure 9.34. Wideband power-based and throughput-based load estimations

The problem with wideband power-based load estimation is that the measured wideband power can include interference from adjacent frequencies. This could originate from another operator's UE located very close to the Node B antenna. Therefore, the interference-based method may overestimate the load of own carrier because of any external interference. The Node B receiver cannot separate the interference from the own carrier and from other carriers by the wideband power measurements.

Throughput-based load estimation does not take interference from adjacent cells or adjacent carriers directly into account. If soft capacity is required, information about the adjacent cell loading can be obtained within RNC. The throughput-based RRM keeps the throughput of the cell at the planned level. If the loading in the adjacent cells is high, this affects the coverage area of the cell.

The third load estimation method in Table 9.10, in the right-hand column, is based simply on the number of connections in the Node Bs. This approach can be used in second generation networks where all connections use fairly similar low bit rates and no high bit rate connections are possible. In third generation networks the mix of different bit rates, services and quality requirements prevents the use of this approach. It is unreasonable to assume that the load caused by one 2-Mbps UE is the same as that caused by one speech UE.

9.4.2 Downlink Load

9.4.2.1 Power-Based Load Estimation

The downlink load of the cell can be determined by the total downlink transmission power, P_{total}. The downlink load factor, η_{DL}, can be defined to be the ratio of the current total transmission power divided by the maximum Node B transmission power P_{max}:

$$\eta_{DL} = \frac{P_{total}}{P_{max}} \tag{9.7}$$

Note that in this load estimation approach the total Node B transmission power P_{total} does not give accurate information concerning how close to the downlink air interface pole capacity the system is operating. In a small cell the same P_{total} corresponds to a higher air interface loading than in a large cell.

9.4.2.2 Throughput-Based Load Estimation

In the downlink, throughput-based load estimation can be effected by using the sum of the downlink allocated bit rates as the downlink load factor, η_{DL}, as follows:

$$\eta_{DL} = \frac{\sum_{j=1}^{N} R_i}{R_{max}} \tag{9.8}$$

where N is the number of downlink connections, including the common channels, R_j is the bit rate of the jth UE, and R_{max} is the maximum allowed throughput of the cell.

It is also possible to weight the UE bit rates with E_b/N_0 values as follows:

$$\eta_{DL} = \sum_{j=1}^{N} R_j \cdot \frac{v_j (E_b/N_0)_j}{W} \cdot [(1 - \bar{\alpha}) + \bar{i}] \tag{9.9}$$

where W is the chip rate, $(E_b/N_0)_j$ is the E_b/N_0 of the jth UE, v_j is the voice activity factor of the jth UE, $\bar{\alpha}$ is the average orthogonality of the cell, and $\bar{i}$ is the average downlink other-to-own cell interference ratio of the cell. Note that Equation (9.9) is similar to the downlink radio network dimensioning (see Section 8.2.2.2).

The average downlink orthogonality can be estimated by the Node B based on the multipath propagation in the uplink. The values of E_b/N_0 need to be assumed based on the typical values for that environment. The average interference from other cells can be obtained in RNC based on the adjacent cell loading.

9.5 Admission Control

9.5.1 Admission Control Principle

If the air interface loading is allowed to increase excessively, the coverage area of the cell is reduced below the planned values, and the quality of service of the existing connections cannot be guaranteed. Before admitting a new UE, admission control needs to check that the admittance will not sacrifice the planned coverage area or the quality of the existing connections. Admission control accepts or rejects a request to establish a radio access bearer in the radio access network. The admission control algorithm is executed when a bearer is set up or modified. The admission control functionality is located in RNC where the load information from several cells can be obtained. The admission control algorithm estimates the load increase that the establishment of the bearer would cause in the radio network. This has to be estimated separately for the uplink and downlink directions. The requesting bearer can be admitted only if both uplink and downlink admission control admit it, otherwise it is rejected because of the excessive interference that it would produce in the network. The limits for admission control are set by the radio network planning.

Several admission control schemes have been suggested in [10–15]. In [10, 12 and 13] the use of the total power received by the Node B is supported as the primary uplink admission control decision criterion, relative to the noise level. The ratio between the total received wideband power and the noise level is often referred to as the noise rise. In [10] and [13] a downlink admission control algorithm based on the total downlink transmission power is presented.

9.5.2   Wideband Power-based Admission Control Strategy

In the interference-based admission control strategy the new UE is not admitted by the uplink admission control algorithm if the new resulting total interference level is higher than the threshold value:

$$I_{\text{total\_old}} + \Delta I < I_{\text{threshold}} \tag{9.10}$$

The threshold value $I_{\text{threshold}}$ is the same as the maximum uplink noise rise and can be set by radio network planning. This noise rise must be included in the link budgets as the interference margin: see Section 8.2.1. Wideband power-based admission control is shown in Figure 9.35. The uplink admission control algorithm estimates the load increase by using either of the two methods presented below. The uplink power increase estimation methods take into account the uplink load curve.

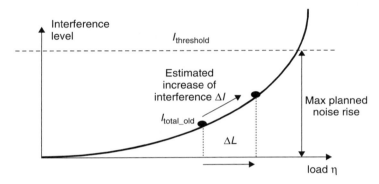

Figure 9.35. Uplink load curve and the estimation of the load increase due to a new UE

Two different uplink power increase estimation methods are shown below. They can be used in the interference-based admission control strategy. The idea is to estimate the increase ΔI of the uplink received wideband interference power I_{total} due to a new UE. The admission of the new UE and the power increase estimation are handled by the admission control functionality.

The first proposed method (the *derivative* method) is presented in Equation (9.13) and the second (the *integral* method) in Equation (9.14). Both take into account the load curve and are based on the derivative of uplink interference with respect to the uplink load factor

$$\frac{dI_{\text{total}}}{d\eta} \tag{9.11}$$

which can be calculated as follows

$$\text{Noise rise} = \frac{I_{\text{total}}}{P_N} = \frac{1}{1 - \eta} \Rightarrow$$

$$I_{\text{total}} = \frac{P_N}{1 - \eta} \Rightarrow \tag{9.12}$$

$$\frac{dI_{\text{total}}}{d\eta} = \frac{P_N}{(1 - \eta)^2}$$

The change in uplink interference power can be obtained by Equation (9.13). This equation is based on the assumption that the power increase is the derivative of the old uplink interference power with respect to the uplink load factor, multiplied by the load factor of the new UE ΔL:

$$\frac{\Delta I}{\Delta L} \approx \frac{dI_{total}}{d\eta} \Leftrightarrow$$

$$\Delta I \approx \frac{dI_{total}}{d\eta} \Delta L \Leftrightarrow$$

$$\Delta I \approx \frac{P_N}{(1-\eta)^2} \Delta L \Leftrightarrow \tag{9.13}$$

$$\Delta I \approx \frac{I_{total}}{1-\eta} \Delta L \Leftrightarrow$$

The second uplink power increase estimation method is based on the integration method, in which the derivative of interference with respect to the load factor is integrated from the old value of the load factor ($\eta_{old} = \eta$) to the new value of the load factor ($\eta_{new} = \eta + \Delta L$) as follows:

$$\Delta I = \int_{I_{total\_old}}^{I_{total\_old}+\Delta I} dI_{total} \Leftrightarrow$$

$$\Delta I = \int_{\eta}^{\eta+\Delta L} \frac{P_N}{(1-\eta)^2} d\eta \Leftrightarrow$$

$$\Delta I = \frac{P_N}{1-\eta-\Delta L} - \frac{P_N}{1-\eta} \Leftrightarrow \tag{9.14}$$

$$\Delta I = \frac{\Delta L}{1-\eta-\Delta L} \cdot \frac{P_N}{1-\eta} \Leftrightarrow$$

$$\Delta I = \frac{I_{total}}{1-\eta-\Delta L} \Delta L$$

In Equations (9.13) and (9.14) the load factor of the new UE ΔL is the estimated load factor of the new connection and can be obtained as

$$\Delta L = \frac{1}{1 + \frac{W}{v \cdot E_b/N_0 \cdot R}} \tag{9.15}$$

where W is the chip rate, R is the bit rate of the new UE, E_b/N_0 is the assumed E_b/N_0 of the new connection and v is the assumed voice activity of the new connection.

The downlink admission control strategy is the same as in the uplink, i.e. the UE is admitted if the new total downlink transmission power does not exceed the predefined target value:

$$P_{total\_old} + \Delta P_{total} > P_{threshold} \tag{9.16}$$

The threshold value $P_{threshold}$ is set by radio network planning. Notice that ΔP_{total} both includes the power of the new UE requesting capacity and the additional power rise of the existing UEs in the system due to the additional interference contributed by the new UE. The load increase ΔP_{total} in the downlink can be estimated based on *a priori* knowledge of the required E_b/N_0, the requested bit rate, and the pilot report from the UE. The pilot report implicitly provides information on the path loss towards the new UE as well as the interference level experienced by the UE.

9.5.3   Throughput-Based Admission Control Strategy

In the throughput-based admission control strategy, the new requesting UE is admitted into the radio access network if

$$\eta_{UL} + \Delta L \eta_{UL\_threshold} \tag{9.17}$$

and the same in downlink:

$$\eta_{DL} + \Delta L < \eta_{DL\_threshold} \tag{9.18}$$

where η_{UL} and η_{DL} are the uplink and downlink load factors before the admittance of the new connection and are estimated as shown in Section 9.4. The load factor of the new UE ΔL is calculated as in Equation (9.15).

Finally, we need to note that different admission control strategies can be used in the uplink and in the downlink.

9.6   Load Control (Congestion Control)

One important task of the RRM functionality is to ensure that the system is not overloaded and remains stable. If the system is properly planned, and the admission control and packets scheduler work sufficiently well, overload situations should be exceptional. If overload is encountered, however, the load control functionality returns the system quickly and controllably back to the targeted load, which is defined by the radio network planning.

The possible load control actions in order to reduce load are listed below:

- Downlink fast load control: Deny downlink power-up commands received from the UE.
- Uplink fast load control: Reduce the uplink E_b/N_0 target used by the uplink fast power control.
- Reduce the throughput of packet data traffic.
- Handover to another WCDMA carrier.
- Handover to GSM.
- Decrease bit rates of real time UEs, e.g. AMR speech codec.
- Drop low priority calls in a controlled fashion.

The first two in this list are fast actions that are carried out within a Node B. These actions can take place within one time slot, i.e. with 1.5 kHz frequency, and provide fast prioritisation of the different services. The instantaneous frame error rate of the non-delay-sensitive connections can be allowed to increase in order to maintain the quality of those services that cannot tolerate retransmission. These actions only cause increased delay of packet data services while the quality of the conversational services, such as speech and video telephony, is maintained.

The other load control actions are typically slower. Packet traffic is reduced by the packet scheduler: see Chapter 10. Inter-frequency and inter-system handovers can also be used as load balancing and load control algorithms and they were described in this chapter.

One example of a real time connection whose bit rate can be decreased by the radio access network is Adaptive Multirate (AMR) speech codec: for further information see Section 2.2.

References

[1] Sipilä, K., Laiho-Steffens, J., Wacker, A. and Jäsberg, M., 'Modelling the Impact of the Fast Power Control on the WCDMA Uplink', *Proceedings of VTC'99 Spring*, Houston, TX, 16–19 May 1999, pp. 1266–1270.

[2] 3GPP TS 25.101 UE Radio Transmission and Reception (FDD).

[3] Salonaho, O. and Laakso, J., 'Flexible Power Allocation for Physical Control Channel in Wideband CDMA', *Proceedings of VTC'99 Spring*, Houston, TX, 16–19 May 1999, pp. 1455–1458.

[4] Sampath, A., Kumar, P. and Holtzman, J., 'On Setting Reverse Link Target SIR in a CDMA System', *Proceedings of VTC'97*, Arizona, 4–7 May 1997, Vol. 2, pp. 929–933.

[5] 3GPP, Technical Specification Group RAN, Working Group 2 (WG2), 'Radio Resource Management Strategies', TR 25.922.

[6] 3GPP TS 25.133 'Requirements for Support of Radio Resource Management (FDD)'.

[7] Hiltunen, K., Binucci, N. and Bergström, J., 'Comparison Between the Periodic and Event-Triggered Intra-Frequency Handover Measurement Reporting in WCDMA', *Proceedings of IEEE WCNC 2000*, Chicago, 23–28 September 2000.

[8] Shapira, J. and Padovani, R., 'Spatial Topology and Dynamics in CDMA Cellular Radio', *Proceedings of 42nd IEEE VTS Conference*, Denver, CO, May 1992, pp. 213–216.

[9] Shapira, J., 'Microcell Engineering in CDMA Cellular Networks', *IEEE Transactions on Vehicular Technology*, Vol. 43, No. 4, November 1994, pp. 817–825.

[10] Dahlman, E., Knutsson, J., Ovesjö, F., Persson, M. and Roobol, C., 'WCDMA – The Radio Interface for Future Mobile Multimedia Communications', *IEEE Transactions on Vehicular Technology*, Vol. 47, No. 4, November 1998, pp. 1105–1118.

[11] Huang, C. and Yates, R., 'Call Admission in Power Controlled CDMA Systems', *Proceedings of VTC'96*, Atlanta, GA, May 1996, pp. 1665–1669.

[12] Knutsson, J., Butovitsch, T., Persson, M. and Yates, R., 'Evaluation of Admission Control Algorithms for CDMA System in a Manhattan Environment', *Proceedings of 2nd CDMA International Conference*, CIC '97, Seoul, South Korea, October 1997, pp. 414–418.

[13] Knutsson, J., Butovitsch, P., Persson, M. and Yates, R., 'Downlink Admission Control Strategies for CDMA Systems in a Manhattan Environment', *Proceedings of VTC'98*, Ottawa, Canada, May 1998, pp. 1453–1457.

[14] Liu, Z. and Zarki, M. 'SIR Based Call Admission Control for DS-CDMA Cellular System,' *IEEE Journal on Selected Areas in Communications*, Vol. 12, 1994, pp. 638–644.

[15] Holma, H. and Laakso, J., 'Uplink Admission Control and Soft Capacity with MUD in CDMA', *Proceedings of VTC'99 Fall*, Amsterdam, Netherlands, 19–22 September 1999, pp. 431–435.

10

Packet Scheduling

Jeroen Wigard, Harri Holma, Renaud Cuny, Nina Madsen, Frank Frederiksen
and Martin Kristensson

This chapter presents the radio access algorithms for supporting packet switched services and analyses their performance. Such services are, for example, messaging, email, WAP/web browsing, streaming video or Voice over IP. This chapter is organised as follows. Packet data protocols are first discussed in Section 10.1. The network delay aspects in terms of round trip time are analysed in Section 10.2. The transport channels and the user specific packet scheduler are introduced in Section 10.3. The cell specific packet scheduler algorithms are discussed in Section 10.4. The packet data system performance results are presented in Section 10.5 and the application performance results in Section 10.6. The services in general are introduced in Chapter 2.

All four UMTS traffic classes – from background to conversational – can be supported by WCDMA radio networks. See Chapter 2 for the introduction of the traffic classes. Background and interactive traffic classes do not require any guaranteed minimum bit rate and they are transmitted through the packet scheduler. The streaming class requires a minimum guaranteed bit rate but tolerates some delay, and packet scheduling can be utilised for streaming. The conversational class is transmitted without scheduling on dedicated channel DCH. The traffic classes and their mapping to transport channels are shown in Figure 10.1. The main focus in this chapter is on those services where packet scheduling can be applied.

10.1   Transmission Control Protocol (TCP)

The characteristics of the incoming data depend on the properties of the transport protocol. Non-real time traffic is typically carried using Transmission Control Protocol, TCP, and packet-based real time traffic, like streaming, typically uses the Real Time Protocol, RTP, on top of User Datagram Protocol, UDP. The protocols are shown in Figure 10.2. The characteristics of TCP are described in this section.

The user plane protocol stack for web browsing application is shown in Figure 10.3. It can be seen that the medium access control (MAC), the radio link control (RLC) and the packet data convergence protocol (PDCP) layer are terminated in the RNC, while the Internet

WCDMA for UMTS, third edition.   Edited by Harri Holma and Antti Toskala
© 2004 John Wiley & Sons, Ltd   ISBN: 0-470-87096-6

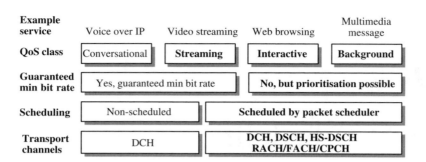

Figure 10.1. Mapping of UMTS traffic classes to scheduling and to transport channels

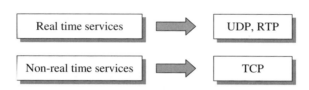

Figure 10.2. Typical packet protocols over WCDMA

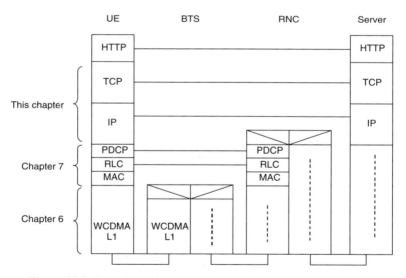

Figure 10.3. Data User Plane Protocol Stack for HTTP web browsing

Protocol (IP), TCP and hypertext transfer protocol (HTTP) are terminated in a server. MAC, RLC and PDCP are covered in Chapter 7. The protocol employed by Web browsing applications is the HTTP, but other protocols and applications such as file transfer protocol (FTP), and mail protocols, like the Internet mail protocol (IMAP) or Telnet would use a similar architecture over TCP/IP. In the protocol structure each entity of a certain layer uses the services offered by the layer just below to communicate with its peer.

A TCP connection is set up between the UE and a server, or intermediate proxies. This server can be located close to the UE or far away. It is possible for a UE in Europe to set-up a TCP connection with a server in the USA. TCP has been designed to provide reliable end-to-end information exchange over an unreliable Internet network since the Internet Protocol does not guarantee any reliability. TCP uses retransmissions and flow control to provide the reliability. TCP is connection oriented between two end points and requires connection establishment and termination. From a wireless performance point of view, one of the most significant characteristics of TCP is the flow control. The receiver end only allows the transmitter to send as much data as the receiver has buffer for. Moreover, the transmitter side adapts the transmission rate to the network capacity/load by means of algorithms such as slow start, and congestion avoidance. The flow control in TCP is accomplished by means of two windows: the congestion window and the offered window. The congestion window is controlled by slow start and congestion avoidance algorithms, whereas the offered window is a window with the size the receiver has buffer for, and the receiver advertises it in every segment transmitted to the server. The offered window is therefore flow control imposed by the receiver. At any time, the minimum of the congestion window and the offered window determines the amount of outstanding data the server can transmit. Therefore, the number of transmitted but unacknowledged data (segments on the fly) can never be larger than the minimum of these two windows. The window, which is the minimum of the two above mentioned windows, is adapted using a sliding window approach:

1. Transmit new segments in the window.

2. Wait for acknowledgements from the receiver.

3. If the acknowledgement is received, slide the window forward and increase the window size, if the offered window is not exceeded.

4. If the acknowledgement is not received before a timer expires, retransmit non-acknowledged segments and decrease the window size.

An example of the progressing of the window, in the case of a positive ACK can be seen in Figure 10.4. In this example the offered window is larger than three segments.

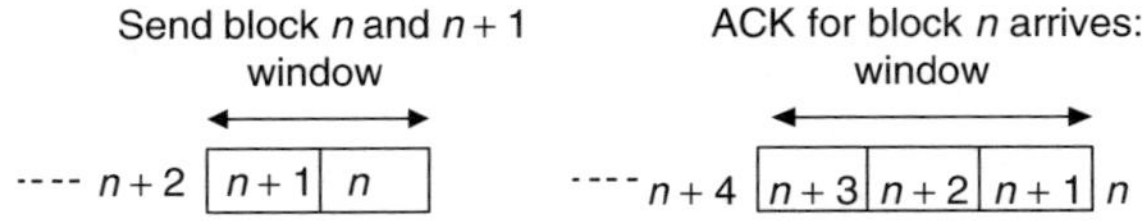

Figure 10.4. Example of TCP window progressing

This flow control approach fully utilises the available bandwidth by adjusting the window size according to the received acknowledgements. The size of the congestion window is adjusted according to the received acknowledgements. Traditionally, the congestion window is initialised to one, two or four times the maximum segment size at the connection establishment. Typical values of maximum segment sizes are 256, 512, 536 or 1460 bytes.

Each time a TCP segment is positively acknowledged, the size of the congestion window is doubled: when the first acknowledgement is received and the initial congestion window was equal to the maximum segment size, the congestion window is increased to two times the maximum segment size, and two segments can be sent. This produces an aggressive exponential increase of the amount of outstanding data the sender can transmit. This approach is called slow start because the starting congestion window is just one maximum segment size. When the congestion window reaches the threshold, called slow start threshold, it is increased linearly at the reception of acknowledgements.

The segment retransmissions in TCP can be triggered by two type of event: reception of several duplicated acknowledgements (usually three duplicated acknowledgements), or timer expiration. The timer is used to detect congestion in the network. Each segment is timed in every cycle, and if its corresponding acknowledgement is not received before the timer expires, the retransmission of non-acknowledged segments is triggered subject to a new window set-up. To avoid further congestion, the slow start threshold is set to half of the current window. The congestion window is set to the maximum segment size and the slow start is started again. This is shown on the right side of Figure 10.5. The Retransmission Time Out (RTO) is calculated from the average and the mean deviation of the round trip time (RTT).

$$RTO = A + 4 \cdot D \qquad (10.1)$$

where A represents the smoothed RTT average and D represents the mean deviation of the RTT. The mean deviation is a good approximation to the standard deviation, and it is included in the RTO computation in order to provide better response to wide fluctuations in the RTT. This means that large variations in the delay of TCP packets over the air interface should be avoided in order to limit the number of TCP timeouts. Details of the RTO calculations are given in [1].

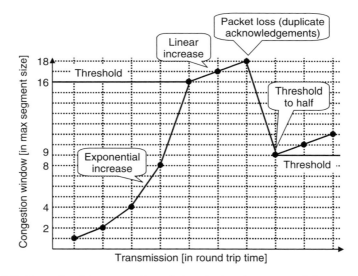

Figure 10.5. Example of the congestion window during slow start and fast retransmission

Upon the reception of a number of duplicated acknowledgements, i.e. multiple acknowledgements with the same next expected segment number n, that exceeds a threshold (usually set to 3), the segment n is assumed to be lost. Note that TCP requires the receiver side to generate an immediate acknowledgement when an out of order segment is received. There are several types of TCP implementation that operate differently in fast retransmission and fast recovery. Here, two of them are mentioned: TCP-Tahoe and TCP-Reno. When the packet loss is detected TCP-Tahoe sets the slow start threshold to half the current window, and the congestion window is set to one maximum segment size which corresponds to what happens at the timer expiration. TCP-Reno sets the slow start threshold and the congestion window to half the current window so the slow start is avoided. While TCP-Reno produces less bursty traffic than TCP-Tahoe, it is much less robust toward phase effects because multiple segment losses of the same congestion episode significantly deteriorate the congestion window size.

For small file sizes the slow start will affect the throughput as the radio link capability is not fully utilised during the slow start. Fundamental to TCP's slow start, timeout and retransmission is the measurement of the round trip time (RTT) experienced on a given connection. This is analysed in the next section.

We further note that, during the TCP connection establishment and termination, typically small IP packets of 40 bytes are transmitted, which can be seen in Figure 10.6. The connection establishment consists of the initiating side sending a SEQ message, indicating it wants to open a TCP connection. The SEQ is used to synchronise the sequence numbers. The responding side simply acknowledges this with an ACK message and sends, at the same time, a SEQ message to open the TCP connection in the return direction. As the last part of the connection establishment, the initiating side sends an ACK message when it has received the SEQ message correctly. The TCP termination is very similar: the terminating side sends a FIN message, indicating that the connection should be closed. To this the responding side answers with an ACK message and a FIN message in order to also close the TCP connection in the other direction. The connection is completely closed when the terminating side has acknowledged the FIN message. The TCP connection establishment and release both take approximately 1.5 x round trip time since three messages are required.

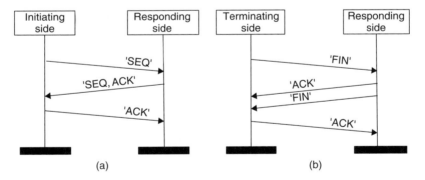

Figure 10.6. TCP connection establishment (a) and termination (b)

TCP is originally designed to work over fixed networks. Using TCP over wireless networks brings some challenges: more narrow bandwidth, longer delay and larger delay variations than in a fixed network. Third generation networks improve the packet data performance compared to second generation systems and bring it closer to fixed line performance. Some enhancements have been considered to optimise TCP, as described in [2]. Some of those solutions may be beneficial for improving the third generation packet data performance as well. These potential TCP optimisation solutions can be classified into standard parameter optimisation, non-standard TCP optimisation and buffer congestion management. The TCP optimisation solutions are briefly described below, but first we list the main TCP parameter optimisation solutions.

- An increase of the initial TCP congestion window. This is especially effective for the transmission of small amounts of data. These amounts of data are commonly seen in such applications as Internet-enabled mobile wireless devices. For large data transfers, on the other hand, the effect of this mechanism is negligible. An initial congestion window size of two segments or larger is recommended in [3].

- An appropriate TCP window size that should be set according to the bandwidth delay product. The bandwidth delay product defines the amount of unacknowledged data that should be transmitted in order to fully utilise the limiting transmission link. The maximum TCP throughput is limited by the following formula:

$$\text{TCP\_throughput} \leq \frac{\text{advertised\_window(bits)}}{\text{round\_trip\_time(s)}} \tag{10.2}$$

Typical TCP window sizes are 16–64 kB and they may be adjusted adaptively. The average WCDMA round trip times are shown in Figure 10.12 to be in the order of 200 ms. Additionally, we need some margin for RLC transmissions. If we assume a maximum round trip time of 400 ms, the achievable TCP throughput is 320–1280 kbps. These numbers indicate that the normal advertised TCP window sizes, that are at least equal to 32 kB, are large enough for TCP connections over WCDMA with 384 kbps data rate.

As non-standard TCP optimisation examples, we can mention:

- Redundancy elimination. This functionality monitors the latest TCP acknowledged sequence number in a router, allowing unnecessary retransmitted TCP segments in that router to be discarded and consequently saving valuable radio capacity. This can be seen in Figure 10.7.

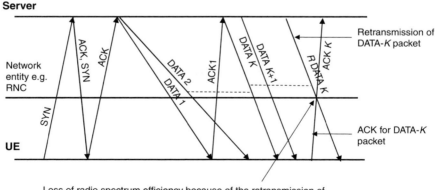

Figure 10.7. Redundancy elimination

- Split TCP. Splitting the TCP connections in to two legs (one leg for the wireless domain and one leg for the fixed Internet) allows a more free optimisation of the transport protocol in the wireless part. Note, however, that this approach breaks the end-to-end design principle of Internet protocols.

This list is not exhaustive and other TCP optimisation methods can be found in the literature. Typically some of these methods (e.g. the split TCP approach), should be used with care and their implications towards end-to-end functionalities such as security should be well understood by network operators. RFC 3135 presents a general IETF view on such network features, underlining the most important issues to consider.

Buffer congestion control mechanisms are used in the case of congestion in shared buffers. The problem arises during buffer overflow since it affects several TCP connections. In the case of user dedicated buffers no such problems exist. The congestion can be generally avoided with large buffers, but long buffering delays are not desirable and should be avoided whenever possible. Buffer congestion most likely happens in the last downlink router buffer before the radio link. Buffer congestion control mechanisms include:

- Random Early Detection (RED), the most well known standard buffer management method. The principle of RED is to increase the packet dropping probability of incoming packets at the same time as the buffer occupancy grows. Ultimately, if the buffer length exceeds a predefined threshold, every incoming packet will be dropped. The advantage of dropping packets before the buffer is completely full is to force some of the TCP sources to reduce their sending rate before congestion actually happens. Another advantage is to avoid global synchronisation in the routers. Global synchronisation happens when the buffer load oscillates from almost empty to congested, and is caused mostly by the TCP congestion control mechanism of multiple connections acting at the same time.

- Fast TCP. An algorithm delays (slightly) TCP acknowledgments if congestion is detected in the reverse direction. As a result, TCP senders will send later their next window of data, giving, in many cases, enough time for the router buffers to recover from congestion. Furthermore, delaying the TCP acknowledgement will slow down the growth of the sending window in the TCP sender, which is also beneficial in the case of congestion. Fast TCP is illustrated in Figure 10.8.

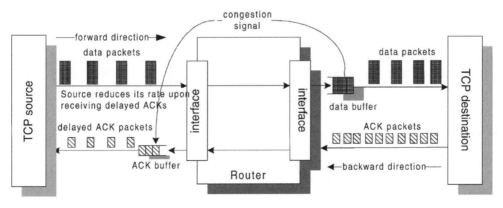

Figure 10.8. Fast TCP

- Window pacing. This scheme reduces the advertised window of TCP acknowledgements if congestion is detected in the reverse direction. The aim is to force the TCP senders to decrease their sending rate. The advantage compared to RED is that packets are actually not dropped in the downlink direction, preventing throughput degradation for the end user. The drawback is that modifying the TCP header is a heavier task than just dropping or delaying incoming IP packets. Also, if IP security is used, this feature cannot be applied since the TCP headers cannot be read anymore.

10.2 Round Trip Time

The round trip time in WCDMA is defined as the delay of a small packet travelling from UE to a server behind GGSN and back. The round trip time is illustrated in Figure 10.9. In fixed IP networks, the round trip time is typically a few tens of milliseconds if the server is close, and up to a few hundred milliseconds if the server is far away. Mobile networks like UMTS increase the delay and the round trip time is longer when TCP connection is established over a WCDMA air interface. A shorter round trip times give a benefit in the response time, which is especially advantageous for TCP slow start and for interactive services with small packets, like gaming; see more detailed discussion on the requirements in Chapter 2. In this section we first analyse the WCDMA round trip time delay budget for small packets and then present the effect of the packet size to the round trip time. Small packets are typically transmitted in TCP connection establishment and in games, while larger packets are transmitted during file or web page download.

Delay from UE to server and back for small IP packet of 32 bytes

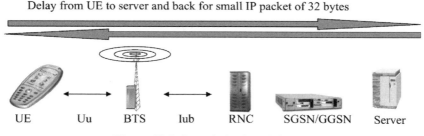

| UE | Uu | BTS | Iub | RNC | SGSN/GGSN | Server |

Figure 10.9. Round trip time definition

The round trip time delay budget for small packets is shown in Figure 10.10, and is typically between 150–200 ms depending on the Transmission Time Interval (TTI = interleaving) used in the air interface. The delay components can be found from [4]. We assume UE delay of 35 ms, Node B delay of 20 ms, Iub delay of 30 ms, including frame protocol and AAL2 processing, RNC 25 ms and air interface 33–66 ms, including buffering and retransmissions. Furthermore, we assume that the core network adds 5 ms and fixed Internet 10 ms delay. Example round trip time measurements are shown in Figure 10.11. The round trip time in this example is approximately 160–200 ms, except for when a block error occurs in the air interface and an RLC retransmission takes place, in which case the round trip time is increased to 320 ms. Since WCDMA round trip time is similar to the fixed Internet round trip time that can be seen for inter-continental connections or for dial-up connections, we may expect that those applications that are designed for fixed Internet

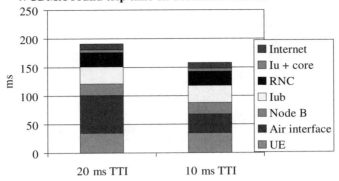

Figure 10.10. WCDMA round trip time for small packets

Pinging 213.161.41.37 with 32 bytes of data:

Reply from 213.161.41.37: bytes = 32 time = 201ms TTL = 255
Reply from 213.161.41.37: bytes = 32 time = 160ms TTL = 255
Reply from 213.161.41.37: bytes = 32 time = 191ms TTL = 255
Reply from 213.161.41.37: bytes = 32 time = 190ms TTL = 255
Reply from 213.161.41.37: bytes = 32 time = 171ms TTL = 255
Reply from 213.161.41.37: bytes = 32 time = 180ms TTL = 255
Reply from 213.161.41.37: bytes = 32 time = 321ms TTL = 255 ⬅ **RLC retransmission**
Reply from 213.161.41.37: bytes = 32 time = 170ms TTL = 255

Reply from 213.161.41.37: bytes = 32 time = 160ms TTL = 255
Reply from 213.161.41.37: bytes = 32 time = 180ms TTL = 255
Reply from 213.161.41.37: bytes = 32 time = 170ms TTL = 255
Reply from 213.161.41.37: bytes = 32 time = 160ms TTL = 255

Ping statistics for 213.161.41.37:
    Packets: Sent = 29, Received = 29, Lost = 0 (0 % loss),
Approximate round trip times in milliseconds:
    Minimum = 160 ms, Maximum = 321 ms, Average = 182 ms ⬅ **Average RTT 182 ms**

Figure 10.11. Example WCDMA round trip time measurements on dedicated channel

typically perform at least satisfactorily when used over the WCDMA air interface. WCDMA round trip time will reduce when product platforms are optimized and when HSDPA with shorter TTI is introduced. It is expected that below 100 ms is feasible with HSDPA.

The round trip time above assumes that the packet fits into one TTI. For example, 64 kbps and 10 ms TTI carries 640 bits = 80 bytes in one TTI. If the packet size is large, multiple TTIs are needed to transmit one packet and the round trip time gets larger. The total average

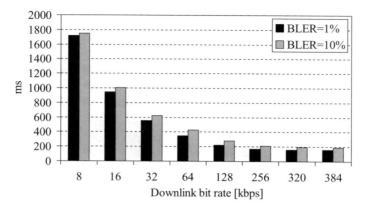

Figure 10.12. WCDMA round trip time for a 1500 B packet in downlink and a 40 B acknowledgement in uplink

round trip times for a 1500 B downlink packet and a 40 B uplink acknowledgement are shown in Figure 10.12 for different downlink bit rates. It is assumed that for uplink acknowledgements an 8 kbps channel is selected when the downlink bit rate is less than or equal to 64 kbps and a 32 kbps uplink channel is used when the downlink bit rate is higher than 64 kbps. The average delay of retransmissions is assumed to be 100 ms. RLC BLERs equal to 1 % and 10 % are shown. It can be noted that using a lower BLER in the radio transmission can somewhat reduce the average round trip time.

It is not only the average round trip time but also the delay jitter that affects the performance. The delay jitter depends on the BLER, the bit rate and the retransmission delay. For low bit rates, one IP packet is typically transmitted over many TTIs, so the probability of at least one of the TTIs failing is quite large, whereas for large bit rates, one IP packet is sent in a few TTIs. Thus, the size of the delay jitter compared to the round trip time is larger for high bit rates.

This section presented the round trip time on a dedicated channel DCH. The following section discusses the common channel state, its round trip time and the delay of the state transitions.

10.3   User-specific Packet Scheduling

The WCDMA packet scheduler is located in RNC. The base station provides the air interface load measurements and the mobile provides uplink traffic volume measurements for the packet scheduler: see Figure 10.13. The user-specific packet scheduler is presented in this section and the cell-specific scheduler in Section 10.4. The user-specific part controls the utilisation of Radio resource control (RRC) states, transport channels and their bit rates according to the traffic volume. The cell-specific part controls the sharing of the radio resources between the simultaneous users.

WCDMA supports three types of transport channel that can be used to transmit packet data: common, dedicated and shared transport channels. These channels are described in Chapter 6 and in this section their properties and feasibility for packet data are discussed.

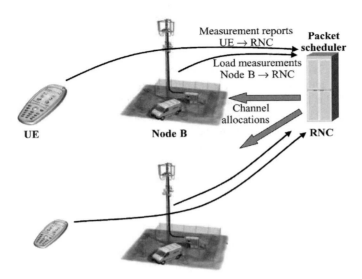

Figure 10.13. The WCDMA packet scheduler, located in RNC

10.3.1 Common Channels (RACH/FACH)

Common channels are the random access channel (RACH) in the uplink and the forward access channel (FACH) in the downlink. Both can carry signalling data but also user data in WCDMA. There are typically only one or a few RACH or FACH channels per sector, for example, 16 kbps RACH with 20 ms TTI and 32 kbps FACH with 10 ms TTI. The advantage of common channels is that no set-up time is needed if the user is in Cell_FACH state. The round trip time on RACH/FACH is illustrated in Figure 10.14: the round trip time is slightly

Pinging www.nokia.com [147.243.3.73] with 32 bytes of data:

Reply from 147.243.3.73: bytes = 32 time = 301ms TTL = 242
Reply from 147.243.3.73: bytes= 32 time = 290ms TTL = 242
Reply from 147.243.3.73: bytes = 32 time = 300ms TTL = 242
Reply from 147.243.3.73: bytes = 32 time = 241ms TTL = 242
Reply from 147.243.3.73: bytes = 32 time = 290ms TTL = 242
Reply from 147.243.3.73: bytes = 32 time = 281ms TTL = 242
Reply from 147.243.3.73: bytes = 32 time = 260ms TTL = 242
Reply from 147.243.3.73: bytes = 32 time = 281ms TTL = 242
Reply from 147.243.3.73: bytes = 32 time = 300ms TTL = 242
Reply from 147.243.3.73: bytes = 32 time = 311ms TTL = 242
Reply from 147.243.3.73: bytes = 32 time = 290ms TTL = 242
Reply from 147.243.3.73: bytes = 32 time = 300ms TTL = 242
Reply from 147.243.3.73: bytes = 32 time = 261ms TTL = 242

Ping statistics for 147.243.3.73:
  Packets: Sent = 13, Received = 13, Lost = 0 (0 % loss),
Approximate round trip times in milliseconds:
  Minimum = 241 ms, Maximum = 311 ms, Average = 285 ms ⬅ **Average RTT 285 ms**

Figure 10.14. Example round trip times for 32 B ping on RACH/FACH

longer than on a dedicated channel but even the first packet goes through as quickly as the other ones.

Common channels do not have a feedback channel, and therefore cannot use fast closed loop power control, but only open loop power control or fixed power. Nor can these channels use soft handover. Therefore, the link level performance of the common channels is not as good as that of the dedicated channels, and more interference is generated than with dedicated channels. The gain of fast power control is analysed in Section 9.2 and the gain of soft handover in Section 9.3. Common channels are most suitable for transmitting small IP packets, for example, during the TCP connection establishment, and for infrequent packets for interactive gaming. Since common channels do not use soft handover but cell reselection, there is a longer delay when the cell reselection takes place. The round trip times on RACH/FACH are shown in Figure 10.15 when cell reselections happen due to mobility. The round trip time is increased from below 300 ms to above 1 s during the cell reselection.

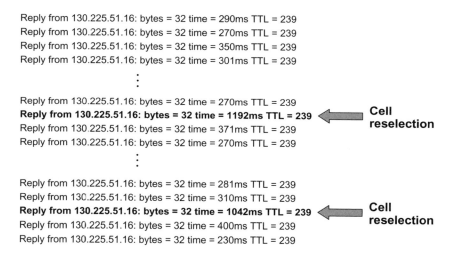

Figure 10.15. Round trip times on RACH/FACH when cell reselections take place

10.3.2   Dedicated Channel (DCH)

The dedicated channel is always a bi-directional channel with both uplink and downlink connections. Because of the feedback channel, fast power control and soft handovers can be used. These features improve their radio performance, and consequently less interference is generated than with common channels. No breaks occur due to mobility since soft handover is used. On the other hand, setting up a dedicated channel takes more time than accessing common channels. The dedicated channel set-up time is illustrated in Figure 10.16. The set-up time is the difference in ping delay between the 1st and 2nd packets, which is 900 ms in this example. The signalling flow chart for this RRC state change is illustrated in Section 7.8.2. The DCH set-up time should preferably be minimised to improve the response times and to avoid the potential TCP timeout risk during DCH set-up. A number of proposals have been discussed in 3GPP to reduce the required signalling messages and to minimise the delay.

Pinging www.nokia.com [147.243.3.73] with 128 bytes of data:

Reply from 147.243.3.73: bytes = 128 time = 1121ms TTL = 242
Reply from 147.243.3.73: bytes = 128 time = 221ms TTL = 242
Reply from 147.243.3.73: bytes = 128 time = 190ms TTL = 242
Reply from 147.243.3.73: bytes = 128 time = 200ms TTL = 242
Reply from 147.243.3.73: bytes = 128 time = 190ms TTL = 242
Reply from 147.243.3.73: bytes = 128 time = 200ms TTL = 242

**DCH allocation 900 ms
(=1121 – 221 ms)**

Figure 10.16. Example DCH set-up time from Cell_FACH state

The dedicated channel can have bit rates from a few kbps up to 384 kbps in the first mobiles, and up to 2 Mbps according to 3GPP. The bit rate can be changed during transmission. When the data transmission is over, the dedicated channel is kept allocated for a few seconds before releasing it and reallocating to another user. During that time the downlink orthogonal code and the network hardware is still dedicated for that connection. Therefore, very bursty traffic on dedicated channels consumes a relatively high number of downlink orthogonal codes and network resources. Also, during TCP slow start the dedicated channel is not fully utilised. The DCH utilisation is illustrated in Figure 10.21. The utilisation for different files sizes and bit rates is calculated in Figure 10.17. The initial

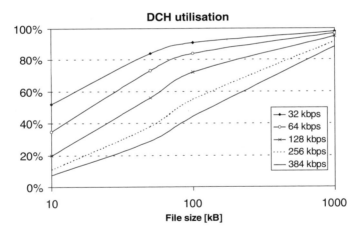

Figure 10.17. Radio bearer utilisation with TCP

congestion window is assumed to be two MSS and the inactivity timer for all bit rates is set to 2 s. With 100 kB file size the channel utilisation is better than 80 % if the bit rate is 64 kbps or below, while the utilisation with 384 kbps is below 50 %. The low utilisation is mainly caused by the inactivity timer of 2 s, since the actual download takes less than 3 s. With 1 MB file size the channel utilisation is approximately 90 % or better for all channel bit rates. It is shown that allocating high bit rates for downloading small files leads to low utilisation factors. This can lead to code shortage and also wastes Iub transmission resources. Therefore, it is beneficial to allocate low bit rates at the start of a download and then increase the bit rate. It can also be noted that the efficiency improves during high load since the data rates are typically decreased from 384 kbps down to 64–128 kbps.

10.3.3   Downlink Shared Channel (DSCH)

The downlink shared channel, DSCH, is targeted to transfer bursty packet data. The idea is to share a single physical channel, i.e. orthogonal code, between many users in a time division manner. This approach saves the limited number of downlink orthogonal codes because several users share the code. If the dedicated channel were used instead, the orthogonal code would be reserved according to the maximum bit rate, and the efficiency of code usage would be lower. Shared channels are used in parallel with a lower bit rate dedicated channel. The dedicated channels carry the physical control channel, including the signalling for fast power control. It should be noted also that shared channels cannot use soft handover. The efficient rate adaptation capability of a shared channel is useful with the TCP slow start and during the end of the file download. With DSCH the resources can be immediately allocated to another user before the DCH inactivity timer expires, improving the resource utilisation compared to DCH. Shared channel is typically not implemented in the first WCDMA networks, nor in the first terminals. The high speed downlink shared channel HS-DSCH is part of the HSDPA concept that is covered in Chapter 11.

10.3.4   Uplink Common Packet Channel (CPCH)

The uplink common packet channel (CPCH) is an extended RACH channel. The CPCH can have fast power control after the access procedure but it cannot use soft handover. The set-up time of the CPCH channel is slightly longer than that of the RACH, but it can be allocated for 64 frames = 640 ms, which means more data can be sent on it than on the RACH. The CPCH is optional for the mobiles. Without wide support of CPCH in mobiles, this channel cannot be efficiently utilised in WCDMA.

10.3.5   Selection of Transport Channel

The transport channels for packet data are summarised in Table 10.1. The High Speed Downlink Shared Channel, HS-DSCH is left out of this description, since the High Speed Downlink Packet Access, HSDPA, is covered in Chapter 11.

Table 10.1. Overview of WCDMA transport channels

	Dedicated channels DCH	Shared channel DSCH	Common channels		
			FACH	RACH	CPCH
RRC state	Cell_DCH	Cell_DCH	Cell_FACH	Cell_FACH	Cell_FACH
Uplink/ Downlink	Both	Downlink	Downlink	Uplink	Uplink
Code usage	According to maximum bit rate	Code shared between users	Fixed codes per cell	Fixed codes per cell	Fixed codes per cell
Fast power control	Yes	Yes	No	No	Yes
Soft handover	Yes	No	No	No	No
Suited for	Medium or large data amounts	Medium or large data amounts	Small data amounts	Small data amounts	Small or medium data amounts
Suited for bursty data	No	Yes	Yes	Yes	Yes
Available in first networks and terminals	Yes	No	Yes	Yes	No

The state transitions can be controlled by the traffic volume threshold from Cell_FACH to Cell_DCH and by the inactivity timer from Cell_DCH to Cell_FACH. If the amount of data in the mobile uplink buffer or in the RNC downlink buffer exceeds the traffic volume threshold, DCH allocation takes place. When DCH is allocated, either for uplink or for downlink, it must be allocated for the other direction at the same time as well, since DCH is a bi-directional channel. The DCH bit rate can be different in uplink and in downlink depending which direction triggers the DCH allocation. The following two examples illustrate the state transitions and transport channel utilisation. Figure 10.18 shows the

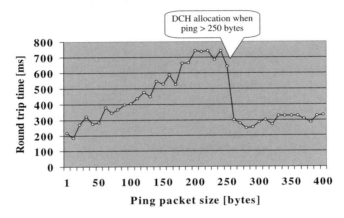

Figure 10.18. Measured round trip time with continuous ping with increasing packet size

average round trip time when ping was performed continuously for packet sizes ranging from 1–400 with 10 byte intervals. DCH was allocated in this case when the data amount exceeded 250 bytes, and kept afterwards. The round trip time drops when DCH is allocated, since DCH clearly has a higher bit rate than RACH/FACH. Figure 10.19 shows an example of getting a directory listing. First, an FTP connection is established on RACH/FACH. The

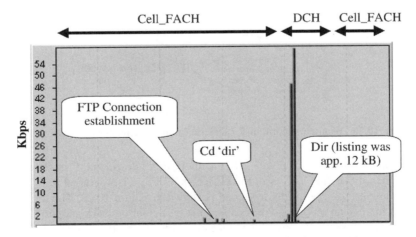

Figure 10.19. Measured transport channel utilisation example

request for directory listing ('dir') is sent on RACH. When the directory listing is transmitted in downlink, the 64 kbps DCH is allocated since the data amount is 12 kB. When the data has been transmitted, the DCH is released after the inactivity timer.

When DCH is allocated, its bit rate can be upgraded or downgraded. The bit rate upgrade takes place if the amount of data in the buffer exceeds a higher threshold. When increasing the bit rate, the maximum link powers and the cell capacity need to be checked. The bit rate downgrade takes place if the DCH capacity is under-utilised. The bit rate downgrade may also be needed if the maximum link power is achieved in the weak coverage area. Figure 10.20 illustrates an algorithm for selecting the transport channel and its bit rate. A bit rate

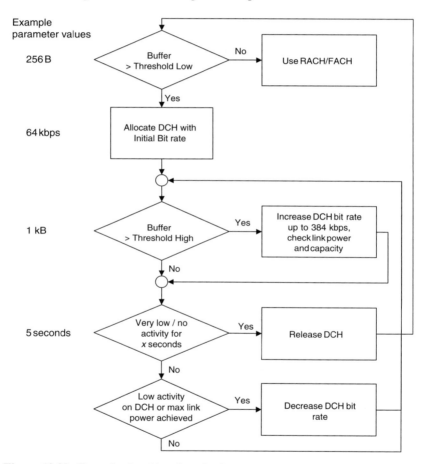

Figure 10.20. Example algorithm for selecting transport channels and their bit rates

allocation example is shown in Figure 10.21 with TCP file download. First, 64 kbps DCH is allocated. When more data arrives in the buffer, the bit rate is upgraded to 384 kbps.

A measurement example of file download is shown in Figure 10.22. The file download is started when there is an on-going voice calls. That particular network supports only 64 kbps with simultaneous voice calls. When the voice call ends, the bit rate is upgraded to 384 kbps. When the second file download starts, 384 kbps is directly allocated.

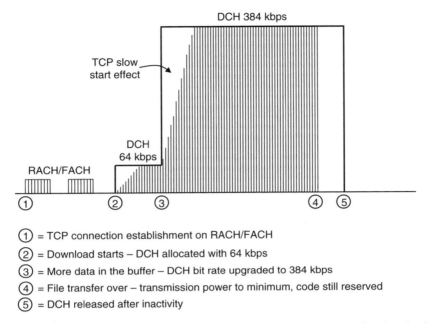

① = TCP connection establishment on RACH/FACH

② = Download starts – DCH allocated with 64 kbps

③ = More data in the buffer – DCH bit rate upgraded to 384 kbps

④ = File transfer over – transmission power to minimum, code still reserved

⑤ = DCH released after inactivity

Figure 10.21. Example usage of transport channels in the case of TCP file download

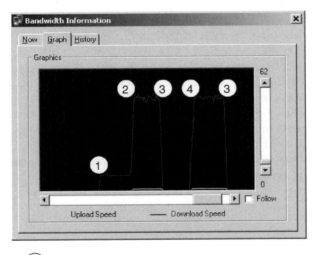

① = File transfer starts, 64 kbps allocated (simultaneous voice call)

② = Bit rate upgraded to 384 kbps due to voice call finish

③ = File transfer over

④ = File transfer starts, 384 kbps allocated

Figure 10.22. Example measured bit rate allocations during TCP download

It is beneficial to allow TCP connection establishment to take place on the common channels. The TCP connection establishment involves transmission of a few small IP packets of 40 bytes. Also, some laptop clients, like VPN clients, tend to send infrequent small or medium size packets. These packets should preferably be transmitted on common channels and no DCH allocation should take place. Suitable traffic volume thresholds are therefore 128–512 bytes.

The DCH inactivity timer depends on the bit rate: the higher the bit rate, the shorter the inactivity timer should be to avoid wasting resources. The resource utilisation was presented in Figure 10.17. The inactivity timer should also be relatively short to save the mobile batteries. However, from the end user point of view, a long inactivity timer would be preferred. If more data arrives after the channel is released, the user will experience another DCH set-up time delay. When setting the value for the inactivity timer, one needs to make a trade-off between end user performance, network resource consumption and mobile battery consumption. Typical values for the inactivity timer are 2–10 seconds.

10.3.6 Paging Channel States

The previous section described the utilisation of Cell_FACH and Cell_DCH states. This section shortly presents Cell_PCH / URA_PCH states. Both the mobile receiver and the transmitter are active in Cell_DCH state, causing relatively high power consumption. Only the mobile receiver is active in Cell_FACH state, reducing power consumption compared to the Cell_DCH state. Cell_FACH state, however, requires continuous reception, while in Cell_PCH state, discontinuous reception can be used, providing very low power consumption and long mobile operation times. Cell_PCH / URA_PCH and Cell_FACH states are similar from the air interface capacity point of view: no interference is caused and no radio resources are reserved. The main difference lies in the mobile power consumption, which is considerably lower in Cell_PCH state than in Cell_FACH state. Typical power consumption values are 200–400 mA in Cell_DCH, 100–200 mA in Cell_FACH and <5 mA in Cell_PCH. These values are naturally affected by the network parameters, like paging cycle, and by the required mobile transmission power in Cell_DCH state. Also, application layer processing, like video encoding and decoding, will affect the power consumption.

The additional delay when moving from Cell_PCH to Cell_FACH state is short if the new capacity request is triggered in uplink. If the new packet arrives in downlink, there is an additional delay caused by the paging cycle to move the mobile from Cell_PCH to Cell_FACH. This extra delay depends on the paging cycle and is, on average, half of the paging period. Paging cycles are normally 320–2560 ms.

The mobile power consumption values are shown in Figure 10.23. These allow roughly four hours of Cell_DCH operation time and hundreds of hours in the Cell_PCH state with a 1000 mAh mobile battery.

10.4 Cell-specific Packet Scheduling

The cell-specific packet scheduler divides the non-real time capacity between simultaneous users. The cell-specific scheduler operates periodically. This period is a configuration parameter and its value typically ranges from 100 ms to 1 s. If the load exceeds the target, the packet scheduler can decrease the load by decreasing the bit rates of packet bearers; if the load is less than the target, it can increase the load by allocating higher bit rates, as shown

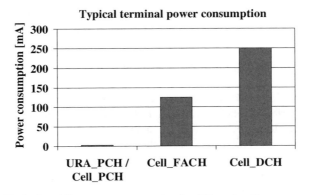

Figure 10.23. Typical mobile power consumption in different radio resource control states

in Figure 10.24. The target of the scheduling is to use efficiently all remaining cell capacity for non-real time connections but also maintain interference levels within planned values so that real time connections are not affected. This section presents the main principles of how the cell-specific packet scheduler operates.

The cell-specific packet scheduler uses the following input information:

- Total Node B power. The total load is estimated using power-based load estimation, as described in Section 9.4.

- Capacity used by non-real time bearers. This capacity can be estimated using throughput-based load estimation.

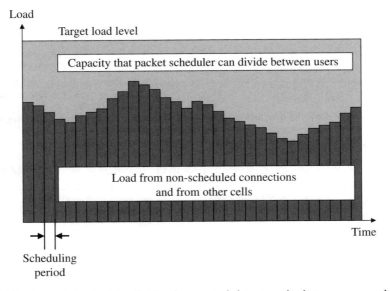

Figure 10.24. The packet scheduler divides the non-real time capacity between non-real time data users

- Target load level from network planning parameters. This parameter defines the maximum interference level that can be tolerated in the cell without affecting the real time connection.

- Bit rate upgrade requests from the user-specific packet scheduler.

The input information and the calculation principles are illustrated in Figure 10.25. Node B provides periodic total power and link power measurements in radio resource reporting to RNC. The total power measurements normally have a faster reporting cycle than link power measurements. The packet scheduler can estimate the total power used by non-controllable traffic, which consists of real time connections and inter-cell interference. This part of the interference cannot be affected by the packet scheduler. The remaining capacity can be divided by the packet scheduler between the simultaneous users. The bit rate upgrade information from the user-specific scheduler is taken into account when changing the bit rates.

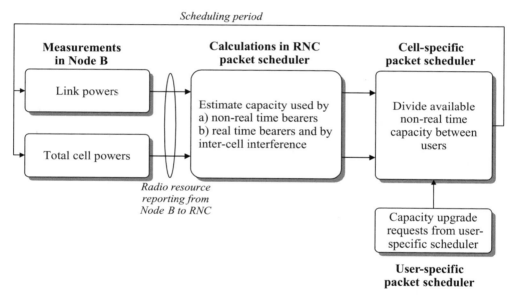

Figure 10.25. Input information and calculation principles of the cell-specific packet scheduler

10.4.1   Priorities

The core network provides QoS parameters over the Iu interface to WCDMA RAN; for more details see Chapter 2. The information includes traffic class, allocation/retention priority and traffic handling priority for the interactive class, as shown in Table 10.2.

Table 10.2. Main QoS parameters for packet scheduling

Traffic class	Interactive class	Background class
Allocation/retention priority	1,2,3	1,2,3
Traffic handling priority	1,2,3	—

The target of these QoS parameters is to share the radio resources more efficiently between the users. When the radio resources are scarce, these parameters help the packet scheduler to decide how to allocate the capacity for the different users. This feature is illustrated in Figure 10.26. The simplest algorithm is one where the higher priority bearers will always get the capacity before the lower priority bearers. Another approach is to give a minimum bit rate, such as 32 kbps, for all users and then allocate the remaining capacity for the high priority users to increase their bit rate.

Packets arriving with different QoS requirements

Scarce radio resources shared according to QoS

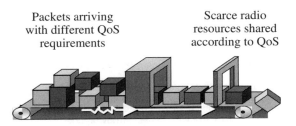

Figure 10.26. QoS priorities are targeted for efficient radio utilisation

QoS parameters can also be used to differentiate the services for performance monitoring. If, for example, MMS, WAP and streaming services are allocated different allocation/ retention priorities, those services can be separated in the radio network performance counters. This allows the operator to have better visibility of the end user performance of the different services. If the services use the same QoS parameters, their radio performances cannot be separately monitored. QoS parameters can also be used to provide better service quality for operator-hosted services compared to other services.

10.4.2 Scheduling Algorithms

The cell-based packet scheduler uses a number of different inputs to select the radio bearer bit rates. An example DCH bit rate allocation is illustrated in Figure 10.27 where new capacity requests keep arriving from user-specific schedulers. The total cell capacity is assumed to be 900 kbps and no real time connections are present. The packet scheduler can divide that 900 kbps capacity between the users. The first and the second users can be allocated 384 kbps. When there is no more room for the maximum 384 kbps capability, one of the highest bit rate bearers is downgraded to make room for the new request. If one of the bearers is released, some of the other allocations can be upgraded correspondingly. Figure 10.27 shows a simplified case. In real life, the maximum load level is not constant but depends on the user locations, on the simultaneous real time connections and on the other cell interference.

10.4.3 Packet Scheduler in Soft Handover

If the mobile is in soft handover, the packet scheduler must take into account the air interface load and the physical resources in all base stations of the active set. The dedicated channels are the only transport channels that can use soft handover. When the mobile is in

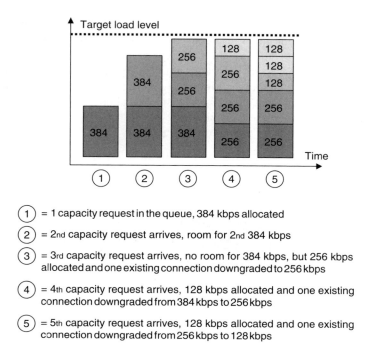

① = 1 capacity request in the queue, 384 kbps allocated

② = 2nd capacity request arrives, room for 2nd 384 kbps

③ = 3rd capacity request arrives, no room for 384 kbps, but 256 kbps allocated and one existing connection downgraded to 256 kbps

④ = 4th capacity request arrives, 128 kbps allocated and one existing connection downgraded from 384 kbps to 256 kbps

⑤ = 5th capacity request arrives, 128 kbps allocated and one existing connection downgraded from 256 kbps to 128 kbps

Figure 10.27. Example DCH bit rate allocation when new requests arrive, no QoS differences

CELL_DCH state and in soft handover, the packet scheduling can be done for each cell separately. Therefore, the responses (scheduled bit rates) of different packet schedulers may differ. The final selection of the bit rate is made according to the most heavily loaded cell in the active set, which has scheduled the lowest bit rate. The interactions are illustrated in Figure 10.28.

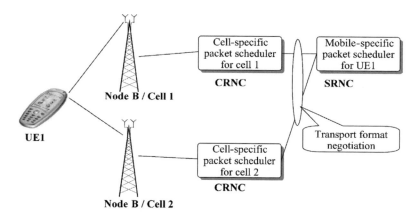

Figure 10.28. Packet scheduling in soft handover

10.5   Packet Data System Performance

In this section, packet scheduling performance is analysed. First, link level performance of packet data is discussed in Section 10.5.1, while system level performance is the topic of Section 10.5.2.

10.5.1   Link Level Performance

The effect of the block error rate, BLER, and retransmissions on throughput is studied at the link level, and an optimal BLER target level is proposed in this section. Packet data performance is studied in the ITU Vehicular A multipath channel using a mobile speed of 3 km/h in the downlink with 64 kbps DCH. The BLER as a function of E_b/N_0 is shown in Figure 10.29.

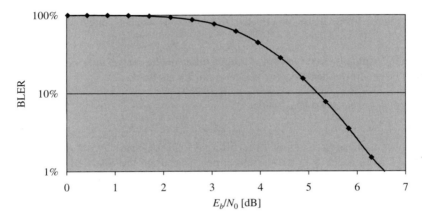

Figure 10.29. BLER as a function of E_b/N_0 for 64 kbps DCH in the downlink

The higher the BLER, the more retransmissions are needed to deliver error-free data. On the other hand, less power, or lower E_b/N_0, is needed for higher BLER levels. What is the optimal BLER operation point that requires the lowest energy per correctly received bit when the retransmissions are taken into account? To find out the optimal BLER point, we use the definition of effective E_b/N_0 that should be minimised to maximise the capacity:

$$\left(E_b/N_0\right)_{\text{effective}} = \frac{E_b/N_0}{1 - BLER} \tag{10.3}$$

The BLER vs. effective E_b/N_0 is shown in Figure 10.30. The relationship between E_b/N_0 and BLER is taken from Figure 10.29. The optimal BLER operation point is around 10 %. If BLER is lower, capacity is wasted because the retransmissions are not efficiently utilised to gain from the additional time diversity. The effective E_b/N_0 is 0.8 dB higher with BLER 1 % than with BLER 10 %. If BLER is higher than 10 %, there are too many retransmissions, causing additional interference. With higher BLER, the average delay will also be longer due to retransmissions, and the quality of the signalling will be reduced. A higher BLER also consumes more downlink orthogonal codes, Node B hardware and Iub resources, because

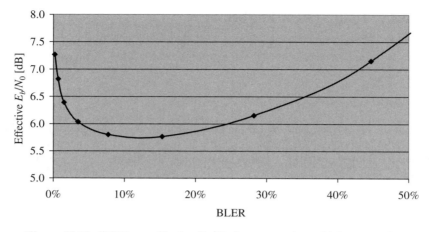

Figure 10.30. BLER vs. effective E_b/N_0 (lowest number = highest capacity)

those resources must be reserved for a longer time for the retransmissions. The BLER level
of 1–10 % is generally assumed for packet data in this book.

10.5.2   System Level Performance

In this section, the system level performance of packet scheduling is analysed. The user bit
rates and system capacity are shown for different load scenarios. A network with 18 cells in
Figure 10.31 is considered in these simulations. The users are distributed uniformly in the

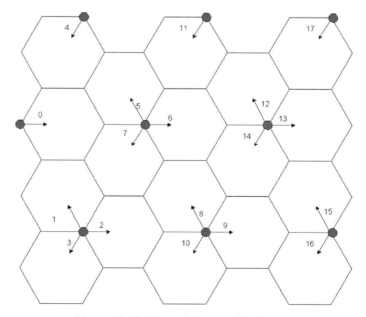

Figure 10.31. Network layout with 18 cells

Table 10.3. Simulation assumptions

Parameter	Value
BLER target	5 %
Packet scheduling period	200 ms
Minimum and initial bit rate	32 kbps
Maximum bit rate	384 kbps
Maximum link power	2 W
Inactivity timer	2–5 seconds

network area. Only the statistics from the four cells in the middle are considered in order to avoid border effects. The main parameters can be seen in Table 10.3.

Users arrive according to a Poisson arrival process. Each user performs the downloading of one packet call, where the packet call distribution is given in Figure 10.32. The minimum packet is 500 B, median 30 kB and the maximum equals 1 MB.

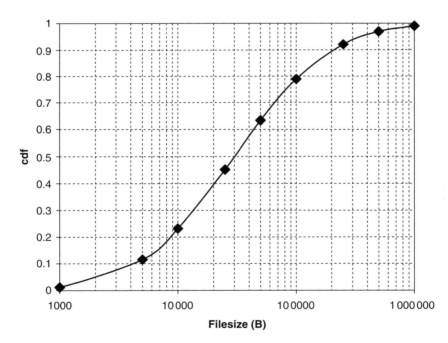

Figure 10.32. Cumulative density function of the packet call distribution (simulation input assumption)

The packet scheduler estimates the bit rate that can be given to a user, according to the available air interface capacity and the maximum link power. First, a 32 kbps is allocated, which is upgraded after 4 s if a bit rate upgrade is possible. The maximum possible bit rate will be given to this user and this bit rate will be kept until the file download is finished. The resources will be kept reserved after the download has finished for 2–5 s in order to avoid repeating releasing and setting up of the dedicated channel.

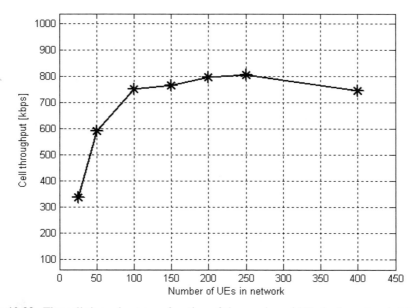

Figure 10.33. The cell throughput as a function of the number of UEs in the network (simulation output result)

Figure 10.33 shows the cell throughput as a function of the load. It can be seen that the throughput initially increases with the number of users. The reason is that the low number of users do not download enough data to fill the whole cell capacity. The maximum capacity is approximately 800 kbps per cell. When there are more than 300 active UEs in the network, the cell throughput starts decreasing. The reason for this is that the users' bit rates decrease with an increasing number of users, while the signalling overhead stays identical per user. Thus, the cell throughput starts to suffer from the large number of signalling channels.

The transmitted bit rate distribution can be seen in Figure 10.34 for three different numbers of users. It can clearly be seen that in the case of a low loaded network, the high bit rates are allocated: nearly 60 % of the time the users get 384 kbps, even though they all start out with 32 kbps for the first four seconds. When the load increases, a shift to the lower bit rates can be seen: in the case of 250 users, only 10% get 384 kbps while 40% of the users get 32 kbps.

10.6   Packet Data Application Performance

The focus in this chapter has, up to now, been on basic packet data performance items, such as packet round trip times and bit rate throughput on the radio, TCP/IP and UDP/IP layers. What has not been discussed thus far is how the applications on top of the TCP/IP and UDP/IP layers over WCDMA are dependent on the underlying end-to-end performance characteristics. This dependence and end-to-end performance is the topic of the following sections.

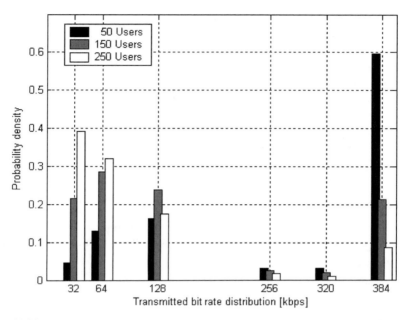

Figure 10.34. The probability density of the transmitted user bit rates at three different loads

10.6.1   Introduction to Application Performance

Applications utilise the packet data transport connections with performance properties such as throughput, delay, jitter, residual bit errors and connection set-up times. See Figure 10.35 for an illustration of protocol layers and their corresponding key performance characteristics. One delay component that is usually included in the initial connection set-up time is the activation of the PDP (Packet Data Protocol) context in the GGSN (Gateway GPRS Support Node). The activation or modification of the PDP context is needed as soon as a connection with a new quality of service traffic class is set up to/from the mobile station. Because one application may use multiple quality of service classes, e.g. the interactive traffic class for signalling and the real time traffic class for user plane data, it is possible that one PDP context is always on and one PDP context is set up on a need basis only.

When designing the applications, both for fixed and wireless networks, it is possible to improve the end user performance by utilising user and control plane compression and

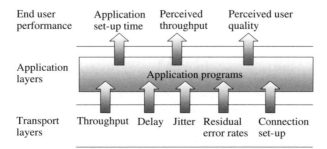

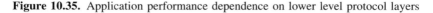

Figure 10.35. Application performance dependence on lower level protocol layers

receiver buffering on the application layer. By doing this the perceived end user throughput and outage times are improved from an end user point of view, even though the underlying transport quality is still the same.

The end user performance measures for applications that we look at are service set-up times, response times during service usage and perceived user plane quality of information presentation (throughput, picture quality, outage, availability).

The target is to illustrate the performance with example packet-based applications, i.e. the different performance enhancement methods are not illustrated in more detail. Depending on the application type, the end user performance expectations are different. The applications are therefore grouped into person-to-person applications, content-to-person applications and business connectivity.

10.6.2   Person-to-person Applications

The performance of the following person-to-person applications is evaluated: push-to-talk over cellular (PoC), real time video sharing, voice over IP (VoIP) and real time games.

10.6.2.1   Push-to-talk over Cellular (PoC)

The PoC application [5] performance is obviously dependent on the speech quality, but since the speech quality already has been studied extensively for circuit switched connections, this section only includes a study of the response times while using PoC. Two central performance measures related to the response times for the push-to-talk service are the start-to-talk delay in Figure 10.36 and the voice-through delay in Figure 10.37. The start-to-talk

Start-to-talk delay
= delay from user A pressing button until
until he gets floor grant message = beep

Figure 10.36. Start-to-talk delay in push-to-talk

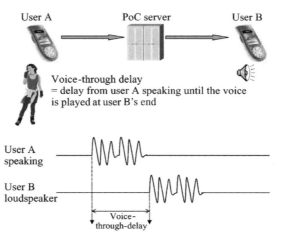

Figure 10.37. Voice-through delay in push-to-talk

delay is the time between user A pressing a button to enable a connection until he/she gets an indication (for example a beep) that the floor is granted. The voice-through delay is the time from when user A starts speaking until the voice starts playing in user B's terminal.

The delay calculation results presented herein include effects like:

- Typical packet round trip times;
- Representative radio bearer and radio access bearer set-up times;
- Paging delays;
- Voice encoding delays in the mobile station;
- Typical PoC server delays.

The receiving terminal may additionally have a receiver buffer of a few hundred ms to compensate for the jitter delays. The jitter buffer delay is, however, not included in the presented calculations.

The voice-through delay and the start-to-talk delay are both affected by what RRC state the mobile station is in when either pushing the start-to-talk button for the first time or when receiving the first speech burst from its peer. It is assumed that both mobile stations start from the Cell_PCH state and that the floor request messages are small enough for the Cell_FACH state. That is, the transition to the Cell_DCH state takes place first when there are speech packets to send. Start-to-talk I in Figure 10.38 and voice-through delay I in

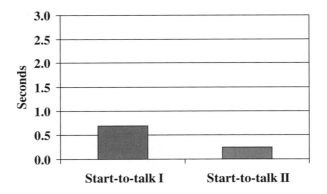

Figure 10.38. Start-to-talk delay in push-to-talk application

Figure 10.39 refer to the case when the mobiles are initially in Cell_PCH state. Note that once a mobile station has transmitted or received its first speech burst it will be in the Cell_DCH state for some time and subsequent start-to-talk and voice-through delays are hence smaller. The transition back from Cell_DCH to Cell_FACH or Cell_PCH takes place first after an inactivity of several seconds, so it is reasonable to assume that during active use of the application, the mobile stations will stay in the Cell_DCH state. Start-to-talk II and voice through delay II refer to the case when the mobiles are already in Cell_DCH state. The DCH allocation delays and round trip times are assumed to be similar to those presented earlier in this chapter. Note that the start-to-talk delay values are short in all cases, since common channels (RACH/FACH) are used to carry the corresponding packets. The voice-through delay is affected by the RRC state since DCH allocation increases the end user

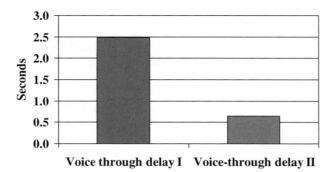

Figure 10.39. Voice-through delay in push-to-talk application

experienced delay. The paging delay does not increase voice-through delay since user B paging is triggered already from the floor request message. For reference: the normal voice-through delay in circuit switched cellular telephony is approx 200 ms. The higher delay in PoC comes mainly from the fact that the voice packets are larger in PoC than in ordinary voice connections.

The push-to-talk application can be used over GPRS as well. The start-to-talk delay in GPRS is slightly longer than in WCDMA. Also, the GPRS voice-through delay is longer when compared to WCDMA DCH. On the other hand, WCDMA voice-through delay is slightly longer than in GPRS if DCH set-up is required in WCDMA. We may note from the PoC performance calculations, that common channels are handy in WCDMA for sending small packets without DCH allocation, and low round trip time in WCDMA is useful in reducing the voice-through delay. We can also note that minimisation of the DCH set-up time is important for optimising the end user performance.

10.6.2.2   Real Time Video Sharing

Two performance metrics for real time video sharing are the set-up delay of the streaming connection and the streaming video delay from user A recording to user B display. The streaming video delay is illustrated in Figure 10.40. The two-way voice can be carried on circuit switched channel in this application. It is desirable to minimise the delay of the streaming video so that the video picture and the voice discussion can be kept reasonably well in synchronisation.

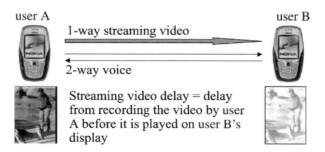

Figure 10.40. Streaming video delay in real time video sharing application

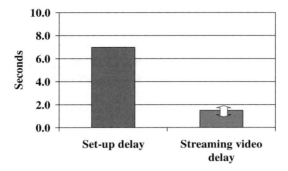

Figure 10.41. Typical delays in real time video sharing

The set-up delay for real time video sharing (shown in Figure 10.41) consists of two main components: session initiation and receiver buffering. The session initiation consists of SIP (Session Initiation Protocol) signalling to set up the streaming connection from user A to user B. By looking at the main delays of the SIP signalling the service set-up should not take longer time than setting up a circuit switched connection today between two mobile stations, i.e. below five to seven seconds. This assumes that efficient SIP compression schemes are used to minimise the SIP message lengths. The streaming delay is, to some extent, a trade-off of picture quality and delay that is caused by receiver buffers. Because of seamless soft handover, the need to compensate for jitter delays is low in WCDMA. Because of this it should be possible to keep the streaming video delay down to two seconds and below.

10.6.2.3 Voice over IP (VoIP)

Voice over IP, video conferencing and other conversational packet services set high requirements on the one-way end-to-end delay: the preferred one-way delay is <150 ms and the limit 400 ms [4]. The short delay is required to enable fluent communication between live end users. The round trip time for WCDMA shown in Figure 10.10 shows that one-way end-to-end delays of approx 200 ms are feasible in WCDMA. This enables good quality conversational packet switched services like, for example, VoIP.

The set-up time is another performance metric that should be minimised to improve end user perceived performance. The estimate of the set-up time consists of RRC state change (DCH allocation), PDP context activation for conversational QoS, paging delay for user B and SIP signalling. The total set-up time is approximately 8 seconds in Figure 10.42. This delay is of the same order as the circuit switched voice call set-up time.

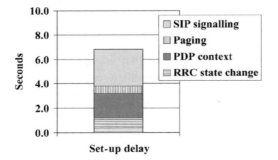

Figure 10.42. Set-up time for VoIP connection

10.6.2.4   Multiplayer Games

Real time network games may require very short delays: action games typically require 200 ms or even shorter round trip time from the UE to the server and back for good quality games. An example action game is Quake II. Real time strategy games, like Age of Kings, typically require a round trip time of around 1 second. Achievable round trip times in GPRS and WCDMA are shown in Figure 10.43, together with the gaming performance requirements. The current GPRS system allows a number of network games, including real time strategy games, from the delay point of view. The current WCDMA delay is short enough even for real time action games with acceptable quality.

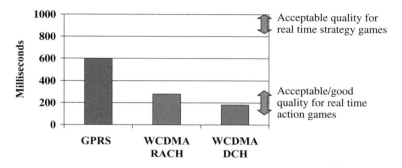

Figure 10.43. Round trip time and delay requirements for real time games

The bit rate requirements for real time games are quite low, mainly <16 kbps. The required data rate is not challenging for WCDMA nor for the current second generation systems. It is mainly the end-to-end delay that is the challenge for real time gaming over cellular systems.

10.6.3   Content-to-person Applications

The following content-to-person applications are considered: WAP browsing, video streaming and content download.

10.6.3.1   WAP Browsing

WAP pages are typically quite small and the download times are fast. The download time of the first WAP page, typically the operator's home page, may be longer. The optimisation of first page download time is considered in this section. Before the UE is able to establish a connection to the WAP gateway, it must have RRC connection to RNC, it must be GPRS attached to SGSN and it must have an active PDP context. If all those procedures need to be done before the download, the first page download can take up to 8–10 seconds, even if the actual download of a small WAP page takes only 2–3 seconds. The optimisation steps to reduce the download time are illustrated in Figure 10.44. The first step is to keep UEs GPRS attached all the time. The second step is to keep UEs RRC connected to RNC so that RRC connection establishment is not required, but only an RRC state change from PCH state to Cell_DCH state. The RRC connection can be maintained if there is an active PDP context for any application, e.g. for presence. The third optimisation step is to use an existing PDP

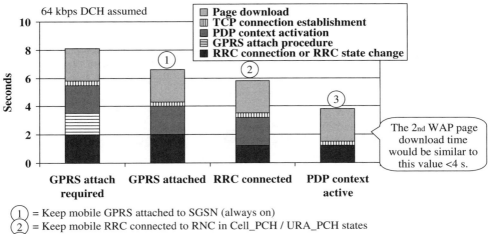

1 = Keep mobile GPRS attached to SGSN (always on)
2 = Keep mobile RRC connected to RNC in Cell_PCH / URA_PCH states
3 = Use the existing PDP context for WAP

Figure 10.44. Download time of first WAP page (home page)

context also for WAP. In this case, only an RRC state change is required before establishing TCP connection and downloading the WAP page. The download time is reduced from the original 8 seconds to below 4 seconds with these optimisation steps. The download time of following WAP pages will be similar to that 4 seconds. If the user already has DCH allocated, the download time is even faster, since no RRC state change is required.

10.6.3.2   Video Streaming

Person-to-person real time video sharing and content-to-person video streaming have a few differences in terms of end user performance. The low delay is important for real time video sharing, while the delay is not as important for content-to-person streaming. The content-to-person streaming is less challenging also because only a downlink connection is required to carry the data, while in real time video sharing, both uplink and downlink connections are needed. Most importantly, video streaming requires a high enough bit rate to carry good quality video and voice. The relationship between bit rate and video quality is illustrated in Figure 10.45. The video streaming bit rates of 20–25 kbps are too low for good quality video streaming, while the bit rates of 60–120 kbps provide a clear improvement in quality. Such bit rates are feasible in WCDMA.

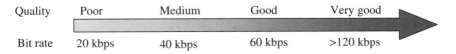
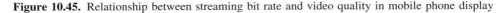

Figure 10.45. Relationship between streaming bit rate and video quality in mobile phone display

10.6.3.3   Content Download

The content download times using TCP are shown in Figure 10.46. Three different content sizes are considered: 100 kB file that could be a short video clip or picture, a 300 kB high

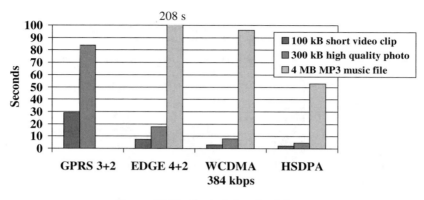

Figure 10.46. Content download times

quality picture and a 4 MB music file. GPRS offers reasonably low response times for file sizes <100 kB which are typical for MMS (Multimedia service). WCDMA offers low download times for high quality photos as well. A 4 MB MP3 music file download time in WCDMA would be less than 100 s. With HSDPA, that music file could be downloaded in less than 1 minute.

10.6.4   Business Connectivity

Business connectivity considers applications running mainly on laptops using a cellular system as a radio modem. The considered applications are Web browsing, Outlook email and Netmeeting. The performance of these applications over WCDMA should preferably be similar to the performance provided by dial-up modems or by broadband DSL and cable modem connections, which set tough requirements on WCDMA performance.

10.6.4.1   Browsing

Web page download times are analysed in this section. The web page download times include RRC state change, Domain Name Server (DNS) query, TCP connection establishment and download of text and graphics in one TCP connection using HTTP1.1. The signalling flow chart is shown in Figure 10.47.

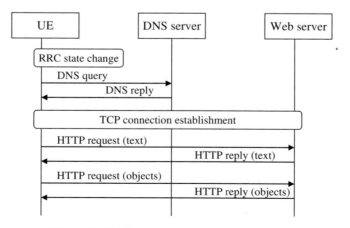

Figure 10.47. Flow chart for web page download

Typical average Internet web page sizes are 100–200 kB today and the size of the pages keeps increasing when more pictures and graphics are introduced to the pages. Measured download times are shown in Figure 10.48 and calculated times in Figure 10.49. Most of the download time comes from the HTTP replies, i.e. downloading the content. There are some differences between the measurement set-up and the calculation assumptions, e.g. Virtual private network (VPN) was used in the measurements.

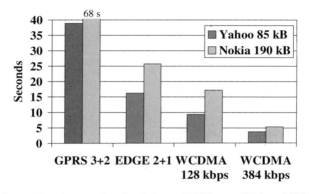

Figure 10.48. Measured web page download times. GPRS uses CS-2 and EDGE uses MCS-7

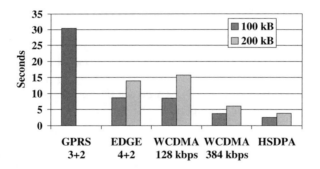

Figure 10.49. Estimated web page download times in average loaded networks

The measurements are done with a three time slot GPRS mobile, like Nokia 6600, with a two time slot EDGE mobile, like Nokia 6220, and with WCDMA bit rates of 128 kbps and 384 kbps, as in Nokia 7600. EDGE modulation and coding scheme were fixed to MCS-7, providing approximately 90 kbps throughput with two time slots. The calculations assume a four time slot EDGE mobile, like Nokia 6230, with an average 40 kbps per time slot and WCDMA 384 kbps connection. The average assumed data rate for HSDPA is 700 kbps.

EDGE provides a major improvement in web browsing performance compared to GPRS. EDGE is >150% faster than GPRS and faster than using dial-up modem connection. WCDMA download times are 40–60 % faster than with EDGE. WCDMA performance in general is similar to low-end DSL connections. HSDPA brings a further 30–40 % reduction in download times compared to WCDMA. HSDPA performance is already close to public WLAN performance.

The web browsing response times could be improved by using Performance Enhancement Proxies, PEPs. These proxies are able to reduce the size of the web page, e.g. by reducing the picture resolution.

10.6.4.2 Email

Two performance measures are shown for Windows 2000 Outlook: time for receiving an email and time for connecting to a mail server. The measured email reception times for two different sizes of an email, 45 kB and 215 kB, are shown in Figure 10.50. Both EDGE and

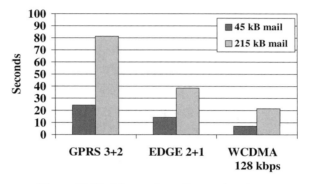

Figure 10.50. Measured email reception times

WCDMA improve the performance compared to GPRS. EDGE is roughly 100 % faster than GPRS, and WCDMA is 100 % faster than EDGE. The reception times with PEP are shown in Figure 10.51. PEP improves the performance for GPRS by 15 % for small emails and up to 50 % for large emails. EDGE and WCDMA do not benefit from PEP for small emails but they also gain roughly 30 % for large mails.

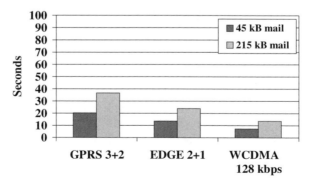

Figure 10.51. Measured email reception times with Performance Enhancement Proxy (PEP)

The measured connection to mail server is illustrated in Figure 10.52. This procedure includes a large number of signalling messages and therefore, a small round trip time is important for the performance. Without PEP, EDGE provides only a minor gain over GPRS,

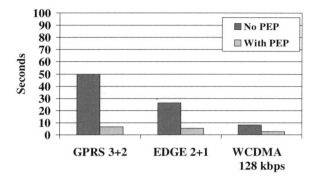

Figure 10.52. Measured connection times to mail server

while WCDMA with lower round trip time provides a major improvement. PEP can improve the connection times by a factor of 3–5 and it is important especially for GPRS and EDGE.

10.6.4.3 Netmeeting

The delay of sharing a slide from laptop to LAN is shown in Figure 10.53. Sharing a simple slide is relatively fast <5 seconds in GPRS. Sharing a heavy slide takes 11 seconds in WCDMA using 64 kbps uplink connection, and 22–33 seconds in GPRS/EDGE.

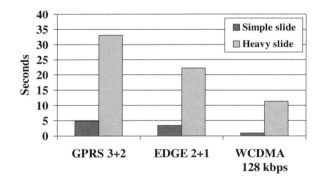

Figure 10.53. Measured time of sharing a slide from cellular system to LAN

10.6.5 Conclusions on Application Performance

The performance of the considered person-to-person applications is mainly defined by the delay: a short delay is required for these applications. WCDMA is able to support these applications by providing a lower delay than second generation systems. The content-to-person applications benefit both from high bit rate capability and from low delay. The presented business applications are sensitive to the delay because of the application signalling involved. The business applications also benefit from the high bit rate capabilities to provide close to DSL/WLAN levels of performance.

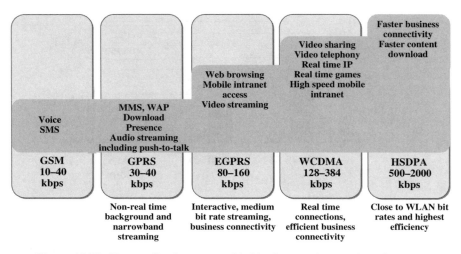

Figure 10.54. New applications are enabled by improved network performance

The performance estimates in this chapter show that GPRS is well suited for background downloads without strict delay requirements, and for downloads of small WAP/web pages. GPRS can also be used for narrowband streaming, like audio streaming. EDGE brings a clear improvement in performance for content-to-person and for business applications. EDGE performance is better than dial-up connection and it allows video streaming bit rates. WCDMA enables a number of person-to-person applications with its short delay. WCDMA also improves business applications and download performance. High Speed Downlink Packet Access, HSDPA, brings a further improvement in end user performance for downlink packet data. The application areas are shown in Figure 10.54.

References

[1] Stevens, W.R., '*TCP/IP Illustrated, Volume 1: The Protocols*' Addison-Wesley Professional Computing Series, 1994.
[2] Inamura, H., Montenegro, G., Ludwig, R., Gurtov, A. and Khafizov, F., '*TCP over 2.5G and 3G Wireless Networks*', internet draft, http://www.ietf.org/internet-drafts/draft-ietf-pilc-2.5g3g-06.txt
[3] Allman, M, Floyd, S. and Partridge, C., '*Increasing TCP's Initial Window*', experimental protocol RFC2414, http://www.faqs.org/rfcs/rfc2414.html
[4] 3GPP Technical Specification 25.853 'Delay Budget within the Access Stratum', 2003.
[5] Industry Specification for PoC Submitted to OMA (Open Mobile Alliance) MAG PoC SWG, Doc# OMA-MAG-POC-2003-0007, August 25, 2003.

11

High-speed Downlink Packet Access

Antti Toskala, Harri Holma, Troels Kolding, Preben Mogensen,
Klaus Pedersen and Karri Ranta-aho

This chapter presents High-speed Downlink Packet Access (HSDPA) for WCDMA – the key new feature included in the Release 5 specifications. The HSDPA concept has been designed to increase downlink packet data throughput by means of fast physical layer (L1) retransmission and transmission combining, as well as fast link adaptation controlled by the Node B (Base Transceiver Station (BTS)). This chapter is organised as follows: First, HSDPA key aspects are presented and a comparison to Release '99 downlink packet access possibilities is made. Next, the impact of HSDPA on the terminal uplink (user equipment (UE)) capability classes is summarised and an HSDPA performance analysis is presented, including a comparison to Release '99 packet data capabilities, as well as performance in the case of a shared carrier between HSDPA and non-HSDPA traffic. The chapter is concluded with a short discussion of evolution possibilities of HSDPA, including a description of the on-going work on uplink improvements in 3GPP.

11.1 Release '99 WCDMA Downlink Packet Data Capabilities

Various methods for packet data transmission in WCDMA downlink already exist in Release '99. As described in Chapter 10, the three different channels in Release '99/ Release 4 WCDMA specifications that can be used for downlink packet data are

- Dedicated Channel (DCH);
- Downlink-shared Channel (DSCH);
- Forward Access Channel (FACH).

The DCH can be used basically for any type of service, and it has a fixed spreading factor (SF) in the downlink. Thus, it reserves the code space capacity according to the peak data

---

WCDMA for UMTS, third edition.   Edited by Harri Holma and Antti Toskala
© 2004 John Wiley & Sons, Ltd   ISBN: 0-470-87096-6

rate for the connection. For example, with Adaptive Multirate (AMR) speech service and packet data, the DCH capacity reserved is equal to the sum of the highest rate used for the AMR speech and the highest rate allowed to be sent simultaneously with full rate AMR. This can be used even up to 2 Mbps, but reserving the code tree for a very high peak rate with low actual duty cycle is obviously not a very efficient use of code resources. The DCH is power-controlled and may be operated in soft handover as well. Further details of the downlink DCH can be found in Section 6.4.5.

The DSCH has been developed to operate always together with a DCH. This way, channel properties can be defined to best suit packet data needs, while leaving the data with tight delay budget, such as speech or video, to be carried by the DCH. The DSCH, in contrast to DCH (or FACH), has a dynamically varying SF informed on a 10 ms frame-by-frame basis with the Transport Format Combination Indicator (TFCI) signalling carried on the associated DCH. The DSCH code resources can be shared between several users and the channel may employ either single code or multicode transmission. The DSCH may be fast power controlled with the associated DCH but does not support soft handover. The associated DCH can be in soft handover, for example speech is provided on DCH if present with packet data. The DSCH operation is described further in Section 6.4.7.

The FACH, carried on the secondary common control physical channel (S-CCPCH) can be used for downlink packet data as well. The FACH is operated normally on its own, and it is sent with a fixed SF and typically at rather high power level to reach all users in the cell, owing to the lack of physical layer feedback in the uplink. There is no fast power control or soft handover for FACH. The S-CCPCH physical layer properties are described in Section 6.5.4. FACH cannot be used in cases in which simultaneous speech and packet data service is required.

11.2 HSDPA Concept

The key idea of the HSDPA concept is to increase packet data throughput with methods known already from Global System for Mobile Communications (GSM)/Enhanced Data rates for Global Evolution (EDGE) standards, including link adaptation and fast physical layer (L1) retransmission combining. The physical layer retransmission handling has been discussed earlier but the inherent large delays of the existing Radio Network Controller (RNC)-based Automatic Repeat reQuest ARQ architecture would result in unrealistic amounts of memory on the terminal side. Thus, architectural changes are needed to arrive at feasible memory requirements, as well as to bring the control for link adaptation closer to the air interface. The transport channel carrying the user data with HSDPA operation is denoted as the High-speed Downlink Shared Channel (HS-DSCH). A comparison of the basic properties and components of HS-DSCH and DSCH is conducted in Table 11.1.

A simple illustration of the general functionality of HSDPA is provided in Figure 11.1. The Node B estimates the channel quality of each active HSDPA user on the basis of, for instance, power control, ACK/NACK ratio, and HSDPA-specific user feedback. Scheduling and link adaptation are then conducted at a fast pace depending on the active scheduling algorithm and the user prioritisation scheme. The channels needed to carry data and downlink/uplink control signalling are described later in this chapter.

With HSDPA, two of the most fundamental features of WCDMA, variable SF and fast power control, are disabled and replaced by means of *adaptive modulation and coding*

Table 11.1. Comparison of fundamental properties of DSCH and HS-DSCH

Feature	DSCH	HS-DSCH
Variable spreading factor	Yes	No
Fast power control	Yes	No
Adaptive modulation and coding (AMC)	No	Yes
Multicode operation	Yes	Yes, extended
Fast L1 HARQ	No	Yes

Note: HARQ: Hybrid Automatic Repeat reQuest.

(AMC), extensive multicode operation and a fast and spectrally efficient retransmission strategy. In the downlink, WCDMA power control dynamics is in the order of 20 dB, compared to the uplink power control dynamics of 70 dB. The downlink dynamics are limited by the intra-cell interference (interference between users on parallel code channels) and by the Node B implementation. This means that for a user close to the Node B, the power control cannot reduce power maximally, and on the other hand reducing the power to beyond 20 dB dynamics would have only marginal impact on the capacity. With HSDPA, this property is now utilised by the link adaptation function and AMC to select a coding and modulation combination that requires higher E_c/I_{0r}, which is available for the user close to the Node B (or with good interference/channel conditions in the short-term sense). This leads to additional user throughput, basically for free. To enable a large dynamic range of the HSDPA link adaptation and to maintain a good spectral efficiency, a user may simultaneously utilise up to 15 multicodes in parallel. The use of more robust coding, fast Hybrid Automatic Repeat Request (HARQ) and multicode operation removes the need for variable SF.

To allow the system to benefit from the short-term variations, the scheduling decisions are done in the Node B. The idea in HSDPA is to enable a scheduling such that, if desired, most of the cell capacity may be allocated to one user for a very short time, when conditions are

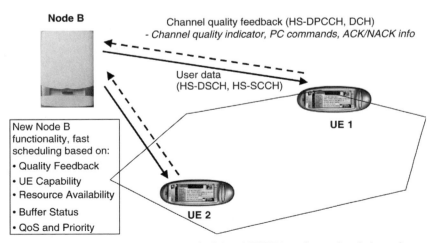

Figure 11.1. General operation principle of HSDPA and associated channels

favourable. In the optimum scenario, the scheduling is able to track the fast fading of the users.

The physical layer packet combining basically means that the terminal stores the received data packets in soft memory and if decoding has failed, the new transmission is combined with the old one before channel decoding. The retransmission can be either identical to the first transmission or contain different bits compared with the channel encoder output that was received during the last transmission. With this incremental redundancy strategy, one can achieve a diversity gain as well as improved decoding efficiency.

11.3   HSDPA Impact on Radio Access Network Architecture

All Release '99 transport channels presented earlier in this book are terminated at the RNC. Hence, the retransmission procedure for the packet data is located in the serving RNC, which also handles the connection for the particular user to the core network. With the introduction of HS-DSCH, additional intelligence in the form of an HSDPA Medium Access Control (MAC) layer is installed in the Node B. This way, retransmissions can be controlled directly by the Node B, leading to faster retransmission and thus shorter delay with packet data operation when retransmissions are needed. Figure 11.2 presents the difference between

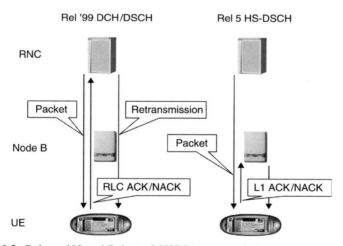

Figure 11.2. Release '99 and Release 5 HSDPA retransmission control in the network

retransmission handling with HSDPA and Release '99 in the case in which the serving and controlling RNCs are the same. In the case where no relocation procedure is used in the network, the actual termination point could be several RNCs further into the network. With HSDPA, the Iub interface between Node B and RNC requires a flow control mechanism to ensure that Node B buffers are used properly and that there is no data loss due to Node B buffer overflow.

The MAC layer protocol in the architecture of HSDPA can be seen in Figure 11.3, showing the different protocol layers for the HS-DSCH. The RNC still retains the functionalities of the Radio Link Control (RLC), such as taking care of the retransmission

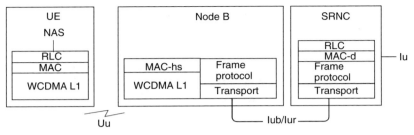

Figure 11.3. HSDPA protocol architecture

in case the HS-DSCH transmission from the Node B fails after, for instance, exceeding the maximum number of physical layer retransmissions. Although there is a new MAC functionality added in the Node B, the RNC still retains the Release '99/Release 4 functionalities. The key functionality of the new Node B MAC functionality (MAC-hs) is to handle the Automatic Repeat Request (ARQ) functionality and scheduling as well as priority handling. Ciphering is done in any case in the RLC layer to ensure that the ciphering mask stays identical for each retransmission to enable physical layer combining of retransmissions.

The type of scheduling to be carried out in Node B is not defined in 3GPP standardisation, only some parameters, such as discard timer or scheduling priority indication, that can be used by RNC to control the handling of an individual user. As the scheduler type has a big impact on the resulting performance and QoS, example packet scheduler types are presented in this chapter in the performance section.

11.4   Release 4 HSDPA Feasibility Study Phase

During Release 4 work, an extensive feasibility study was performed on the HSDPA feature to investigate the gains achievable with different methods and the resulting complexity of various alternatives. The items of particular interest were obviously the relative capacity improvement and the resulting increases in the terminal complexity with physical layer ARQ processing, as well as backwards compatibility and coexistence with Release '99 terminals and infrastructure. The results presented in [1] compared the HSDPA cell packet data throughput against Release '99 DSCH performance as presented, and the conclusions drawn were that HSDPA increased the cell throughput up to 100 % compared to Release '99.

The evaluation was conducted for a one-path Rayleigh fading channel environment using *C/I* scheduling. The results from the feasibility study phase were produced for relative comparison purposes only. The HSDPA performance with more elaborate analysis is discussed later in this chapter.

11.5   HSDPA Physical Layer Structure

The HSDPA is operated similarly to DSCH together with DCH, which carries the services with tighter delay constraints, such as AMR speech. To implement the HSDPA feature, three new channels are introduced in the physical layer specifications [2]:

- HS-DSCH carries the user data in the downlink direction, with the peak rate reaching up to 10 Mbps range with 16 QAM (quadrature amplitude modulation).

- High-speed Shared Control Channel (HS-SCCH) carries the necessary physical layer control information to enable decoding of the data on HS-DSCH and to perform the possible physical layer combining of the data sent on HS-DSCH in the case of retransmission of an erroneous packet.

- Uplink High-Speed Dedicated Physical Control Channel (HS-DPCCH) carries the necessary control information in the uplink, namely, ARQ acknowledgements (both positive and negative ones) and downlink quality feedback information.

These three channel types are discussed in the following sections.

11.5.1 High-speed Downlink Shared Channel (HS-DSCH)

The HS-DSCH has specific characteristics in many ways compared with existing Release '99 channels. The Transmission Time Interval (TTI) or interleaving period has been defined to be 2 ms (three slots) to achieve a short round trip delay for the operation between the terminal and Node B for retransmissions. The HS-DSCH 2 ms TTI is short compared to the 10, 20, 40 or 80 ms TTI sizes supported in Release '99. Adding a higher order modulation scheme, 16 QAM, as well as lower encoding redundancy has increased the instantaneous peak data rate. In the code domain perspective, the SF is fixed; it is always 16, and multicode transmission as well as code multiplexing of different users can take place. The maximum number of codes that can be allocated is 15, but depending on the terminal (UE) capability, individual terminals may receive a maximum of 5, 10 or 15 codes. The total number of channelisation codes with spreading factor 16 is 16 (under the same scrambling code), but as there is a need to have code space available for common channels, HS-SCCHs and for the associated DCH, the maximum usable number of codes was set to 15. A simple scenario is illustrated in Figure 11.4, where two users are using the same HS-DSCH. Both users check

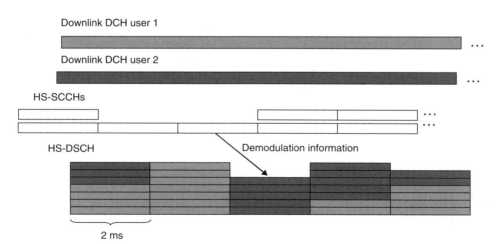

Figure 11.4. Code multiplexing example with two active users

the information from the HS-SCCHs to determine which HS-DSCH codes to despread, as well as other parameters necessary for correct detection.

11.5.1.1   HS-DSCH Modulation

As stated earlier, 16 QAM modulation was introduced in addition to Release '99 Quadrature Phase Shift Keying (QPSK) modulation. Even during the feasibility study phase, 8 PSK and 64 QAM were considered, but eventually these schemes were discarded for performance and complexity reasons. 16 QAM, with the constellation example shown in Figure 11.5, doubles the peak data rate compared to QPSK and allows up to 10 Mbps peak data rate with 15 codes of SF 16. However, the use of higher order modulation is not without cost in the mobile radio environment. With Release '99 channels, only a phase estimate is necessary for the demodulation process. Even when 16 QAM is used, amplitude estimation is required to separate the constellation points. Further, more accurate phase information is needed since constellation points have smaller differences in phase domain compared to QPSK. The HS-DSCH capable terminal needs to obtain an estimate of the relative amplitude ratio of the DSCH power level compared to the pilot power level, and this requires that Node B should not adjust the HS-DSCH power between slots if 16 QAM is used in the frame. Otherwise, the performance is degraded as the validity of an amplitude estimate obtained from Common Pilot Channel (CPICH) and estimated power difference between CPICH and HS-DSCH would no longer be valid.

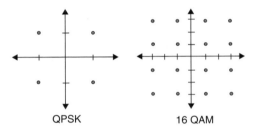

QPSK   16 QAM

Figure 11.5. QPSK and 16 QAM constellations

11.5.1.2   HS-DSCH Channel Coding

The HS-DSCH channel coding has some simplifications when compared to Release '99. As there is only one transport channel active on the HS-DSCH, the blocks related to the channel multiplexing for the same users can be left out. Further, the interleaving only spans over a single 2 ms period and there is no separate intra-frame or inter-frame interleaving. Finally, turbo coding is the only coding scheme used. However, by varying the transport block size, the modulation scheme and a number of multicodes, other effective code rates other than 1/3 become available. In this manner, code rates within the range 0.15–0.98 can be achieved. By varying the code rate, the number of bits per code can be increased at the expense of reduced coding gain. The major difference is the addition of the hybrid ARQ (HARQ) functionality as shown in Figure 11.6. When using QPSK, the Release '99 channel interleaver is used and when using 16 QAM, two parallel (identical) channel interleavers are applied. As discussed earlier, the HSDPA-capable Node B has the responsibility of selecting the transport format to be used along with the modulation and number of codes on the basis of the information available at the Node B scheduler.

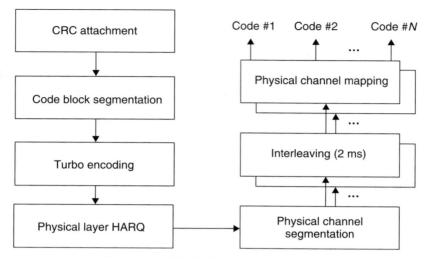

Figure 11.6. HS-DSCH channel coding chain

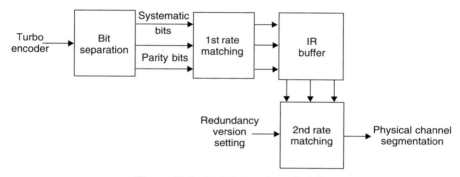

Figure 11.7. HARQ function principle

The HARQ functionality is implemented by means of a two-stage rate-matching functionality, with the principle illustrated in Figure 11.7. The principle shown in Figure 11.7 contains a buffer between the rate-matching stages to allow tuning of the redundancy settings for different retransmissions between the rate-matching stages. The buffer shown should be considered only as a virtual buffer as the obvious practical rate-matching implementation would consist of a single rate-matching block without buffering any blocks after the first rate-matching stage. The HARQ functionality is basically operated in two different ways. It is possible to send identical retransmissions, which is often referred to as chase or soft combining. With different parameters, the transmissions will not be identical and then the principle of incremental redundancy is used. In this case, for example, the first transmission could consist of systematic bits, while the second transmission would consist of only parity bits. The latter method has a slightly better performance but it also needs more memory in the receiver, as the individual retransmissions cannot be just added.

The terminal default memory requirements are set on the basis of soft combining and at maximum data rate (supported by the terminal). Hence, at the highest data rate, only soft combining may be used, while with lower data rates, also incremental redundancy can be used.

With a 16 QAM constellation, the different bits mapped to the 16 QAM symbols have different reliability. This is compensated in connection with the ARQ process with a method called *constellation rearrangement*. With constellation rearrangement, the different retransmissions use slightly different mapping of the bits to 16 QAM symbols to improve the performance. Further details on the HS-DSCH channel coding can be found from [3].

11.5.1.3 HS-DSCH Versus Other Downlink Channel Types for Packet Data

In Table 11.2, a comparison of different channel types is presented with respect to the key physical layer properties. In all cases except for the DCH, the packet data itself is not operated in soft handover. The HARQ operation with HS-DSCH will also be employed at the RLC level if the physical layer ARQ timers or the maximum number of retransmissions are exceeded.

Table 11.2. Comparison of different channel types

Channel	HS-DSCH	DSCH	Downlink DCH	FACH
Spreading factor	Fixed, 16	Variable (256-4) frame-by-frame	Fixed, (512-4)	Fixed (256-4)
Modulation	QPSK/16 QAM	QPSK	QPSK	QPSK
Power control	Fixed/slow power setting	Fast, based on the associated DCH	Fast with 1500 kHz	Fixed/slow power setting
HARQ	Packet combining at L1	RLC level	RLC level	RLC level
Interleaving	2 ms	10–80 ms	10–80 ms	10–80 ms
Channel coding schemes	Turbo coding	Turbo and convolutional coding	Turbo and convolutional coding	Turbo and convolutional coding
Transport channel multiplexing	No	Yes	Yes	Yes
Soft handover	For associated DCH	For associated DCH	Yes	No
Inclusion in specification	Release 5	Release '99	Release '99	Release '99

11.5.2 High-speed Shared Control Channel (HS-SCCH)

The high-speed shared control channel (HS-SCCH) carries the key information necessary for HS-DSCH demodulation. The UTRAN needs to allocate a number of HS-SCCHs that correspond to the maximum number of users that will be code-multiplexed. If there is no data on the HS-DSCH, then there is no need to transmit the HS-SCCH either. From the network point of view, there may be a high number of HS-SCCHs allocated, but each terminal will only need to consider a maximum of four HS-SCCHs at a given time. The HS-SCCHs that are to be considered are signalled to the terminal by the network. In reality, the need for more than four HS-SCCHs is very unlikely. However, more than one HS-SCCH may be needed to better match the available codes to the terminals with limited HSDPA capability.

Each HS-SCCH block has a three-slot duration that is divided into two functional parts. The first slot (first part) carries the time-critical information that is needed to start the demodulation process in due time to avoid chip level buffering. The next two slots (second part) contain less time-critical parameters including Cyclic Redundancy Check (CRC) to check the validity of the HS-SCCH information and HARQ process information. For protection, both HS-SCCH parts employ terminal-specific masking to allow the terminal to decide whether the detected control channel is actually intended for the particular terminal.

The HS-SCCH uses SF 128 that can accommodate 40 bits per slot (after channel encoding) because there are no pilot or Transmit Power Control TPC bits on HS-SCCH. The HS-SCCH uses half rate convolution coding with both parts encoded separately from each other because the time-critical information is required to be available immediately after the first slot and thus cannot be interleaved together with Part 2.

The HS-SCCH Part 1 parameters indicate the following:

- Codes to despread. This also relates to the terminal capability in which each terminal category indicates whether the current terminal can despread a maximum of 5, 10 or 15 codes.

- Modulation to indicate if QPSK or 16 QAM is used.

The HS-SCCH Part 2 parameters indicate the following:

- Redundancy version information to allow proper decoding and combining with the possible earlier transmissions.

- ARQ process number to show which ARQ process the data belongs to.

- First transmission or retransmission indicator to indicate whether the transmission is to be combined with the existing data in the buffer (if not successfully decoded earlier) or whether the buffer should be flushed and filled with new data.

Parameters such as actual channel coding rate are not signalled but can be derived from the transport block size and other transport format parameters.

As illustrated in Figure 11.8, the terminal has a single slot duration to determine which codes to despread from the HS-DSCH. The use of terminal-specific masking allows the

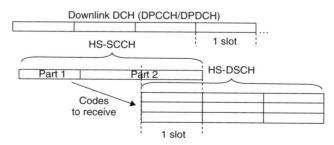

Figure 11.8. HS-SCCH and HS-DSCH timing relationship

terminal to check whether data was intended for it. The total number of HS-SCCHs that a single terminal monitors (the Part 1 of each channel) is a maximum of four, but in case there is data for the terminal in consecutive TTIs, then the HS-SCCH shall be the same for that terminal between TTIs to increase signalling reliability. This kind of approach is also necessary not only to avoid the terminal having to buffer data not necessarily intended for it, but also as there could be more codes in use than supported by the terminal capability. The downlink DCH timing is not tied to the HS-SCCH (or consequently HS-DSCH) timing.

11.5.3 Uplink High-speed Dedicated Physical Control Channel (HS-DPCCH)

The uplink direction has to carry both ACK/NACK information for the physical layer retransmissions and the quality feedback information to be used in the Node B scheduler to determine to which terminal to transmit and at which data rate. It was required to ensure operation in soft handover in the case that not all Node Bs have been upgraded to support HSDPA. Thus, it was decided to leave the existing uplink channel structure unchanged and add the needed new information elements on a parallel code channel that is named the *Uplink High-speed Dedicated Physical Control Channel* (HS-DPCCH). The HS-DPCCH is divided into two parts as shown in Figure 11.9 and carries the following information:

- ACK/NACK transmission, to reflect the results of the CRC check after the packet decoding and combining.

- Downlink Channel Quality Indicator (CQI) to indicate which estimated transport block size, modulation type and number of parallel codes could be received correctly (with reasonable BLER) in the downlink direction.

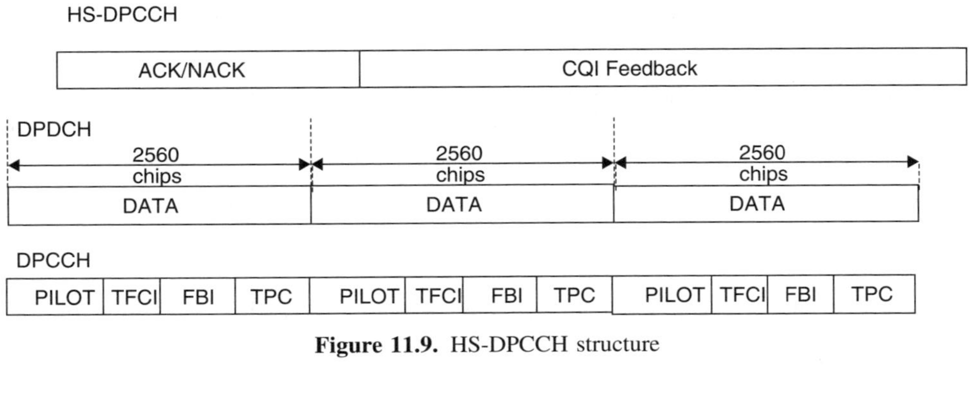

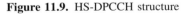

Figure 11.9. HS-DPCCH structure

In 3rd Generation Partnership Project (3GPP) standardisation, there was a lively discussion on this aspect, as it is not a trivial issue to define a feedback method that (1) takes into account different receiver implementations and so forth and (2) simultaneously, is easy to convert to suitable scheduler information in the Node B side. In any case, the feedback information consists of 5 bits that carry quality-related information. One signalling state is reserved for the state 'do not bother to transmit' and other states represent the transmission that the terminal can receive at the current time. Hence, these states range in quality from single code QPSK transmission up to 15 codes 16 QAM transmission (including various coding rates). Obviously, the terminal capability restrictions need to be taken into account in addition to the feedback signalling, and thus, the terminals that do not support a certain number of codes in part of the Channel Quality Indicator (CQI) feedback table shall signal the value for power-reduction factor related to the most demanding combination supported from the CQI table. The CQI table consists of roughly evenly spaced reference transport block size, number of codes and modulation combination that also define the resulting coding rate.

The HS-DPCCH needs some part of the uplink transmission power, which has an impact on the link budget for the uplink. The resulting uplink coverage impact is discussed later in connection with performance.

11.5.4   HSDPA Physical Layer Operation Procedure

The HSDPA physical layer operation goes through the following steps:

- The scheduler in the Node B evaluates for different users what the channel conditions are, how much data is pending in the buffer for each user, how much time has elapsed since a particular user was last served, for which users retransmissions are pending and so forth. Deciding the exact criteria that have to be taken into account in the scheduler is naturally a vendor-specific implementation issue.

- Once a terminal has been determined to be served in a particular TTI, the Node B identifies the necessary HS-DSCH parameters, for instance, how many codes are available or can be filled, can 16 QAM be used and what are the terminal capability limitations? The terminal soft memory capability also defines which kind of HARQ can be used.

- The Node B starts to transmit the HS-SCCH two slots before the corresponding HS-DSCH TTI to inform the terminal of the necessary parameters. The HS-SCCH selection is free (from the set of maximum four channels) assuming there was no data for the terminal in the previous HS-DSCH frame.

- The terminal monitors the HS-SCCHs given by the network and once the terminal has decoded Part 1 from an HS-SCCH intended for that terminal, it will start to decode the rest of that HS-SCCH and will buffer the necessary codes from the HS-DSCH.

- Upon having the HS-SCCH parameters decoded from Part 2, the terminal can determine to which ARQ process the data belongs and whether it needs to be combined with data already in the soft buffer.

- Upon decoding the potentially combined data, the terminal sends in the uplink direction an ACK/NACK indicator, depending on the outcome of the CRC check conducted on the HS-DSCH data.

- If the network continues to transmit data for the same terminal in consecutive TTIs, the terminal will stay on the same HS-SCCH that was used during the previous TTI.

The HSDPA operation procedure has strictly specified timing values for the terminal operation from the HS-SCCH reception via HS-DSCH decoding to the uplink ACK/NACK transmission. The key timing value from the terminal point of view is the 7.5 slots from the end of the HS-DSCH TTI to the start of the ACK/NACK transmission in the HS-DPCCH in the uplink. The timing relationship between downlink, DL and uplink, UL is illustrated in

Figure 11.10. The network side is asynchronous in terms of when to send a retransmission in the downlink. Therefore, depending on the implementation, different amounts of time can be spent on the scheduling process in the network side.

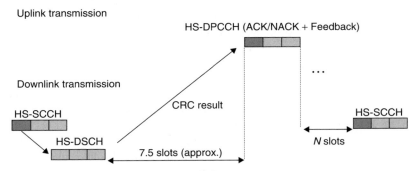

Figure 11.10. Terminal timing with respect to one HARQ process

Terminal capabilities do not impact the timing of an individual TTI transmission but do define how often one can transmit to the terminal. The capabilities include information of the minimum inter-TTI interval that tells whether consecutive TTIs may be used or not. Value 1 indicates that consecutive TTIs may be used, while values 2 and 3 correspond to leaving a minimum of one or two empty TTIs between packet transmissions.

Since downlink DCH, and consecutively uplink DCH, are not slot-aligned to the HSDPA transport channels, the uplink HS-DPCCH may start in the middle of the uplink slot as well, and this needs to be taken into account in the uplink power setting process. The uplink timing is thus quantised to 256 chips (symbol-aligned) and minimum values to 7.5 slots $-$ 128 chips, 7.5 slots $+$ 128 chips. This is illustrated in Figure 11.11.

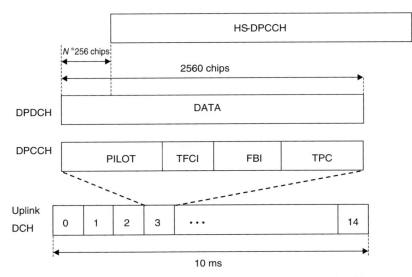

Figure 11.11. Uplink DPCH and HS-SCCH timing relationship

11.6   HSDPA Terminal Capability and Achievable Data Rates

The HSDPA feature is optional for terminals in Release 5 with a total of 12 different categories of terminal (from a physical layer point of view) with resulting maximum data rates ranging between 0.9 and 14.4 Mbps. The HSDPA capability is otherwise independent from Release '99-based capabilities, but if HS-DSCH has been configured for the terminal, then DCH capability in the downlink is limited to the value given by the terminal. A terminal can indicate 32, 64, 128 or 384 kbps DCH capability, as described in Chapter 6.

The terminal capability classes are shown in Table 11.3. The first ten HSDPA terminal capability categories need to support 16 QAM, but the last two, categories 11 and 12, support only QPSK modulation. The differences between classes lie in the maximum number of parallel codes that must be supported and whether the reception in every 2 ms TTI is required. The highest HSDPA class supports 10 Mbps. Besides the values indicated in Table 11.3, there is the soft buffer capability with two principles used for determining the value for soft buffer capability. The specifications indicate the absolute values, which should be understood in the way that a higher value means support for incremental redundancy at maximum data rate, while a lower value permits only soft combining at full rate. While determining when incremental redundancy can be applied also, one needs to observe the memory partitioning per ARQ process defined by the SRNC. There is a maximum of eight ARQ processes per terminal.

Table 11.3. HSDPA terminal capability categories

Category	Maximum number of parallel codes HS-DSCH	Minimum inter-TTI interval	Transport channel bits per TTI	ARQ type at maximum data rate	Achievable maximum data rate (Mbps)
1	5	3	7298	Soft	1.2
2	5	3	7298	IR	1.2
3	5	2	7298	Soft	1.8
4	5	2	7298	IR	1.8
5	5	1	7298	Soft	3.6
6	5	1	7298	IR	3.6
7	10	1	14 411	Soft	7.2
8	10	1	14 411	IR	7.2
9	15	1	20 251	Soft	10.2
10	15	1	27 952	IR	14.4
11	5	2	3630	Soft	0.9
12	5	1	3630	Soft	1.8

Category number 10 is intended to allow the theoretical maximum data rate of 14.4 Mbps, permitting basically the data rate that is achievable with rate 1/3 turbo coding and significant puncturing, resulting in the code rate close to 1. For category 9, the maximum turbo-encoding block size (from Release '99) has been taken into account when calculating the values, thus resulting in the 10.2 Mbps peak user data rate value with four turbo-encoding blocks. It should be noted that, for HSDPA operation, the terminal will not report individual values but only the category. The classes shown in Table 11.3 are as included in [4] with

12 distinct terminal classes. From a Layer 2/3 point of view, the important terminal capability parameter to note is the RLC reordering buffer size that basically determines the window length of the packets that can be 'in the pipeline' to ensure in-sequence delivery of data to higher layers in the terminal. The minimum values range from 50 to 150 kB, depending on the UE category.

Besides the parameter part of the UE capability, the terminal data rate can be largely varied by changing the coding rate as well. Table 11.4 shows the achievable data rates when keeping the number of codes constant (15) and changing the coding rate as well as the modulation. Table 11.4 shows some example bit rates without overhead considerations for different transport format and resource combinations (TFRCs).

Table 11.4. Theoretical bit rates with 15 multicodes for different TFRCs

TFRC	Modulation	Effective code rate	Max. throughput (Mbps)
1	QPSK	$\frac{1}{4}$	1.8
2	QPSK	2/4	3.6
3	QPSK	$\frac{3}{4}$	5.3
4	16 QAM	2/4	7.2
5	16 QAM	$\frac{3}{4}$	10.7

These theoretical data rates can be allocated for a single user or divided between several users. This way, the network can match the allocated power/code resources to the terminal capabilities and data requirements of the active terminals. In contrast to Release '99 operation, it is worth noting that the data rate negotiated with the core network is typically smaller than the peak data rate used in the air interface. Thus, even if the maximum data rate negotiated with the core network was, e.g., 1 Mbps or 2 Mbps, the physical layer would use (if conditions permit) a peak data rate of, e.g., 3.6 Mbps.

11.7 Mobility with HSDPA

The mobility procedures for HSDPA users are affected by the fact that transmission of the HS-PDSCH and the HS-SCCH to a user belongs to only one of the radio links assigned to the UE, namely the serving HS-DSCH cell. UTRAN determines the serving HS-DSCH cell for an HSDPA-capable UE, just as it is UTRAN that selects the cells in a certain user's active set for DCH transmission/reception. Synchronised change of the serving HS-DSCH cell is supported between UTRAN and the UE, so that connectivity on HSDPA is achieved if the UE moves from one cell to another, so that start and stop of transmission and reception of the HS-PDSCH and the HS-SCCH is done at a certain time dictated by UTRAN. This allows implementation of HSDPA with full mobility and coverage to fully exploit the advantages of this scheme over Release '99 channels. The serving HS-DSCH cell may be changed without updating the user's active set for the Release '99 dedicated channels, or in combination with establishment, release, or reconfiguration of the dedicated channels. In order to enable such procedures, a new measurement event from the user is included in Release 5 to inform UTRAN of the best serving HS-DSCH cell.

In the following sub-sections we will briefly discuss the new UE measurement event for support of mobility for HSDPA users, as well as outlining the procedures for intra- and inter-Node B HS-DSCH to HS-DSCH handover. Finally, in Section 11.7.4 we address handover from HS-DSCH to DCH. To further narrow the scope of this section, we only address intra-frequency handovers for HSDPA users, even though inter-frequency handovers are also applicable for HSDPA users, triggered by, for instance, compressed mode measurements from the user, as discussed in Chapter 9.

11.7.1 Measurement Event for Best Serving HS-DSCH Cell

As discussed in Section 9.3, it is the user's serving RNC that determines the cells that should belong to the user's active set for transmission of dedicated channels. The serving RNC typically bases its decisions on requests received from the user that are triggered by measurements on the P-CPICH from the cells in the user's candidate set. Similarly, for HSDPA, a measurement event 1d has been defined, which is called the measurement event for best serving HS-DSCH cell [5]. This measurement basically reports the best serving HS-DSCH cell to the serving RNC based on a measurement of the P-CPICH E_c/I_0 or the P-CPICH received signal code power (RSCP) measurements for the potential candidate cells for the serving HS-DSCH cell, as illustrated in Figure 11.12. It is possible to configure this measurement event so that all cells in the user's candidate set are taken into account, or to restrict the measurement event so that only the current cells in the user's active set for dedicated channels are considered. Usage of a hysteresis margin to avoid fast change of the serving HS-DSCH cell is also possible for this measurement event, as well as specification of a cell individual offset (CIO) to favour certain cells, i.e. for instance, to extend their HSDPA coverage area.

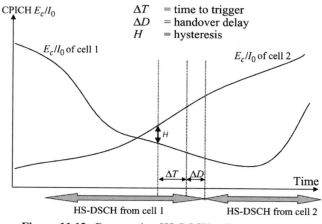

Figure 11.12. Best serving HS-DSCH cell measurement

11.7.2 Intra-Node B HS-DSCH to HS-DSCH Handover

Once the serving RNC decides to make an intra-Node B handover from a source HS-DSCH cell to a new target HS-DSCH cell under the same Node B, as illustrated in Figure 11.13, the

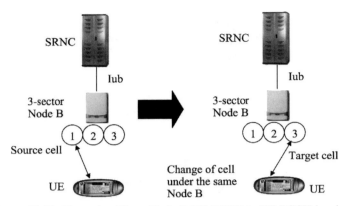

Figure 11.13. Example of intra-Node B HS-DSCH to HS-DSCH handover

serving RNC sends a synchronised radio link reconfiguration prepare message to the Node B, as well as a radio resource control (RRC) physical channel reconfiguration message to the user. At a specified time index where the handover from the source cell to the new target cell is carried out, the source cell stops transmitting to the user, and the MAC-hs packet scheduler in the target cell is thereafter allowed to control transmission to the user. Similarly, the terminal starts to listen to the HS-SCCH (or several HS-SCCHs depending on the MAC-hs configuration) from the new target cell, i.e. the new serving HS-DSCH cell. This also implies that the CQI reports from the user are measured from the channel quality corresponding to the new target cell. It is typically recommended that the MAC-hs in the target cell does not start transmitting to the user until it has received the first CQI report that is measured from the target cell.

Prior to the HS-DSCH handover from the source cell to the new target cell, there are likely to be several PDUs buffered in the source cell's MAC-hs for the user, both PDUs that have never been transmitted to the user and pending PDUs in the Hybrid ARQ manager that are either awaiting Ack/Nack on the uplink HS-DPCCH or PDUs that are waiting to be retransmitted to the user. Assuming that the Node B supports MAC-hs preservation, all the PDUs for the user are moved from the MAC-hs in the source cell to the MAC-hs in the target cell during the HS-DSCH handover. This means that the status of the Hybrid ARQ manager is also preserved without triggering any higher layer retransmission such as, for instance, RLC retransmissions during intra-Node B HS-DSCH to HS-DSCH handover. If the Node B does not support MAC-hs preservation, then handling of the not completed PDU is the same as in the inter-Node B handover case.

During intra-Node B HS-DSCH to HS-DSCH handover, it is likely with a fairly high probability that the user's associated DPCH is potentially in two-way softer handover. Under such conditions, the uplink HS-DPCCH may also be regarded as being in two-way softer handover, so Rake fingers for demodulation of the HS-DPCCH are allocated to both cells in the user's active set. This implies that uplink coverage of the HS-DPCCH is improved for users in softer handover and no power control problems are expected.

11.7.3 Inter-Node B HS-DSCH to HS-DSCH Handover

Inter-Node B HS-DSCH to HS-DSCH handover is also supported by the 3GPP specifications, where the serving HS-DSCH source cell is under one Node B, while the new target

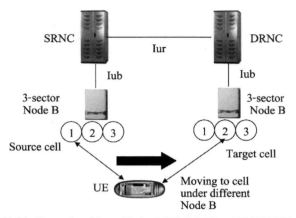

Figure 11.14. Example of inter-Node B HS-DSCH to HS-DSCH handover

cell is under another Node B, and potentially also under another RNC, as illustrated in Figure 11.14. Once the serving RNC decides to initiate such a handover, a synchronised radio link reconfiguration prepare message is sent to the drifting RNC and the Node B that controls the target cell, as well as a radio resource control (RRC) physical channel reconfiguration message to the user. At the time, the cell change is implemented, the MAC-hs for the user in the source cell is reset, which basically means that all buffered PDUs for the user are deleted, including the pending PDUs in the hybrid ARQ manager. At the same time index, the flow control unit in the MAC-hs in the target cell starts to request PDUs from the MAC-d in the serving RNC, so that it can start to transmit data on the HS-DSCH to the user.

As the PDUs that were buffered in the source cell prior to the handover are deleted, these PDUs must be recovered by higher layer retransmissions such as, for instance, RLC retransmissions. When the RLC protocol realises that the PDUs it has originally forwarded to the source cell are not acknowledged, it will initiate retransmissions, which basically implies forwarding the same PDUs to the new target cell that were deleted in the source cell. In order to reduce the potential PDU transmission delays during this recovery phase, the RLC protocol at the user end can be configured to send an RLC status report to the UTRAN at the first time incident after the serving HS-DSCH cell has been changed [6]. This implies that the RLC protocol in the RNC can immediately start to forward the PDUs that were deleted in the source cell prior to the HS-DSCH cell change.

For user applications that do not include any higher layer retransmission mechanisms such as, for instance, applications running over UDP (User Datagram Protocol) and RLC transparent or unacknowledged mode, the PDUs that are deleted in the source cell's MAC-hs prior to the handover are lost forever. For such applications, having large data amounts (many PDUs) buffered in the MAC-hs should therefore be avoided, as these may be lost if an inter-Node B HS-DSCH to HS-DSCH handover is suddenly initiated.

11.7.4   HS-DSCH to DCH Handover

Handover from an HS-DSCH to DCH may potentially be needed for HSDPA users that are moving from a cell with HSDPA to a cell without HSDPA (Release '99 compliant only cell),

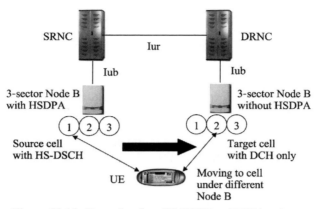

Figure 11.15. Example of an HS-DSCH to DCH handover

as illustrated in Figure 11.15. Once the serving RNC decides to initiate such a handover, a synchronised radio link reconfiguration prepare message is sent to the involved Node Bs, as well as a radio resource control (RRC) physical channel reconfiguration message to the user. In a similar way to the inter-Node B HS-DSCH to HS-DSCH handover, the HS-DSCH to DCH handover results in a reset of the PDUs in MAC-hs in the source cell, which subsequently requires recovery via higher layer retransmissions such as, for instance, RLC retransmissions.

The Release 5 specifications also support implementation of handover from DCH to HS-DSCH. This handover type may, for instance, be used if a user is moving from a non-HSDPA-capable cell into an HSDPA-capable cell, or to optimise the load balance between HSDPA and DCH use in a cell.

Table 11.5 presents a summary of the different handover modes and their characteristics. Notice that the handover delay is estimated to be below 500 ms. The actual handover delay will, in practice, depend on the RNC implementation and the size of the RRC message that is

Table 11.5. Summary of HSDPA handover types and their characteristics

	Intra-Node B HS-DSCH to HS-DSCH	Inter-Node B HS-DSCH to HS-DSCH	HS-DSCH to DCH
Handover measurement		By UE	
Handover decision		By serving RNC (SRNC)	
Packet retransmissions	Packets forwarded from source MAC-hs to target MAC-hs	Packets not forwarded. RLC retransmissions used from SRNC	RLC retransmissions used from SRNC
Packet losses	No	No, when RLC acknowledged mode used.	No, when RLC acknowledged mode used.
Uplink HS-DPCCH	Softer handover can be used for HS-DPCCH	HS-DPCCH received by one cell.	—

sent to the user during the handover phase and the data rate on the Layer 3 signalling channel on the associated DPCH.

11.8 HSDPA Performance

In this section, different performance aspects related to HSDPA are discussed. Since the two most basic features of WCDMA, fast power control and variable SF, have been disabled, a performance evaluation of HSDPA involves considerations that differ somewhat from the general WCDMA analysis. For packet data traffic, HSDPA offers a significant gain over the existing Release '99 DCH and DSCH bearers. It facilitates very fast per-2 ms switching among users, which gives high trunking efficiency and code utilisation for bursty packet services. Further, with the introduction of higher order modulation and reduced channel encoding, even very high radio quality conditions can be mapped into increased user throughput and cell capacity. Finally, advanced packet scheduling, which considers the user's instantaneous radio channel conditions, can produce a very high cell capacity while maintaining tight end-to-end QoS control. In the following sub-sections, single user and multiuser issues are discussed separately. After this description, some examples of HSDPA system performance are given, looking first at the system performance in the 'all HSDPA users' scenario and then looking at the situation when operating the system in emigration phase, where a large number of terminals do not yet have HSDPA capability.

11.8.1 Factors Governing Performance

The HSDPA mode of operation encounters a change in environment and channel performance by fast adaptation of modulation, coding and code resource settings. The performance of HSDPA depends on a number of factors that include the following:

- *Channel conditions*: Time dispersion, cell environment, terminal velocity as well as experienced own cell interference to other cell interference ratio (I_{or}/I_{oc}). Compared to the DCHs, the average I_{or}/I_{oc} ratio at the cell edge is reduced for HSDPA owing to a lack of soft handover gain. Macro cell network measurements indicate typical values down to -5 dB compared to approximately -2 to 0 dB for DCH.

- *Terminal performance*: Basic detector performance (e.g. sensitivity and interference suppression capability) and HSDPA capability level, including supported peak data rates and number of multicodes.

- *Nature and accuracy of radio resource management* (*RRM*): Power and code resources allocated to the HSDPA channel and accuracy/philosophy of Signal to Interference power ratio (SIR) estimation and packet scheduling algorithms.

For a terminal with high detection performance, some experienced SIR would potentially map into a higher throughput performance experienced directly by the HSDPA user.

11.8.2 Spectral Efficiency, Code Efficiency and Dynamic Range

In WCDMA, both spectral efficiency and code efficiency are important optimisation criteria to accommodate code-limited and power-limited system states. In this respect, HSDPA provides some important improvements over Release '99 DCH and DSCH:

- Spectral efficiency is improved at lower SIR ranges (medium to long distance from Node B) by introducing more efficient coding and fast HARQ with redundancy combining. HARQ combines each packet retransmission with earlier transmissions, such that no transmissions are wasted. Further, extensive multicode operations offer high spectral efficiency, similar to variable SF but with higher resolution. At very good SIR conditions (vicinity of Node B), HSDPA offers higher peak data rates and thus better channel utilisation and spectral efficiency.

- Code efficiency is obtained by offering more user bits per symbol and thus more data per channelisation code. This is obtained through higher order modulation and reduced coding. Further, the use of time multiplexing and shared channels generally leads to better code utilisation for bursty traffic, as described in Chapter 10.

The principle of HSDPA is to adapt to the current channel conditions by selecting the most suitable modulation and coding scheme, leading to the highest throughput level. In reality, the available data rate range may be slightly limited at both ends due to packet header overhead and practical detection limitations. The maximum peak data rate is thus often described to be of the order of 11–12 Mbps. The key measure for describing the link performance is the narrowband *signal to interference and noise ratio* (SINR) as experienced by the UE detector (e.g. the received E_s/N_0). In hostile environments, the availability of high SINR is limited, which reduces the link and cell throughput capabilities.

An example SINR-to-throughput mapping function is illustrated in Figure 11.16 for a Pedestrian-A profile with a Rake receiver moving at 3 kmph. The curve includes the first transmission *block error rate* (BLER) and thus considers the basic HARQ mechanism.

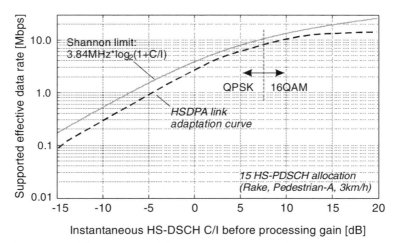

Figure 11.16. SINR to throughput mapping table with a single HS-PDSCH

The HARQ mechanism provides some additional data rate coverage in the lower end, and provides a smoother transmission between the different transport block size settings. On the curve, the operating regions for the two modulation options are also illustrated. As QPSK requires less power per user bit to be received correctly, the available options of higher code rate and multiple HS-PDSCHs are used before switching to 16 QAM. Measured in the SINR domain, the total link adaptation dynamic range is of the order of 30–35 dB. It is comparable to the dynamic range of power control with variable spreading factor, but is shifted in order to work at higher SINR and throughput values. When only a single HS-PDSCH code is employed, the transition curve saturates earlier, at a maximum peak data rate value around 900 kbps. For reference, Figure 11.16 also illustrates the theoretical Shannon capacity for a 3.84 MHz bandwidth. There is an approximate 2 dB difference, due to mainly decoder limitations and receiver estimation inaccuracies.

The single user link adaptation performance depends on other issues, such as CQI measuring, transmission, and processing delays. This adds to the inherent delay associated with the two-slot time difference between the HS-SCCH frame and the corresponding HS-DSCH packet. The minimum total delay is around 6 ms between the time of estimation of the CQI report and the time when the first packet based on this report can be received by the UE. If the UE employs CQI repetition to gain in uplink coverage, this delay increases further. As mentioned earlier, the target BLER for the CQI report is 10 %, but even higher spectral efficiency can be achieved by operating the system at a 1st transmission BLER level of 15–40 %. However, operation of the system at a lower target BLER may be attractive from delay and hardware utilisation considerations, thus, in the simulation in this chapter 10 % is chosen as the target value for 1st transmission BLER. The link adaptation performance when only a single user is being scheduled with a certain average G-factor is depicted in Figure 11.17 for the 15 code case as well (G-factor is the ratio between

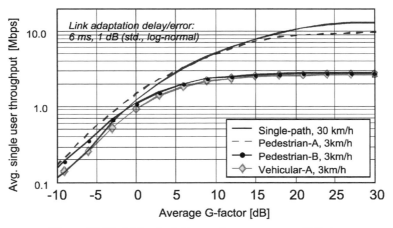

Figure 11.17. Link adaptation performance versus G-factor

wideband received own cell power and other cell interference plus noise). Figure 11.17 assumes the use of non-identical retransmissions and 75 % power allocation for HSDPA use. For a typical macro cell environment, the G-factor near the cell edge is approximately −3 dB, while the median G-factor is around 2 dB. For users in good conditions, the G-factor may be of the order of 12–15 dB. A Rake receiver is assumed and it is seen that this receiver type is limited at low interference levels by the lack of orthogonality in the Pedestrian-B and Vehicular-A environments.

While the HS-DSCH offers high spectral efficiency, it should be noted that at least one (non-power controlled) HS-SCCH is needed to operate the system. This also implies that the data rate carried on the HS-DSCH should be sufficient to compensate for the interference due to the relative HS-SCCH overhead. As mentioned previously, code multiplexing can be used to send HSDPA data to several users within the same TTI by sharing the HS-PDSCH code set between them. Code multiplexing is useful when a single user cannot utilise the total power and/or code resources due to lack of buffered data, or due to the network being able to transmit more codes than the UE supports. Considering the overhead of having multiple parallel HS-SCCHs and the fact that all UEs support a minimum of five codes, it is not expected that more than three users need be code multiplexed in practice, even if all the cell traffic would be using HSDPA. In general, HSDPA offers the best potential for large packet sizes and bit rates. Services resulting in small packet sizes at low data rates, as for instance gaming applications, may therefore be best served using other channel types.

The dependence between the average user throughput per code and the code power is shown in Figure 11.18 for different I_{or}/I_{oc} conditions and different channel profiles using HARQ with soft combining. Owing to the code efficiency inherent in the higher order TFRCs, HS-DSCH supports higher data rates when more power is allocated to the code. However, by noting that the slopes of the curves in Figure 11.18 generally decrease, it is clear that the spectral efficiency degrades as the power is increased. However, if only limited code resources are available, the available power can be utilised better compared to, for instance, DSCH, which is hard-limited to 128 kbps per code at an SF 16 level. To achieve 384 kbps with DSCH, the code resources must be doubled (SF 8). Comparing the difference between the Pedestrian-A and Vehicular-A channel profiles, it is evident that the gain achieved by increasing the power is higher when the terminal is limited by time dispersion. At low values of I_{or}/I_{oc}, the terminal is mainly interference-limited and the two cases become similar.

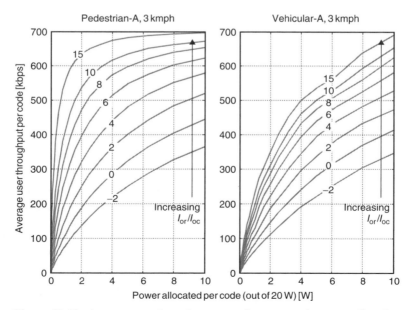

Figure 11.18. Average user throughput per code versus code power allocation

11.8.3   User Scheduling, Cell Throughput and Coverage

The HSDPA cell throughput depends significantly on the interference distribution across the cell, the time dispersion, and the multicode and power resources allocated to HSDPA. In Figure 11.19, the *cumulative distribution function* (CDF) of instantaneous user throughput for both macro cell outdoor and micro cell outdoor–indoor scenarios is considered. The shown CDFs correspond to the case in which fair time scheduling is employed. Fair time scheduling means that the same power is allocated to all users such that users with better channel conditions experience a higher throughput. Figure 11.19 assumes that the available capacity of the cell is allocated to the studied user and that other cells are fully loaded. Note that in the micro cell case, 30 % of the users have sufficient channel quality to support peak data rates exceeding 10 Mbps due to limited time dispersion and high cell isolation. The mean bit rate that can be obtained is more than 5 Mbps. For the macro cell case, the presence of time dispersion and high levels of other cell interference widely limits the available peak data rates. Nevertheless, peak data rates of more than 512 kbps are supported 70 % of the time and the mean bit rate is more than 1 Mbps. For users located in the vicinity of the Node B, time dispersion limits the maximum peak data rate to around 6 to 7 Mbps.

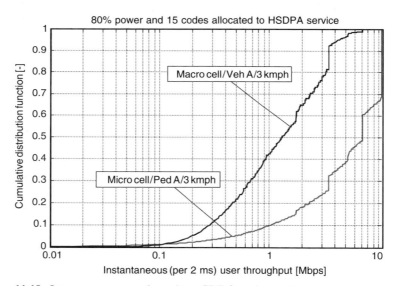

Figure 11.19. Instantaneous user throughput CDF for micro cell and macro cell scenarios

As discussed earlier, with 16 QAM, the channel estimation is more challenging than with QPSK and thus it is not usable in all cell locations. With a macro cell environment (with Vehicular-A channel model) the probability for using 16 QAM is between 5 % and 10 %, assuming the terminal has a normal Rake receiver. When the delay profile is more favourable and cell isolation is higher with a micro cell environment, then the probability increases to approximately 25 % with the Pedestrian-A environment. The value of 30 % for the cell area in Figure 11.19 lacks some imperfections, such as the interference between code channels due to hardware imperfections, which shows more in the Pedestrian-A type environment, where the orthogonality is well preserved by the channel itself.

The chosen packet scheduling method has a significant impact on the overall cell throughput and the end-user perceived QoS. This aspect is related to the gain by *multiuser diversity*. With fast scheduling and multiple users it is possible at any given time to pick the 'best' user in the cell, e.g. a selection diversity mechanism that may be of very high order. The concept of multiuser diversity is illustrated in Figure 11.20(a). Such a means of scheduling is denoted advanced or opportunistic packet scheduling, as opposed to blind packet scheduling methods that do not consider the radio conditions. Examples of the latter type are the round robin in time and fair throughput packet schedulers. Probably the most straightforward and aggressive advanced packet scheduler is the maximum-throughput or maximum-C/I packet scheduler, which always schedules the user with the best instantaneous channel quality. Its principle is depicted in Figure 11.20(b). The main drawbacks of this scheduler are its inherent unfairness and coverage limitations. Several publications list different HSDPA packet scheduler options, including [7,8].

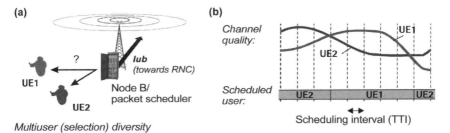

Figure 11.20. Illustration of (a) the multiuser diversity principle and (b) scheduling to the user with the highest instantaneous radio channel quality

One of the often referred to fast scheduling methods is the *proportional fair* algorithm [9,10], which offers an attractive trade-off between user fairness and cell capacity. The absolute performance of proportional fair scheduling with WCDMA/HSDPA has been studied in several references, including [11–13]. The proportional fair scheduling idea is to schedule users only when they experience good instantaneous channel conditions (e.g. they experience constructive fading); thereby improving both the user throughput and cell throughput for time-shared channels. To identify the best user for scheduling, a relative channel quality indicator is calculated for each user as the ratio:

$$\text{Scheduling metric} = \frac{\text{User's instantaneously supported data rate}}{\text{User's average served throughput}} \qquad (11.1)$$

Thus, a user is prioritised either if (1) he/she has good instantaneous conditions compared to the average level or (2) the user has been served with little throughput in the past. The latter ensures scheduling robustness such that users with static channel conditions are also supported. To compute this scheduling metric, the packet scheduler utilises the CQI information as well as the information from the previous transmissions. In deploying the proportional fair packet scheduler, the averaging function must be designed to take the service requirements into account to establish the right trade-off between delays and convergence of the algorithm [11].

The proportional fair scheduling method results in all users getting approximately an equal probability of becoming active, even though they may experience very different average channel quality [14]. The performance of advanced scheduling can be modified to meet applicable QoS requirements, see e.g. [8,15, 16]. While the proportional fair method in Equation (11.1) places high emphasis on the users near the cell edge, it still offers an uneven data distribution across the cell.

The HSDPA bearer capacity gain for proportional fair scheduling over simple round robin in time scheduling is of the order of 40–60% for macro cell environments, and can be theoretically higher for certain environments and operating conditions [17]. These gain figures assume that the user selection diversity order is higher than, e.g., 6–10. If users have low service activity cycles, this means that the physical number of users needs to be larger for a high scheduling gain. Another fundamental requirement for the proportional fair method to give a significant system gain is that the channel variations must be slow enough such that the scheduler can track the channel conditions when considering inherent link adaptation and packet scheduling delays. Previous studies indicate that significant performance is achieved as long as the UE velocity is less than around 25–30 km/h [12]. Beyond this point, the proportional fair scheduler gives a performance similar to the traditional round robin in time scheduler. Further, the user's channel conditions should be changing fast enough that packet delay requirements do not prevent the scheduler from waiting for the following constructive fade for the user.

Figure 11.21 presents the relative performance between Release '99 and HSDPA in two different environments. As seen from the numbers, HSDPA increases the cell throughput more than 100 % compared to the Release '99 in the macro cell case and in the micro cell case, the gain of HSDPA exceeds 200 % (even up to 300 %) owing to the availability of very high user peak data rates. However, for the most extreme cases, the practical imperfections associated with the terminal and Node B hardware, link adaptation and packet scheduling may limit the achievable cell throughput in practice. Further, it is assumed that in favourable conditions, a user will always utilise the available throughput. The application level impacts with HSDPA are contained in Chapter 10.

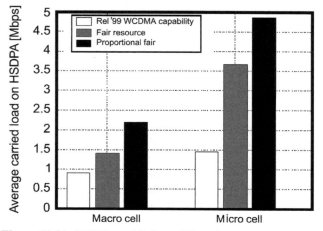

Figure 11.21. HSDPA and Release '99 performance comparison

Another important area for observation is the coverage with HSDPA. The downlink coverage is of interest in terms of what kind of data rate can be offered at the cell edge. As such, the downlink data rate will adapt automatically to the coverage situation, based on the CQI feedback from the UE.

As the HS-DSCH does not employ a fast power control, the coverage is defined as the area over which the average user throughput is of some value. The average user data rate coverage follows the I_{or}/I_{oc} distribution of the cell and the amount of time dispersion. The user data rate downlink coverage for a macro cell scenario, including significant AMC errors, is illustrated in Figure 11.22. Compared to cell throughput capacity, the single user data rate coverage is significantly lower, since there is no gain of switching between users with favourable channel conditions, however, the total cell capacity can still benefit from the operation in a soft handover area, assuming reasonable scheduling and not too tight timing constraints for the scheduler operation. The flexible support for different handover types for HSDPA-capable users makes it possible to obtain full coverage and mobility for HSDPA users receiving data on the HS-DSCH within an area that is covered by HSDPA-capable Node Bs. This implies that even users with an active set size larger than one (i.e. the associated DPCH is in soft handover) can receive data on the HS-DSCH, and thereby benefit from the higher data rate supported for this channel type compared to DCH. The HS-DSCH is still more spectral efficient than the DCH in soft handover (benefits from soft handover macro and micro diversity gain), since it benefits from fast link adaptation, effective time diversity and soft combining from the Layer 1 Hybrid ARQ scheme, and multiuser diversity from using fast Node B scheduling.

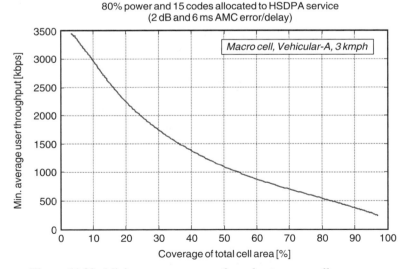

Figure 11.22. Minimum average user throughput versus cell coverage

The uplink data coverage as such is not directly impacted by HSDPA operation, but there needs to be sufficient power margin available in the uplink for the signalling in the HS-DPCCH, as well as for the associated DCH. In the case of TCP/IP-based traffic, e.g. web browsing, the uplink traffic consists, in addition to the application data, of the TCP/IP acknowledgements. Depending on the TCP/IP block size, the resulting uplink data rate will vary, e.g. from 16 kbps onwards for the 500 kbps downlink data rate. These acknowledgements need to be carried by the uplink as well as the minimum necessary Layer 2/3 signalling (e.g. handover-related measurements), thus, uplink planning should have coverage roughly equal to 64 kbps data rate. This ensures that downlink throughput is not compromised by the poor uplink performance due to missing or delayed TCP/IP acknowledgements. On top of this, possible service multiplexing (e.g. with AMR speech service, as shown in Figure 11.23) needs may also need to be accommodated. Note that Layer 2/3 signalling and the HS-SCCH are not shown in the figure. The exact value to be used in cell planning will depend on many parameters, including the power offsets and repetition factors for the HS-DPCCH ACK/NACK and CQI fields. With the 3.6 Mbps or higher peak rates there is obviously going to be a need for more TCP/IP acknowledgements, but those data rates are not expected to be available at the cell edge in any case.

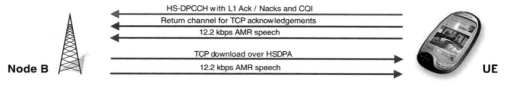

Figure 11.23. Different data flows between the UE and Node B

Besides the transmission itself, the addition of a new code channel will increase the peak-to-average ratio of the terminal transmission when HS-DPCCH is present. This causes the terminal to use more back-off to maintain the required spectrum mask for the transmission. The specifications are expected (the topic is still being addressed in 3GPP) to allow the terminal transmission power to be reduced by at most 1 dB in cases when the DPCCH/DPDCH power ratio is reasonable, e.g with user data rates around 32 kbps. With the high-power DPDCH with higher data rates, the transmission power is not reduced at all. This allows the configurations with DPDCH (user) data rate in the order of 64 kbps or higher to have no additional impact on the link budget, except the actual power needed for HS-DPCCH and for DCH transmission. For very low data rates such as 16 or 32 kbps with the 1 dB reduction, the uplink connection will not suffer range problems if the network was dimensioned to enable an uplink transmission rate of 64 kbps or more in the whole network.

11.8.4 HSDPA Network Performance with Mixed Non-HSDPA and HSDPA Terminals

Typically the WCDMA networks shall start using HSDPA when there is a large existing user base in place. Thus, it becomes essential to understand the HSDPA performance in the case of mixed non-HSDPA mobiles and HSDPA mobiles. System level simulation results are studied for the example case where traffic is carried on both Release '99 dedicated channels and over Release 5 HSDPA on the same carrier, and 5 HS-PDSCH codes are allocated for the HSDPA user with a single HS-SCCH. Release '99 channels can use the remaining code resources. This provides a maximum peak data rate of 3.6 Mbps with 16 QAM and allows one user to be scheduled at a time.

The simulation results are obtained from dynamic cellular network simulations, where users are moving within an area covered by many three-sector Node Bs, as illustrated in Figure 11.24. Dynamic models for the user mobility, traffic models, variations of the radio propagation conditions, etc, are used. The ITU Pedestrian-A delay profile is used, while otherwise the setting is closer to a macro cell environment.

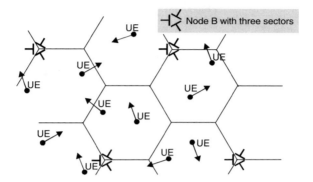

Figure 11.24. Network topology for the reference simulation set-up

All traffic on DCHs uses a constant data rate of 64 kbps, with power control. The data rate on the HS-DSCH is adjusted for every TTI as a function of the CQI received. The proportional fair scheduler, as discussed earlier, is used with HARQ, assuming soft combining. The available Node B transmit power for HS-PDSCH and HS-SCCH codes is fixed in each simulation to a value in the range from 3 W to 9 W, with 20 W total power.

Let us first consider the total average cell capacity that can be achieved with such a system configuration, assuming that the offered traffic in the network is sufficiently high that the HS-DSCH is utilised in every TTI, and the average power allocated to transmission of DCH is used. Figure 11.25 shows the average cell throughput for the Release '99 DCH and Release 5 HS-DSCH as a function of the power that is allocated to HSDPA transmission. The total cell throughput (i.e. the sum of the Release '99 DCH and Release 5 HS-DSCH throughput) is also plotted. It is observed that the HS-DSCH throughput increases when the HSDPA power is increased, while the DCH throughput decreases as less and less power becomes available for transmission of such channels.

At 7 W HSDPA power, we can achieve an average cell throughput of 1.4 Mbps on the HS-DSCH, and an average cell throughput of approximately 440 kbps on the Release '99 DCHs. With only non-HSDPA terminals active in the cell and no power/codes reserved for HS-PDSCH/HS-SCCH transmission, we are able to achieve an average cell throughput of 1.0 Mbps. This implies that with HSDPA enabled, the cell throughput is increased by a factor of 1.7, which is basically equivalent to an average gain in cell throughput of 70 %. The capacity gain is achieved mainly from the multiuser diversity gain offered by the fast MAC-hs proportional fair scheduler and the higher spectral efficiency on the HS-DSCH by using fast link adaptation with adaptive modulation and coding, as well as the improved Layer 1 Hybrid ARQ scheme with soft combining of retransmissions. If the radio channel power delay profile is more challenging, such as Vehicular-A, a similar gain is also observed, though the absolute values are lower.

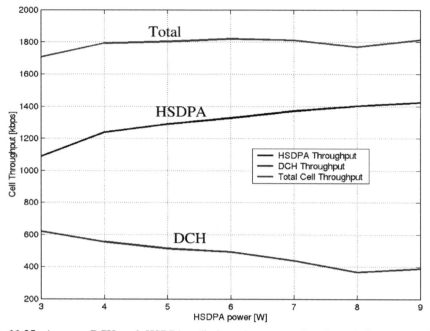

Figure 11.25. Average DCH and HSDPA cell throughput as a function of the power allocated to HSDPA

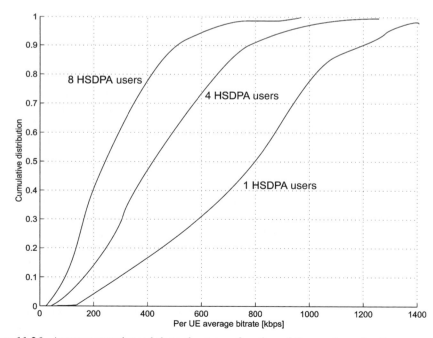

Figure 11.26. Average experienced throughput as a function of the number of active users per cell

The average throughput that the HSDPA users experience depends on the number of simultaneous users that are sharing the HS-DSCH channel, as well as their relative experienced signal quality, i.e. symbol energy to noise plus interference ratio (E_sN_0).

Figure 11.26 shows the cumulative distribution function (cdf) of the average experienced throughput per HSDPA user, depending on the number of simultaneous active users sharing the HS-DSCH. These results are obtained for the case where 7 W and 5 HS-PDSCH codes are allocated to HSDPA transmission. For the case with only one active HSDPA user in the cell, the average experienced per user throughput is 800 kbps at the median, and with a 10 % probability it is higher than 1.3 Mbps (typically observed for those users that are close to the Node B). When increasing the number of simultaneously active HSDPA users to four, the median per user throughput is decreased to approximately 400 kbps, because more users have to share the available capacity on the HS-DSCH. However, notice that the median throughput is only decreased by a factor of two when increasing the number of users from one to four. This behaviour is observed because HSDPA benefits from fast scheduling multiuser diversity gain when four users are present, while there is, of course, no such gain available for the single user scenario. For eight simultaneous active HSDPA users, the achievable median per user throughput is of the order of 220 kbps. Hence, the experienced per user throughput depends strongly on the number of simultaneously active HSDPA users that are sharing the HS-DSCH.

11.9   Terminal Receiver Aspects

The terminal receiver aspects were discussed earlier in the chapter, since one of the new challenges is the need for amplitude estimates for 16 QAM detection. However, there are other challenges coming from the use of 16 QAM as well. A good quality voice call in WCDMA typically requires a C/I of -20 dB compared to 10 dB for GSM. Since the interference, including the inter-symbol interference, can be 20 dB above the signal level, the WCDMA voice signal is very robust against interference and does not benefit significantly from equalisers. However, for the high peak data rates provided with HSDPA service, higher C/I (E_b/N_0) values above 0 dB are required and, consequently, the signal becomes less robust against inter-symbol interference.

Hence, the HSDPA concept with 16 QAM transmission potentially benefits from equaliser concepts that reduce the interference from multipath components. The multipath interference cancellation receiver shown in Figure 11.27 was discussed and analysed in [1]. The same receiver front-end as employed in the Rake receiver is used as a pre-stage to provide draft symbol estimates. Those estimates are then used to remove the multipath interference from the received signal, and new symbol estimates can be obtained with the same

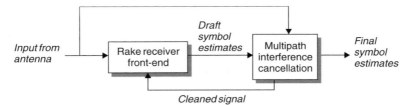

Figure 11.27. Example multipath interference cancellation

matched filter. After a few iterations, the final symbol estimates are calculated. Another type of advanced receiver is a linear equaliser. The advanced receiver algorithms (with uplink focus) are discussed in more detail in Section 12.6.

Advanced receivers make it possible to provide higher bit rates in multipath channels compared to what is achievable with normal Rake receivers. On the other hand, the complexity of such receivers is significantly higher than for the standard Rake receiver. In 3GPP standardisation there is no intention to specify any receiver solutions, but only performance requirements in particular cases.

During 2004, work on the improved HSDPA performance requirements is to be started in 3GPP, with the focus being on two technology directions, advanced receivers and RX diversity. RX diversity can improve the performance of HSDPA when diversity is small, but the link level improvements with RX diversity (or additional diversity in general) are not necessarily additive with the gains from the scheduling. The advanced receiver battles the inter-symbol interference and thus makes the 16 QAM especially usable more often by enabling higher data rates, especially in the vehicular type of environment.

11.10   Evolution Beyond Release 5

As described previously, the HSDPA concept of Release 5 is able to provide a clear increase in the WCDMA downlink packet data throughput. It is obvious that further enhancements on top of the HSDPA feature can be considered for increased user bit rates and cell throughput. Possible techniques raised previously include further improvements in the downlink with advanced antenna techniques and applying similar techniques to HSDPA also for the uplink direction, which are briefly discussed in this section. The second edition of this book also contained fast cell selection (FCS) which was then determined in 3GPP not worth adding to the specifications.

11.10.1   Multiple Receiver and Transmit Antenna Techniques

Using several transmitter antennas in the Node B and several receiver antennas in the terminal can increase the HSDPA bit rates. Such approaches are commonly denoted as multiple input multiple output (MIMO) techniques. Higher data rates can be achieved either (1) by an improved antenna transmit and receive diversity leading to better channel quality or (2) by reusing the spreading code on different antennas (higher throughput per code due to data layering). To distinguish between several sub-streams sharing the same code, the terminal uses multiple antennas and spatial signal processing. An example of a MIMO receiver with two antennas is shown in Figure 11.28. The space–time Rake combiner is the multiple antenna generalisation of the conventional Rake combiner. As seen from the micro

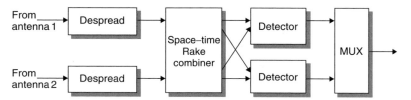

Figure 11.28. Example MIMO receiver

cell results that were shown earlier, up to 20 to 30 % of the users may have a channel quality that exceeds the requirements for 10 Mbps. For this scenario, MIMO schemes could potentially increase the system performance. However, inherent complexity and sensitivity issues must be considered in this context as well. As such, the MIMO technology will be studied further for future releases.

11.10.2 High Speed Uplink Packet Access (HSUPA)

Following the creation of the Release 5 specifications with HSDPA during the first half of 2002, the next natural direction of interest was to see if something similar could be done for the uplink. The 3GPP Release 6 feasibility study for Enhanced Uplink for UTRA FDD was started in Autumn 2002, with the focus on evaluating potential performance enhancements for the uplink dedicated transport channels. The scope of the study was either to enhance the uplink performance in general, or to enhance the uplink performance for background, interactive and streaming traffic classes. More concretely, the target was to improve the uplink air interface capacity utilisation and end user experience by increasing both the cell throughput and the coverage of the higher bit rates in the uplink. For the uplink, increasing the coverage of higher data rates, already now possible, was of greater interest than increasing the theoretical peak data rates. Possibilities for delay reductions in the packet data transmission and in setting up the dedicated connections were also studied. 3GPP concluded the studies in March 2004 and, consequently, a work item was initiated.

The techniques that were studied for enhancing the uplink dedicated transport channels were, for the most part, the familiar ones discussed earlier in this chapter for HSDPA, such as:

- Fast HARQ terminated at Node B;
- Fast Node B based uplink scheduling;
- Higher order modulation.

Unfortunately, the fundamental differences between uplink and downlink data transmission make it impossible simply to introduce the HSDPA solutions in the uplink as they are already seen in the downlink. The key difference between uplink and downlink is the handling of the total transmission power resource. In the downlink direction, the power resource is centralised, while in the uplink, the power resource available for an individual user is limited by the terminal power amplifier capabilities. Thus, it can be claimed that a pure time division approach, which is in place with a maximum data rate in HSDPA, would not make sense in the WCDMA uplink. Further with uplink, the power control operation has much larger dynamics compared to the downlink, where the interference between the codes limits the dynamics, thus giving an occasional 'free lunch' for the higher order modulation/coding as the symbol energy cannot be reduced after a certain point. The benefit from higher order modulation in HSDPA also comes from avoiding channelisation code limitations, while uplink, with user-specific scrambling, could utilise more codes and remain with lower order modulation than 16 QAM. Additionally, the power control cannot be abandoned in the case of continuous uplink transmission due to the near–far problem.

Another area requiring specific attention is the operation in soft handover, where the receiving base station in the uplink will vary as a function of the terminal movement and changes in radio conditions.

Thus, it is foreseen that benefits can be obtained but they are not expected to be of same order of magnitude as has been achieved with HSDPA.

In the following sections a more detailed look is taken at the techniques that were studied in 3GPP for enhancing the performance of the uplink DCH. The feasibility study was finalised in March 2004 and a work item was set up. The resulting improvements are to be reflected in forthcoming WCDMA specification Releases, with an expected completion date between the end of 2004 and mid 2005. The work will focus on the HARQ, uplink scheduling and TTI/channel structure definition work. The work has a few different terms associated with it, namely High Speed Uplink Packet Access (HSUPA) or Enhanced Uplink DCH, E-DCH. The conclusions from the 3GPP feasibility study phase can be found in [18].

11.10.2.1 Fast Hybrid ARQ in the Uplink

The simple principle of fast HARQ is to allow the Node B to ask for the UE to retransmit the uplink packet if it was not received correctly. Moreover, the Node B can use different methods for combining the multiple transmissions of a single packet, hence reducing the required received E_c/N_0 of individual transmissions. An N-channel stop-and-wait HARQ protocol, similar to the one used in HSDPA, is also considered for the uplink. Different HARQ protocols and combining methods with HSDPA were covered earlier in this chapter. The operation of uplink HARQ retransmission in comparison to Release '99 RLC level retransmission is depicted in Figure 11.29.

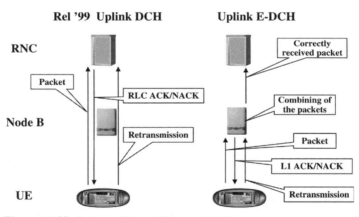

Figure 11.29. Release '99 and Enhanced Uplink retransmission control

With fast HARQ, the BLER target of the first transmission could be significantly higher, as the delay experienced from the retransmission of an erroneously received packet is dramatically reduced when compared to RLC level retransmission. A higher BLER target reduces the UE transmission power required for a given data rate. Hence, for the same cell loading, the cell capacity can be increased. Considering a fixed data rate, lower energy per bit contributes to range improvement as well. There is, however, a penalty in increasing the BLER target too much, since having a large number of retransmissions starts to show in the average delay seen by the UE, even if the peak delay values are not seen very often due to avoidance of RLC retransmission. The effective throughput for a fixed data rate is also reduced with high BLER, as an increasing number of packets are transmitted more than once.

Allowing a large number of physical layer retransmissions makes the RLC level retransmission probability close to zero. This would benefit especially delay and delay variance sensitive services, as the slow RLC retransmissions would be eliminated. If the initial BLER target was already low, e.g. below 10 %, the average delay is not reduced that drastically, as most of the transmissions will anyway go through at the first attempt. However, the maximum delay seen from the case when one or more RLC retransmissions are needed reduces significantly. Hence, maximum delay and delay variance are reduced but the cell capacity and data rate coverage are not affected. Finding a balance between the different parameterisation of the HARQ, penalties in the system level and gains in the link level requires careful study, and it could prove beneficial to set the operating point for the HARQ separately for different kinds of service.

How the uplink HARQ should operate when the UE is in SHO is currently under discussion, but a gain from HARQ in SHO has been shown in [18]. HARQ operation in SHO introduces one additional complexity component not present in HSDPA HARQ. In a CDMA system, the gain from soft handover comes from one Node B receiving a packet correctly, while another Node B is failing in decoding. Hence, one Node B sends a positive acknowledgement to the UE and the other sends a negative acknowledgement. In such a case, the network has received the packet and the UE should not send the same packet again, as shown in Figure 11.30, and respectively, the HARQ procedure in the Node B with failed

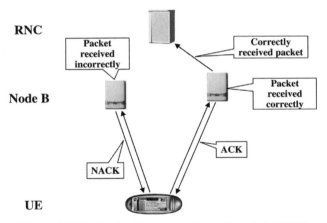

Figure 11.30. Soft handover operation with uplink HARQ

packet should recover from this. The RNC needs to ensure the in-sequence delivery to higher layers and to make the selection combining for the packets received from different Node Bs. An additional problem due to the limited UE power is the possibility of the UE not having enough transmission power to maintain the same data rate for retransmission that was used for the initial transmission.

11.10.2.2 Fast Packet Scheduling in the Uplink

Currently, the uplink scheduling is RNC based, and thus, due to signalling delays, its nature is more statistical than dynamic. The RNC gives the UEs a set of data rates (TFCs) based on the uplink load measurement reports from the Node Bs and traffic volume measurement reports from the UEs. Due to the higher layer signalling required between the nodes, the scheduling delay and scheduling period cannot be very short. To ensure that the uplink remains non-congested all the time, a relatively conservative data rate allocation approach must be applied. Otherwise, the probability of uplink overload becomes too high.

With the physical layer scheduling in Node B, the scheduling period could be shorter and the physical layer measurement information readily available in Node B could be used as a basis for scheduling. This enables more up-to-date scheduling decisions and, thus, better use of the available uplink air interface capacity.

Significant shortening of the scheduling periods allows for more dynamic control over the uplink air interface capacity. Whenever one UE stops transmitting or reduces its transmission data rate, the freed capacity can quickly and efficiently be allocated to other UEs. The side benefit of this is that the target operating point of the uplink load can be closer to the maximum load level without increasing the probability of overloading the uplink, as the uplink noise rise probability distribution can be made narrower, as shown in Figure 11.31.

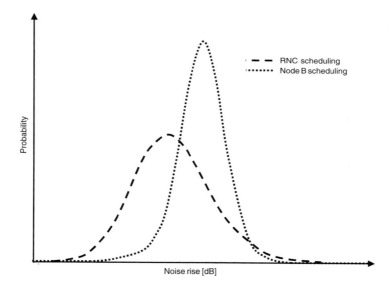

Figure 11.31. Probability distribution functions of the uplink noise rise

Figure 11.32 illustrates the possibility of operating closer to the maximum load limit due to the smaller marginal load area. When faster scheduling also covers uplink congestion control, the Prx_target can be moved closer to Prx_threshold, which is considered to be the overload limit. Node B based scheduling needs to have information on the UE uplink transmission needs, as well as having (fast) signalling to inform the UE of the allowed data rate at a given time. For the RNC, these means exist via RRC signalling, which obviously has limitations for the reaction speed. Figure 11.33 illustrates the current scheduling and the proposed concepts for E-DCH operation.

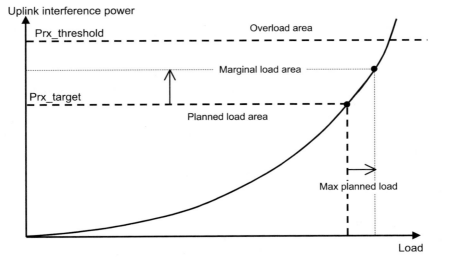

Figure 11.32. Uplink load curve and fast scheduling impact on maximum planned load

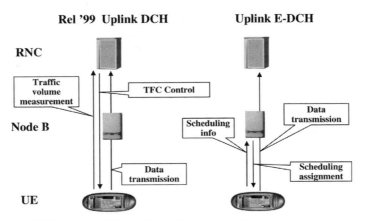

Figure 11.33. Release '99 and Node B schedulers have the same components

11.10.2.3 Changes to the Physical Channel Structure

The fundamental question in introducing an enhanced dedicated channel in the uplink is, how will the Release '99 channel structures be affected? The uplink data transmission may remain unchanged and only the signalling may give new requirements to Layer 1. However, it is just as possible that an enhanced dedicated physical data channel will be introduced parallel to existing DPDCH.

Similarly, several options on how to convey the new signalling over the air interface are being investigated, as follows:

- Uplink signalling. The alternatives on the table are to have the signalling embedded on the uplink DPDCH or DPCCH, or to have a separate code channel added, similar to HS-DPCCH with HSDPA.

- Downlink signalling. Similarly, using the DPDCH and DPCCH can be considered, as well as a shared (e.g. similar to HS-SCCH) or separate new dedicated code channel for signalling.

For the downlink, a dedicated code channel is practically out of the question due to code limitations in the downlink. Also, the modification of HS-SCCH needed for HSDPA transmission would not be backwards compatible with Release 5 HSDPA UEs and is most probably out of the question. The more uplink code channels are introduced, the worse the UE peak to average power ratio will grow, so there is a motivation not to let the number of uplink code channels grow without any limits.

11.10.2.4 Transmission Time Interval Duration

Finally, introducing a shorter Transmission Time Interval (TTI) for uplink is also under discussion. The motivation for the 2 ms TTI introduced for HSDPA downlink was due to the lack of power control being employed for HS-PDSCH transmission. Such motivation is absent in the uplink as disabling uplink power control would reduce the system capacity. The motivation for a shorter TTI in the uplink is mainly due to the possibility of reducing HARQ retransmission delay. The challenge for the 2 ms TTI comes from the resulting uplink range with limited power resources in the terminal. With an equal amount of data in the TTI, it is possible to transmit less energy during 2 ms than 10 ms. Additionally, the interleaving gain is reduced when moving to 2 ms. Thus, even if shorter TTIs were used, for the cell edge operation, the use of 10 ms TTI will be part of the specification.

11.11 Conclusion

In this chapter the HSDPA concept was introduced and its performance was considered. The main aspects discussed can be summarised as follows:

- The HSDPA concept utilises a distributed architecture in which the processing is closer to the air interface at Node B for low delay link adaptation.

- The HSDPA concept provides a 50–100 % higher cell throughput than the Release '99 DCH/DSCH in macro cell and more than 100 % gain in micro cell scenarios. For micro cell, the HS-DSCH can support up to 5 Mbps per sector per carrier, that is, 1 bit/s/Hz/cell.

- The HSDPA concept offers more than 100 % higher peak user bit rates than Release '99, and the difference is even larger if observing the maximum downlink DCH data rate supported by the networks as being 384 kbps. HS-DSCH bit rates are comparable to Digital Subscriber Line (DSL) modem bit rates. The mean user bit rates in a large macro cell environment can exceed 1 Mbps and in small micro cells 5 Mbps.

- The HSDPA concept is able to support efficiently not only non-real time UMTS QoS classes, but also real time streaming UMTS QoS classes with guaranteed bit rates.

- The use of HSDPA also provides significant benefits in the case of mixed terminal deployment, with cell code and power resources shared between HSDPA and non-HSDPA users.

- The applicability of HSDPA techniques for the uplink direction is under investigation in 3GPP for inclusion in forthcoming Releases, though the expected gain is not foreseen to be as big an improvement as was achieved with HSDPA in the downlink. A work item is ongoing to specify the HSUPA improvements.

References

[1] 3GPP Technical Report 25.848, Physical layer aspects of UTRA High Speed Downlink Packet Access, version 4.0.0, March 2001.

[2] 3GPP Technical Specification 25.211, Physical Channels and Mapping of Transport Channels onto Physical Channels (FDD), version 5.0.0, March 2002.

[3] 3GPP Technical Specification 25.212, Multiplexing and Channel Coding (FDD), version 5.0.0, March 2002.

[4] 3GPP Technical Specification 25.306, UE Radio Access Capabilities, version 5.1.0, June 2002.

[5] 3GPP Technical Specification 25.331, Radio Resource Control (RRC), Release 5, December 2003

[6] 3GPP Technical Specification 25.322, Radio Link Control (RLC), December 2003.

[7] Elliot, R. C. and Krzymien, W. A. 'Scheduling Algorithms for the cdma2000 Packet Data Evolution', *Proceedings of the IEEE Vehicular Technology Conference (VTC)*, Vancouver, Canada, September 2002, vol. 1, pp. 304–310.

[8] Ameigeiras, P. '*Packet Scheduling and Quality of Service in HSDPA*', Ph.D. thesis, Department of Communication Technology, Aalborg University, Denmark, October 2003.

[9] Kelly, F. 'Charging and Rate Control for Elastic Traffic,' *European Transactions on Telecommunications*, 1997, vol. 8, pp. 33–37.

[10] Jalali, A., Padovani, R. and Pankaj, R. 'Data Throughput of CDMA–HDR High Efficiency–High Data Rate Personal Communication Wireless System,' *Proceedings of Vehicular Technology Conference (VTC)*, Tokyo, Japan, May 2000, vol. 3, pp. 1854–1858.

[11] Kolding, T. E. 'Link and System Performance Aspects of Proportional Fair Scheduling in WCDMA/HSDPA', *Proceedings of 58th IEEE Vehicular Technology Conference (VTC)*, Florida, USA, October 2003, vol. 2, pp. 1454–1458.

[12] Ramiro-Moreno, J., Pedersen, K. I. and Mogensen, P. E. 'Network Performance of Transmit and Receive Antenna Diversity in HSDPA under Different Packet Scheduling Strategies,' *Proceedings of 57th IEEE Vehicular Technology Conference (VTC)*, Jeju, South Korea, April 2003, vol. 2, pp. 1454–1458.

[13] Parkvall, S., Dahlman, E., Frenger, P., Beming, P. and Persson, M. 'The High Speed Packet Data Evolution of WCDMA,' *Proceedings of the 12th IEEE Symposium of Personal, Indoor, and Mobile Radio Communications (PIMRC)*, San Diego, California, USA, September 2001, vol. 2, pp. G27–G31.

[14] Holtzman, J. M. 'Asymptotic Analysis of Proportional Fair Algorithm', *IEEE Proc. Personal Indoor Mobile Radio Communications (PIMRC)*, Septermber, 2001, pp. F33–F37.

[15] Andrews, M., Kumaran, K., Ramanan, K., Stolyar, A. and Whiting, P. 'Providing Quality of Service over a Shared Wireless Link,' IEEE Communications Magazine, February 2001, vol. 39, no. 2, pp. 150–154.

[16] Kolding, T. E., Pedersen, K. I., Wigard, J., Frederiksen, F. and Mogensen, P. E. 'High Speed Downlink Packet Access: WCDMA Evolution', IEEE Vehicular Technology Society (VTS) News, February 2003, vol. 50, no. 1, pp. 4–10.

[17] Hosein, P. A. 'QoS Control for WCDMA High Speed Packet Data', International Workshop on Mobile and Wireless Communications Networks, Stockholm, Sweden, September 2002, pp. 169–173.

[18] 3GPP Technical Report, TR 25.896, Feasibility Study for Enhanced Uplink for UTRA FDD, Release 6, version 6.0.0, March 2004.

12

Physical Layer Performance

Harri Holma, Jussi Reunanen, Leo Chan, Preben Mogensen, Klaus Pedersen,
Kari Horneman, Jaakko Vihriälä and Markku Juntti

12.1 Introduction

This chapter presents coverage and capacity results and investigates the impact of the radio
propagation environment, advanced base station (Node B) solutions, and WCDMA physical
layer parameters. The advanced base station solutions include both baseband and antenna
techniques. Cell coverage is especially important in terms of the initial network deployment
investment, where the offered traffic load does not set demands for a fine grid of cells;
WCDMA network coverage is analysed in Section 12.2. The importance of capacity will
increase after the initial coverage deployment when the amount of traffic increases;
WCDMA capacity is presented in Section 12.3. In this chapter, we also present the
WCDMA air interface capacity, which is limited by interference. We here assume that
there are sufficient baseband hardware resources in Node B, transmission network capacity,
and RNC resources to support the maximum air interface capacity. The results from field
capacity trials are presented in Section 12.4. The 3GPP performance requirements are
analysed in Section 12.5, and Section 12.6 presents the possible performance enhancements
that are supported by the 3GPP standard, including adaptive antenna structures and multiuser
detection with advanced baseband processing. The focus areas of this chapter are illustrated
in Figure 12.1.

The radio network planning and the optimisation also have a major impact on coverage
and capacity. Their effects are presented in Chapters 8 and 9. The sharing of the capacity
between simultaneous users by the packet scheduler is discussed in Chapter 10.

12.2 Cell Coverage

Cell coverage is important when the offered traffic to a cell is insufficient to fully utilise the
operator's available spectrum; this is typically the case even for urban areas at initial network
deployment phase, and remains the case for rural areas with low traffic density. For
conversational class (i.e. symmetrical bit rate allocation) the macro cell coverage is

WCDMA for UMTS, third edition. Edited by Harri Holma and Antti Toskala
© 2004 John Wiley & Sons, Ltd ISBN: 0-470-87096-6

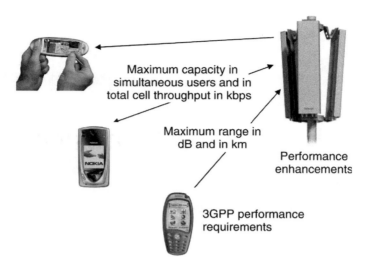

Figure 12.1. Focus areas of this chapter

determined by the uplink direction, because the transmission power of the UE is much lower than that of the macro cell Node B. The output power of the UE is typically 21 dBm (125 mW) and that of the macro cell Node B 40–46 dBm (10–40 W) per sector (cell). Hence, in this section we focus on uplink coverage. Also, in Section 8.2.2 the macro cell coverage was shown to be limited by uplink link budget.

Let us first consider how the link budget improvement in dB impacts the maximal cell coverage area [km^2]. We here assume an improvement in link budget of ΔL dB, for example, by deploying advanced base station antenna techniques, mast head amplifier, or improved base station sensitivity by other techniques. The effect of the improvements in the link budget, ΔL, on the relative cell radius increase, $\Delta R/R$, can be calculated assuming a macro cell propagation model, for example the Okumura–Hata model from Section 8.2. R is the default cell range and ΔR is the relative increase in cell range. In this example, the path loss exponent is set to 3.52, which results in

$$\Delta L = 35.2 \log_{10}\left(1 + \frac{\Delta R}{R}\right) \tag{12.1}$$

The relative cell area increase $\Delta A/A$ can be calculated as

$$1 + \frac{\Delta A}{A} = \left(1 + \frac{\Delta R}{R}\right)^2 = \left(10^{\frac{\Delta L}{35.2}}\right)^2 \tag{12.2}$$

The required relative base station site density with a given improvement in the link performance is calculated in Table 12.1. The base station density is inversely proportional to the cell area. For example, with a link performance improvement of 5.3 dB, the base station density can be reduced to approximately 50 %.

Typically, the radio access network represents a major part of the total UMTS network investment, and most of the radio access network costs are base station site related costs.

Table 12.1. Reduction in the required base station site density
with an improved link budget

Improvement in the link budget	Relative number of sites
0.0 dB = Reference case	100 %
1.0 dB	88 %
2.0 dB	77 %
3.0 dB	68 %
4.0 dB	59 %
5.0 dB	52 %
6.0 dB	46 %
10.0 dB	27 %

Therefore, a reduction in the number of cell sites is important in reducing the required investment for the UMTS network.

The factors affecting the maximum path loss can be seen from the link budget – see Section 8.2 – and are shown in Figure 12.2. The effect of the base station solutions, the bit rate and the diversity is described in this chapter. The relationship between uplink loading and coverage has been discussed in Section 8.2.2.

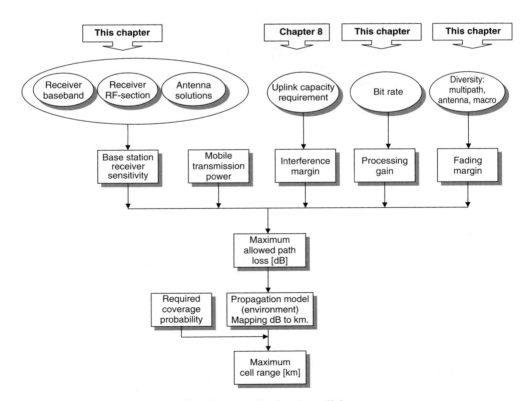

Figure 12.2. Factors affecting the uplink coverage

12.2.1   Uplink Coverage

In this section we evaluate the effect of the physical layer parameters and the base station solutions on the WCDMA uplink coverage.

12.2.1.1   Data Bit Rate

Uplink cell coverage depends on the transmitted bit rate: for higher bit rates, the processing gain is reduced and the coverage range is hence reduced, see the uplink link budgets in the tables in Section 8.2, row k. The relationship between bit rate and coverage is quite different for real time and for non-real time applications. The relative coverage versus bit rate is first shown in Figure 12.3 for identical outage probability for all bit rates. This plot corresponds to real time applications demanding a guaranteed bit rate with a high coverage probability, e.g. 95 %. The same maximum transmit power of the terminal is assumed for all bit rates and a suburban propagation model is assumed. In this example, the uplink range of 384 Mbps is reduced to 62 % of the range of 64 kbps. If the cell layout is planned for 384 kbps uplink coverage instead of 64 kbps, the base station site density must be increased by a factor of $(1/0.62)^2 = 2.6$, which would be quite challenging.

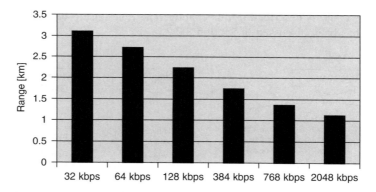

Figure 12.3. Uplink range of real time guaranteed bit rates in suburban area

Non-real time applications do not require a minimum guaranteed bit rate with low outage probability. If we can allow temporarily a lower bit rate, the cell size does not need to be decreased to offer high data rates with high probability. The distribution of the uplink bit rate is shown in Figure 12.4 for a cell that is planned to provide 64 kbps with 95 % probability. Such a cell is able to provide 384 kbps with 80 % coverage probability. The average bit rate in this case is >330 kbps. The underlying reason for the difference between real time and non-real time bit rates is shown in Figure 12.5: the median mobile output power for 64 kbps is 5 dBm and only 8 % of the mobiles are transmitting with full power (5 % of them are in outage). It is possible for most mobiles to increase their transmission power to obtain a higher data rate for non-real time services.

There are a number of solutions to improve uplink coverage, for example, beamforming antennas or higher order antenna diversity reception. If we can obtain improvements in the uplink link budget, this will increase the coverage probability of uplink data rates. The uplink bit rate improvements for non-real time and for guaranteed real time services

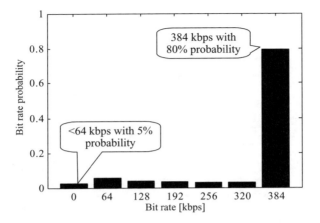

Figure 12.4. Distribution of the uplink non-real time bit rates

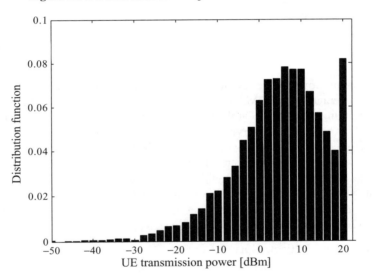

Figure 12.5. Simulated distribution of the mobile transmission power with 64 kbps (5 % outage)

are shown in Figure 12.6. The non-real time bit rate is the average for a 384 kbps capable terminal, while the real time bit rate is the one that can be provided with 95 % probability. The reference case (= 0 dB improvement) provides 64 kbps with 95 % probability. The guaranteed bit rate increases 6-fold to 384 kbps when an 8 dB improvement is obtained in the link budget. The average non-real time bit rate is already >330 kbps in this reference case and hence the improvement in the average bit rates is only moderate with the link budget improvements.

Wide area uplink coverage for high bit rate real time services will be challenging in UMTS, as shown in Figure 12.3, and providing full 384 kbps or higher data rate uplink coverage requires high base station site density. These results also point out the importance of the solutions that improve uplink coverage in third generation systems. In second

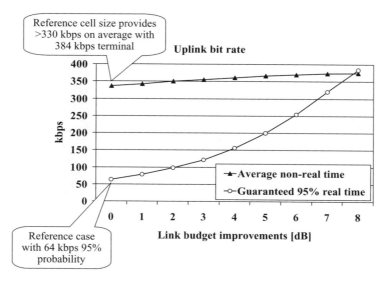

Figure 12.6. Real time and non-real time bit rates as a function of link budget improvements

generation systems, coverage issues are less challenging since only low bit rate services are offered. The coverage of WCDMA data services is compared to GSM900 and GSM1800 speech coverage in Section 8.4. On the other hand, high average bit rates beyond 300 kbps can be offered for non-real time services with GSM site densities. Non-real time applications can benefit from those locations where the mobile has enough transmission power for 384 kbps, but these applications can survive in those locations where only 64 kbps is available. The bit rate adaptation capability of the application is beneficial when running over cellular systems. The reduction of the uplink bit rate is possible for non-real time packet data services and also for AMR speech service, which supports different bit rates from 4.75 kbps to 12.2 kbps. The coverage of AMR speech service is discussed in the next section.

12.2.1.2 Adaptive Multirate Speech Codec

With Adaptive Multirate (AMR) speech codec it is possible to switch to a lower bit rate – and thereby increase the processing gain – if the link budget becomes insufficient to retain a low target BLER due to poor cell coverage. The AMR speech codec is introduced in Chapter 2. The gain in the link budget by reducing the AMR bit rate can be calculated as follows:

$$\text{Coverage\_gain} = 10 \cdot \log_{10}\left(\frac{\text{DPDCH}(12.2\text{kbps}) + \text{DPCCH}}{\text{DPDCH}(\text{AMR\_bit\_rate[kbps]}) + \text{DPCCH}}\right)$$

$$= 10 \cdot \log_{10}\left(\frac{12.2 + 12.2 \cdot 10^{\frac{-3\text{dB}}{10}}}{\text{AMR\_bit\_rate[kbps]} + 12.2 \cdot 10^{\frac{-3\text{dB}}{10}}}\right) \qquad (12.3)$$

where the power difference between DPCCH and DPDCH is assumed to be −3.0 dB for 12.2 kbps AMR speech. DPCCH is the physical layer control channel and DPDCH the physical layer data channel. For different AMR bit rates, the DPCCH power is kept the same, while the power of the DPDCH is changed according to the bit rate. The reduction of the total transmission power is calculated in Equation (12.3) and can be used to provide a larger uplink cell range for AMR speech. The coverage gain by reducing the bit rate from 12.2 kbps to 7.95 kbps is 1.1 dB, and the gain by reducing the bit rate from 12.2 kbps to 4.75 kbps is 2.3 dB. An example of cell range values with different AMR bit rates is shown in Figure 12.7.

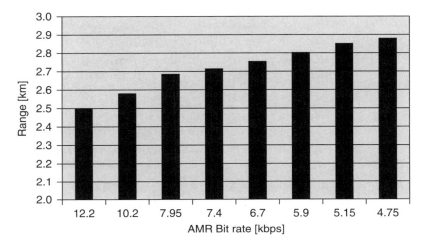

Figure 12.7. Example of uplink cell coverage range for different AMR speech codec bit rates

The link budget gain with lower AMR bit rate can typically not be used in the cell dimensioning to make the cell range larger, since the coverage is often defined and limited by a certain minimum real time data rate, e.g. by 64 kbps video. The link budget gain can rather be used to guarantee a better voice quality under challenging radio conditions, e.g. indoor environments with a high penetration loss, or during compressed mode measurements. The compressed mode is described in Chapter 9.

12.2.1.3 Multipath Diversity

Multipath diversity reduces the fast fading probability of the signal and thus the required fading margin in the link budget. The fast fading margin is on row p in the link budget of Table 8.4. The gain of multipath diversity in terms of coverage is illustrated in Figure 12.8 for the cases of: 1-path, 2-path and 4-path Rayleigh fading. Multipath diversity can be obtained, for example, from Rake reception under time dispersive radio channel conditions or antenna diversity. In this section the effect of the multipath diversity on the uplink coverage is shown in terms of simulation and laboratory measurement results for two different multipath profiles: ITU Pedestrian A – providing only little multipath diversity, and ITU Vehicular A – providing more multipath diversity. The multipath diversity gain is here defined as the reduction of the average mobile transmission power when there is multipath diversity available. Two-branch receive antenna diversity is assumed at the base station.

The multipath diversity gain is shown in Table 12.2 at 3 and 20 km/h [1]. The gain is in the range of 1.0 and 1.6 dB respectively, for both simulation and measurement results. These results are obtained with fast power control active. However, if the terminal is operating at the cell edge and transmitting with constant full power, the fast power control does not

Table 12.2. Multipath diversity gain for AMR voice with fast
power control [1]

	Simulations	Laboratory measurements
3 km/h	1.0 dB	1.3 dB
20 km/h	1.5 dB	1.6 dB

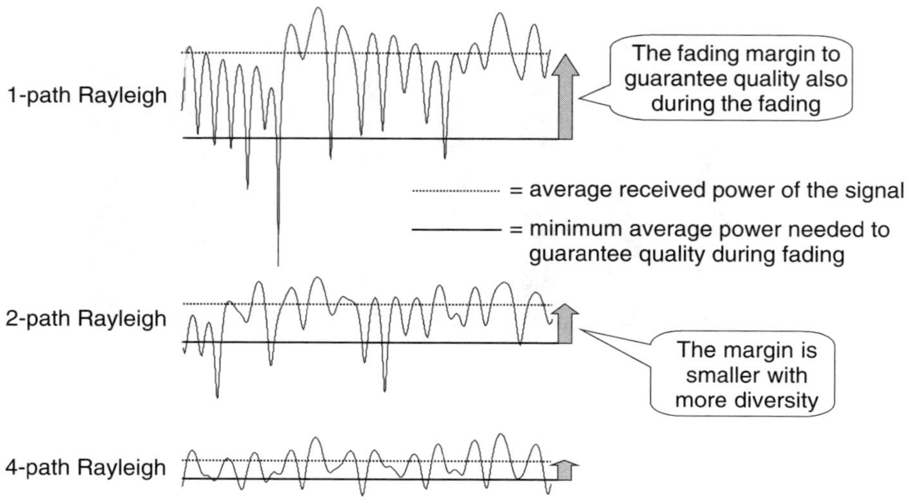

Figure 12.8. Diversity reduces the required fading margin

compensate for the fast fading and the importance of the multipath diversity significantly increases, as shown in Table 12.3. The degree of multipath diversity obtained by a Rake receiver depends on the time dispersion of the radio channel relative to the transmission bandwidth. Hence, the multipath diversity gain is larger for wideband CDMA than for narrowband CDMA for the same radio environment: see Section 3.3.

Table 12.3. Multipath diversity gain for AMR voice with
power control and with constant maximum transmit power

	With power control	With constant maximum transmit power
3 km/h	1.0 dB	2.8 dB

12.2.1.4 Soft Handover

During soft handover the uplink transmission from the terminal is received by two or more base stations. Since during soft handover there are at least two base stations detecting the UE transmitted signal, the probability of a correctly detected signal increases, and thus a soft

Table 12.4. Soft and softer handover gains against fast fading

	ITU Pedestrian A	ITU Vehicular A
Softer handover gain, equal mean path loss to both sectors	5.3 dB	3.1 dB
Soft handover gain, equal mean path loss to both base stations	4.0 dB	2.2 dB
Soft handover gain, 3 dB higher mean path loss to the worst receiving base stations	2.7 dB	0.8 dB

handover gain is obtained. The soft handover gain for uplink coverage is shown in Table 12.4 for the case of 3 km/h, two receiving base stations, and AMR speech. We assume here that the fast fading is uncorrelated between the base stations and sectors. Two cases are shown: when the mean path losses to the two base stations are identical, and when there is a 3 dB mean difference in the path loss. These two cases are illustrated on the upper row of Figure 12.9. The first case gives the highest soft handover gain. When the difference in mean path loss becomes large, the soft handover gain vanishes and at a certain mean power difference the terminal will leave soft handover and only remain connected to the strongest base station. A typical value for the window drop is 2–4 dB, see Chapter 9 for more details. The results show that the lower the multipath diversity, the larger the soft handover gain. For equal mean path loss the soft handover gain is 4 dB for an ITU Pedestrian profile and 2.2 dB for an ITU Vehicular A profile.

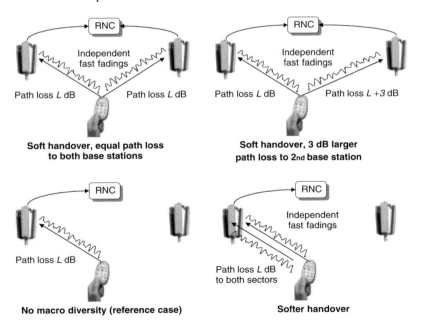

Figure 12.9. Soft and softer handover cases for the soft handover gain evaluation

Uplink soft handover uses selection combining in RNC based on a CRC check, while in softer handover, the uplink transmission from the mobile is received by two sectors of one Node B. In softer handover the signals from two sectors are maximal ratio combined in the baseband Rake receiver unit of the base station, see Section 3.6. The soft and softer handover gains with equal power to both sectors and base stations are shown in Table 12.4. Softer handover provides 0.9–1.3 dB more gain than soft handover.

12.2.1.5 Base Station Receive Antenna Diversity

Ideally, 3 dB coverage gain can be obtained with receive antenna diversity, even if the antenna diversity branches have fully correlated fading. The reason is that the desired signals from two antenna branches can be combined coherently, while the received thermal noises are combined non-coherently. The 3 dB gain assumes ideal channel estimation, but the degradation of non-ideal channel estimation is marginal. Additionally, antenna diversity also provides a significant gain against fast fading for the case of uncorrelated or low correlated antenna branches. Network operators typically select antenna diversity topologies that ensure an envelope correlation of less than 0.7. The Node B receive antenna diversity gains are obtained at the expense of increased or duplicate hardware in the Node B, including RF front-end, baseband hardware, antenna feeders, antennas or antenna ports.

Two different diversity antenna topologies are shown in Figure 12.10. Low correlated antenna branches can be obtained by space or polarisation diversity. The advantage of polarisation diversity is that the diversity branches do not need separate physical antenna structures, see the left picture of Figure 12.10. The performance of polarisation diversity in GSM has been presented in [3], [4] and [5], and for WCDMA in [6].

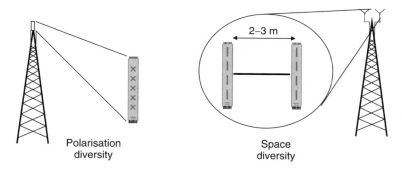

Figure 12.10. Polarisation and space diversity antennas

Simulated and measured antenna diversity gain results are shown in Table 12.5. It can be observed that the gain is higher at low mobile speeds of 3 km/h and 20 km/h than for 120 km/h. The reason is that for high mobile speeds the link performance benefits from time

Table 12.5. Antenna diversity gain for AMR speech with fast power control [1]

	ITU Pedestrian A		ITU Vehicular A	
	Simulations	Laboratory measurements	Simulations	Laboratory measurements
3 km/h	5.5 dB	5.3 dB	3.7 dB	3.3 dB
20 km/h	5.0 dB	5.9 dB	3.5 dB	3.5 dB
120 km/h	4.0 dB	4.4 dB	3.0 dB	3.4 dB

diversity provided by the interleaving, and hence the additional gain from antenna diversity is reduced. We can also note that the gain is higher when the amount of multipath diversity is small as in the ITU Pedestrian A channel. The antenna diversity gain at low mobile speed is up to 5–6 dB for the ITU Pedestrian A profile and 3–4 dB for ITU Vehicular A profile. For the simulated case, the antenna branches are uncorrelated and for the measured case, the branches are practically uncorrelated.

The performance of uplink diversity reception can be further extended by deploying four-branch antenna reception. The four-branch antenna configuration can be obtained using two antennas with polarisation diversity with a separation of 2–3 metres to combine polarisation and space diversity, i.e. obtain four low correlated antenna branches. The two antennas can also be placed very close to each other, even in a single radome, to make the visual impact lower. However, in that case the branch correlation between the two polarisation antenna structures is expected to be high. The two four-branch antenna options are shown in Figure 12.11.

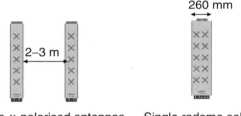

Two *x*-polarised antennas   Single radome solution

Figure 12.11. Four-branch receive antenna configurations

The simulated diversity gains of two- and four-branch diversity are summarised in Figure 12.12. These results assume separate antennas in four-branch reception, i.e. low branch correlation, and constant maximum transmit power of the mobile. Hence, it should be noted

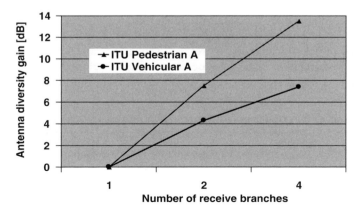

Figure 12.12. Antenna diversity gain with one-, two- and four-branch reception for the case of constant maximum transmit power of the mobile

that the results cannot be directly compared to the results in Table 12.5. The gain of four-branch diversity over two-branch diversity in ITU Vehicular A is 3.1 dB. The gain of the single radome solution is typically 0.2–0.4 dB lower, due to the higher antenna branch correlation shown in the measurement part.

The more diversity already available, the smaller the diversity gain from an additional diversity feature. This rule applies to antenna diversity and to all different kinds of diversity. Therefore, there is no *a priori* value for any diversity gain, because the gains depend on the degree of diversity from other diversity techniques.

Field Measurements of Four-branch Receive Antenna Diversity

The field performance of four-branch reception was tested in the WCDMA network in Espoo, Finland. The measurement area is in the middle of Figure 8.18. The measurement environment is of the urban and sub-urban type. The measurement routes are shown in Table 12.6.

Table 12.6. Measurement routes

Route A	up to 40 km/h in Leppävaara / Lintuvaara
Route B	up to 70 km/h on Ring I
Route C	below 10 km/h in Mäkkylä

In the field measurements the mobile transmission power was recorded slot-by-slot with three different base station antenna configurations:

1. Two-branch reception with one polarisation diversity antenna.

2. Four-branch reception with two polarisation diversity antennas separated by 1 m.

3. Four-branch reception with two polarisation diversity antennas side-by-side (emulates single radome solution).

For each configuration the route was measured several times. The different measurement routes are made comparable using the differential Global Positioning System, GPS. The average transmission power over the measurement route is calculated from dBm values. These measured mobile transmission powers are shown in Table 12.7.

The multipath propagation in the measured environment is closer to ITU Vehicular A than to ITU Pedestrian A. We therefore compare the measurement results to the simulation results

Table 12.7. Measured logarithmic average mobile transmission powers

Route	Antenna separation	2-branch reception	4-branch reception	4-branch gain over 2-branch
Route A	1 m separation	6.95 dBm	4.44 dBm	2.5 dB
	no separation	6.95 dBm	4.83 dBm	2.1 dB
Route B	1 m separation	7.90 dBm	4.59 dBm	3.3 dB
	no separation	7.90 dBm	4.86 dBm	3.1 dB
Route C	1 m separation	5.63 dBm	2.54 dBm	3.1 dB

of the ITU Vehicular A profile. The simulated gain of four-branch reception over two-branch reception in Figure 12.12 is 3.1 dB with separate antennas, and the average measured gain with 1 m separation is 3.0 dB in Table 12.7.

The difference between separate antennas and the single radome solution is 0.2–0.4 dB. The impact of antenna branch correlation for the two spaced antenna structures is small because the diversity order is already large: multipath and polarisation diversity.

It can be concluded that four-branch receive antenna diversity is an effective technique to increase the uplink coverage area. A 3 dB improvement in the uplink performance reduces the required site density by about 30 % according to Table 12.1.

12.2.2 Downlink Coverage

The Node B transmit power is typically 20 W (43 dBm), while the mobile transmit power is only 125 mW (21 dBm). With a low number of simultaneous connections, it is possible to allocate a high power per mobile connection in downlink. Hence, better coverage can be given for high bit rate services in downlink than in uplink. The downlink coverage is affected by the maximum link power that is a network planning parameter. The downlink coverage is also affected by the amount of inter-cell interference. In this example the G factor, i.e. own cell to other cell interference ratio, at the cell edge is assumed to be -2.5 dB, which corresponds to approximately -12 dB CPICH E_c/I_0 with medium base station transmission power in large cells. The calculation assumes that CPICH is allocated 2 W and other common channels 1 W. The other cell transmission power is assumed to be 10 W and the maximum path loss at the cell edge 156 dB. The results are shown in Figure 12.13. 2 W link power provides 384 kbps at 60 % of the maximum cell range and 64 kbps with full coverage. 5 W power allocation gives 384 kbps at 80 % of the maximum cell range, while 10 W power allocation gives practically full 384 kbps coverage.

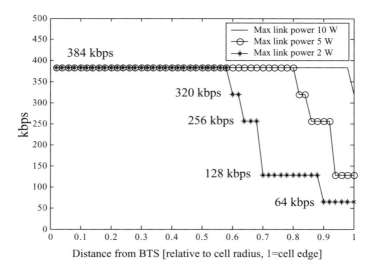

Figure 12.13. Downlink coverage with different maximum link powers

12.3 Downlink Cell Capacity

The WCDMA downlink air interface capacity has been shown to be less than the uplink capacity [7–9]. The main reason is that better receiver techniques can be used in the Node B than in the mobile. These techniques include receiver antenna diversity and multiuser detection. Additionally, in UMTS, the downlink capacity is expected to be more important than the uplink capacity because of the asymmetric downloading type of traffic. In this section the downlink capacity and its performance enhancements are therefore considered. WCDMA capacity evaluation is studied also in [10].

The following sections present two aspects that impact upon the downlink capacity, and which are different from the uplink: The issue of orthogonal codes is described in Section 12.3.1 and the performance gain of downlink transmit diversity in Section 12.3.2. Additionally, we discuss the WCDMA voice capacity with AMR codec and Voice over IP (VoIP) in Section 12.3.3.

12.3.1 Downlink Orthogonal Codes

12.3.1.1 Multipath Diversity Gain in Downlink

The effect of the downlink orthogonal codes on capacity is considered in this section. In downlink, short orthogonal channelisation codes are used to separate users in a cell. Within one scrambling code the channelisation codes are orthogonal, but only in a one-path channel. In the case of a time dispersive multipath channel, the orthogonality is partly lost, and own-cell users sharing one scrambling code also interfere with each other. The downlink performance in the ITU Vehicular A and ITU Pedestrian A multipath profiles is presented below for the case of 8 kbps, 10 ms interleaving, and 1 % BLER. The ITU Pedestrian A channel is close to a single-path channel and does, on one hand, preserve almost full own-cell orthogonality, but does not provide much multipath diversity, while the ITU Vehicular A channel gives a significant degree of multipath diversity but the orthogonality is partly lost. The simulation scenario is shown in Figure 12.14. The required transmission power per

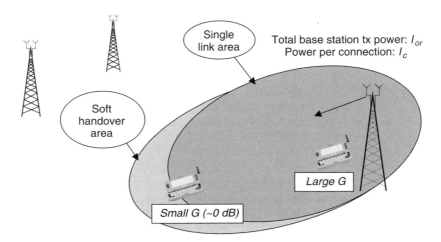

Figure 12.14. Simulation scenario for downlink performance evaluation

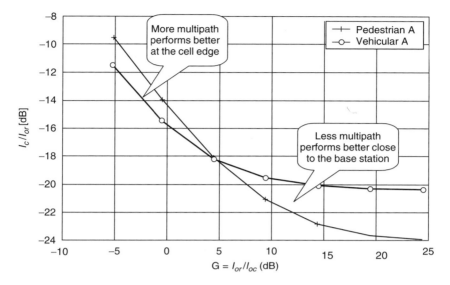

Figure 12.15. Effect of multipath propagation

speech connection ($= I_c$) as compared to the total base station power ($= I_{or}$) is shown on the vertical axis in Figure 12.15. For example, the value of $-20\,\text{dB}$ means that this connection takes $10^{(-20\,\text{dB}/10)} = 1\%$ of the total base station transmission power. The lower the value on the vertical axis, the better the performance. The horizontal axis shows the total transmitted power from this base station divided by the received interference from the other cells, including thermal noise ($= I_{oc}$). This ratio I_{or}/I_{oc} is also known as the geometry factor, G. A high value of G is obtained when the mobile is close to the base station, and a low value, typically $-3\,\text{dB}$, at the cell edge.

We can observe some important issues about downlink performance from Figure 12.15. At the cell edge, i.e. for low values of G, the multipath diversity in the ITU Vehicular A channel gives a better performance compared to less multipath diversity in the ITU Pedestrian A channel. This is because other cell interference dominates over own-cell interference. Close to the base station the performance is better in the ITU Pedestrian A channel because the multipath propagation in the ITU Vehicular A channel reduces the orthogonality of the downlink codes. Furthermore, there is not much need for diversity close to the base station, since the intra-cell interference experiences the same fast fading as the desired user's signal. If signal and interference have the same fading, the signal to interference ratio remains fairly constant despite the fading. The effect of soft handover is not shown in these simulations but it would improve the performance, especially in the ITU Pedestrian A channel at the cell edge by providing extra soft handover diversity – macro diversity. The macro diversity gain is presented in detail in Section 9.3.1.3.

We note that in the downlink the multipath propagation is not clearly beneficial – it gives diversity gain at the cell edge but at the same time reduces orthogonality close to the Node B. Hence, the multipath propagation does not necessarily improve downlink capacity

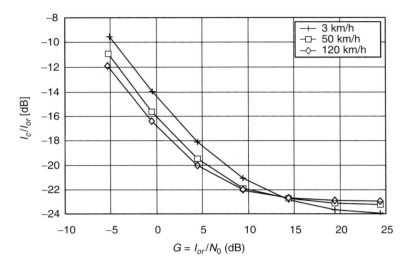

Figure 12.16. Effect of mobile speed in the ITU Pedestrian A channel

because of the loss of orthogonality. The loss of the orthogonality in multipath channel could be improved with interference cancellation receivers or equalisers in the mobile. Such receivers are discussed in Chapter 11 for High Speed Downlink Packet Access, HSDPA.

The effect of the mobile speed on downlink performance in the Pedestrian A channel is shown in Figure 12.16. At the cell edge the best performance is obtained for high mobile speeds, while close to the base station, low mobile speeds perform better. This behaviour can be explained by the fact that for high mobile speeds interleaving and channel coding, here convolutional code, provide time diversity and coding gain. In Figure 12.15 it was shown that diversity is important at the cell edge to improve the performance.

12.3.1.2 Downlink Capacity in Different Environments

In this section the WCDMA capacity formulas from Section 8.2.2 are used to evaluate the effect of orthogonal codes on the downlink capacity in macro and micro cellular environments. The downlink orthogonal codes make the WCDMA downlink more resistant to intra-cell interference than the uplink direction, and the effect of inter-cell interference from adjacent base stations has a large effect on the downlink capacity. The amount of interference from the adjacent cells depends on the propagation environment and the network planning. Here we assume that the amount of inter-cell interference is lower in micro cells where street corners isolate the cells more strictly than in macro cells. This cell isolation is represented in the formula by the other-to-own cell interference ratio i. We also assume that in micro cellular environments there is less multipath propagation, and thus a better orthogonality of the downlink codes. On the other hand, less multipath propagation gives less multipath diversity, and therefore we assume a higher E_b/N_0 requirement in the downlink in micro cells than in macro cells.

The assumed loading in uplink is allowed to be 60 % and in downlink 80 % of WCDMA pole capacity. A lower loading is assumed in uplink than in downlink because the coverage is more challenging in uplink. A higher loading results in smaller coverage, as shown in

Table 12.8. Assumptions in the throughput calculations

	Macro cell	Micro cell
Downlink orthogonality	0.5	0.9
Other-to-own cell interference ratio i	0.65	0.4
Uplink E_b/N_0 with 2-branch diversity	2.0 dB	2.0 dB
Uplink loading	60 %	60 %
Downlink E_b/N_0, no transmit diversity	5.0 dB	6.5 dB
Downlink loading	80 %	80 %
Downlink common channels	15 %	15 %
Block error rate BLER	1 %	1 %

Section 8.2.2. We assume that 15 % of the downlink capacity is allocated for downlink common channels, for more information about these channels see Section 8.2.2. A user bit rate of 64 kbps is assumed in the uplink calculation.

We calculate the example data throughputs in macro and micro cellular environments in both uplink and downlink. The assumptions of the calculations are shown in Table 12.8 and the results in Table 12.9. The capacities in Table 12.9 assume that the users are equally distributed over the cell area and the same bit rate is allocated for all users.

Table 12.9. Data throughput in macro and micro cell environments per sector per carrier

	Macro cell	Micro cell
Uplink	900 kbps	1060 kbps
Downlink	710 kbps	1160 kbps

In macro cells the uplink throughput is higher than the downlink throughput, while in micro cells the uplink and downlink throughputs are very similar. We can note that the downlink capacity is more sensitive to the propagation and multipath environment than the uplink capacity. The reason is the application of the orthogonal codes.

The capacity calculations above assume that all cells are fully loaded. If the adjacent cells have lower loading, it is possible to have an even higher cell capacity. The extreme case is an isolated cell without any inter-cell interference. Figure 12.17 shows three different cell capacities with 384 kbps connections. The first one is the typical multicell capacity, the second one single cell capacity with orthogonality of 0.5 and the third one single cell capacity with orthogonality close to 1, i.e. single path model. In the third case, the capacity is code limited with a maximum seven simultaneous users of 384 kbps. In the case of favourable orthogonality conditions and low other-cell to own-cell interference ratio, the cell capacity can be clearly higher than in the typical multicell case.

12.3.1.3   Number of Orthogonal Codes

The number of downlink orthogonal codes within one scrambling code is limited. With a spreading factor of SF, the maximum number of orthogonal codes is SF. This code limitation

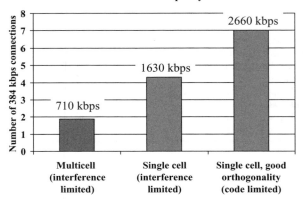

Figure 12.17. 384 kbps data capacity in multicell and single cell cases

can place an upper limit on the downlink capacity if the propagation environment is favourable and the network planning and hardware support such a high capacity. In this section the achievable downlink capacity with one set of orthogonal codes is estimated. The assumptions in these calculations are shown in Table 12.10 and the results in Table 12.11.

Table 12.10. Assumptions in the calculation of Table 12.11

Common channels	10 codes with SF = 128
Soft handover overhead	20%
Spreading factor (SF) for half rate speech	256
Spreading factor (SF) for full rate speech	128
Chip rate	3.84 Mcps
Modulation	QPSK (2 bits per symbol)
Average DPCCH overhead for data	10%
Channel coding rate for data	1/3 with 30% puncturing

Table 12.11. Maximum downlink capacity with one scrambling code per sector

Speech, full rate (AMR 12.2 kbps and 10.2 kbps)	128 channels	Number of codes with spreading factor of 128
	*(128 − 10)/128	Common channel overhead
	/1.2	Soft handover overhead
	= **98 channels**	
Speech, half rate (AMR ≤ 7.95 kbps)	2*98 channels	Spreading factor of 256
	= **196 channels**	
Packet data	3.84e6	Chip rate
	*(128 − 10)/128	Common channel overhead
	/1.2	Soft handover overhead
	*2	QPSK modulation
	*0.9	DPCCH overhead
	/3	1/3 rate channel coding
	/(1 − 0.3)	30% puncturing
	= **2.5 Mbps**	

Part of the downlink orthogonal codes must be reserved for the common channels and for soft and softer handover overhead. These factors are taken into account in Table 12.10 and Table 12.11. The maximum number of full rate speech channels per sector is 98 with these assumptions, and the maximum data throughput is 2.5 Mbps per sector.

The number of orthogonal codes is not a hard-blocking limitation for the downlink capacity. If this number is not large enough, a second (or more) scrambling code can be taken into use in the downlink, which gives a second set of orthogonal short codes: see Section 6.3. These two sets of orthogonal codes are not orthogonal to each other. Hence, if the second scrambling code is used, the code channels under the second scrambling code cause much more interference to those under the first scrambling code than the other code channels under the first scrambling code. A second scrambling code will be needed with downlink smart antenna solutions, but in this case the scrambling codes can be spatially isolated to reduce the non-orthogonal interference from multiple scrambling codes in a cell, see Figure 12.46.

12.3.2   Downlink Transmit Diversity

The downlink capacity could obviously be improved by using receive antenna diversity in the mobile. For small and cheap mobiles it is not, however, feasible to use two antennas and receiver chains. Furthermore, two receiver chains in the mobile will increase power consumption. The WCDMA standard therefore supports the use of Node B transmit diversity. The target of the transmit diversity is to move the complexity of antenna diversity in downlink from the mobile reception to the Node B transmission. The supported downlink transmit diversity modes are described with physical layer procedures in Section 6.6. With transmit diversity, the downlink signal is transmitted via two base station antenna branches. If receive diversity is already deployed in the Node B and we duplex the downlink transmission to the receive antennas, there is no need for extra antennas for downlink diversity. In Figure 12.10 both antennas could be used for reception and for transmission.

In this section we analyse the performance gain from the downlink transmit diversity. The performance gain from transmit diversity can be divided into two parts: (1) coherent combining gain and (2) diversity gain against fast fading. The coherent combining gain can be obtained because the signal is combined coherently, while interference is combined non-coherently. The gain from ideal coherent combining is 3 dB with two antennas. With downlink transmit diversity it is possible to obtain coherent combining in the mobile reception if the phases from the two transmission antennas are adjusted according to the feedback commands (estimated antenna weights) from the mobile in the closed loop transmit diversity. The coherent combining is, however, not perfect because of the discrete values of the antenna weights and delays in the feedback commands. The downlink transmit diversity with feedback is depicted in Figure 12.18.

Both the closed loop and the open loop transmit diversity provide gain against fading because the fast fading is uncorrelated from the two transmit antennas. The gain is larger when there is less multipath diversity. The importance of the diversity is discussed in detail in Section 9.2.1.2. The gains from the downlink transmit diversity are summarised in Table 12.12.

It is important to note the difference between the two sources of diversity in the downlink: multipath and transmit diversity. Multipath diversity reduces the orthogonality of the downlink codes, while transmit diversity keeps the downlink codes orthogonal in flat fading

Table 12.12. Comparison of uplink receive and downlink transmit diversity

	Coherent combining gain	Diversity gain
How to obtain gain	Feedback loop from mobile to base station to control the transmission phases to make received signals to combine coherently in mobile	Uncorrelated fading from the two transmission antennas
Non-idealities in obtaining the gain	Discrete steps in feedback loop Delay in feedback loop Multipath propagation	Correlation between transmission antennas

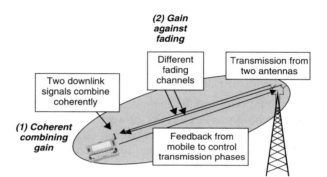

Figure 12.18. Downlink transmit diversity with feedback

channels. In order to maximise the interference-limited downlink capacity, it would be beneficial to avoid multipath propagation to keep the codes orthogonal and to provide the diversity with transmit antenna diversity.

The effect of the downlink transmit diversity gains on downlink capacity and coverage is illustrated in Figure 12.19. The simulation results typically show an average gain of

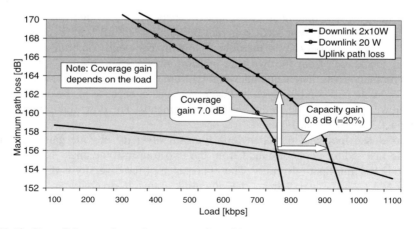

Figure 12.19. Downlink capacity and coverage gains with transmit diversity. A 0.8 dB link level gain from transmit diversity is assumed

0.5–1.0 dB in the macro cell environment. A 0.8 dB gain – including coherent combining gain and diversity gain against fading – is assumed here. This gain implies that the average power of each downlink connection can be reduced by 0.8 dB while maintaining the same quality. At the same time the system can support 0.8 dB, i.e. 20 % (=10^(10.8/10)), more users. If we allow, for example, a maximum path loss of 156 dB, the capacity can be increased by 20 % from 760 kbps to 910 kbps. The transmit diversity gain can be used alternatively to improve the downlink coverage while keeping the load unchanged. In the example in Figure 12.19, the maximum path loss could be increased by 7 dB, from 156 dB to 163 dB, if the load were kept at 760 kbps. The coverage gain is higher than the capacity gain because of the WCDMA load curve. It may not be possible to utilise the downlink coverage gains and extend the cell size with downlink transmit diversity if the uplink is the limiting direction in coverage. The coverage gain could be used alternatively to reduce the required base station transmission power. If we keep the load unchanged at 760 kbps and the maximum path loss unchanged at 156 dB, we could reduce the transmission power by 7 dB, from 20 W to 2 × 2.0 W.

Transmit diversity is also supported with Release 5 High-Speed Downlink Packet Access (HSDPA). As HSDPA uses fast scheduling, there is a conflict with benefits from transmit diversity and HSDPA. The fast scheduling with HSDPA benefits from the wider C/I distribution, which is made narrower by the transmit diversity methods. Especially with the open loop transmit diversity, the HSDPA performance can improve in the link level but in the system level there is no clear gain compared to single antenna transmission. With closed loop mode 1 there are some benefits even in the system level due to the feedback.

12.3.3 Downlink Voice Capacity

WCDMA voice capacity with AMR voice codec is addressed in this section. Both circuit switched voice and Voice over IP (VoIP) are considered. AMR codec is introduced in Chapter 2. The voice capacity numbers in Chapter 8 refer to the full rate AMR 12.2 kbps. With path loss of 156 dB the maximum number of voice users is 66. The voice capacity can be increased by using a lower bit rate AMR mode. We estimate the capacity of the lower AMR modes with the following equation

$$\text{Voice capacity} = 66 \text{ users} \cdot \frac{12.2 \text{ kbps}}{\text{AMR bit rate[kbps]}} \cdot 10^{\left(E_b/N_{0_{12.2\,\text{kbps}}} - E_b/N_{0_\text{AMR}}\right)/10} \quad (12.4)$$

where the capacity increases according to the reduction of the bit rate. Additionally, we take into account that the physical layer overhead increases for the lower AMR rates and that is modelled here as a higher E_b/N_0. Table 12.13 shows the voice capacity with three AMR modes: 12.2 kbps, 7.95 kbps and 4.75 kbps. The voice capacity approximately doubles when

Table 12.13. Voice capacity with different AMR modes

	AMR 12.2 kbps	AMR 7.95 kbps	AMR 4.75 kbps
E_b/N_0	7.0 dB	7.5 dB	8.0 dB
Capacity	66 users	90 users	134 users

the bit rate is reduced from 12.2 kbps to 4.75 kbps. The AMR bit rate can be controlled by the operator and it allows a trade-off to be made between voice capacity and voice quality. The AMR bit rate can be adjusted dynamically according to the instantaneous network load. The AMR voice capacity can be further increased by 15–20 % by using AMR source adaptation, for details see Chapter 2.

The voice over IP (VoIP) uses the same dedicated channels in the air interface as the circuit switched voice. The flexibility of the WCDMA air interface allows the introduction of the VoIP service without any modifications to the physical layer standard. The VoIP call from the packet core network includes IP headers that are considerably large compared to the voice payload. In order to save air interface resources, the IP headers are compressed by Packet Data Convergence Protocol, PDCP, in RNC, which is part of the 3GPP Release 4 standard. For more details see Chapter 7. The compressed IP headers are delivered over the WCDMA air interface. The scenario is shown in Figure 12.20.

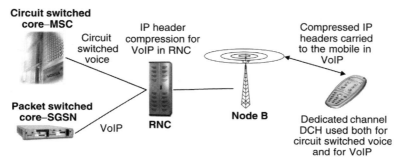

Figure 12.20. Circuit switched voice and voice over IP with WCDMA

The VoIP service will affect the WCDMA voice capacity because of increased overhead, even if IP headers are compressed. That overhead includes compressed IP headers, RLC headers, real time protocol (RTP) payload headers and real time control protocol (RTCP). We assume that the average overhead of compressed IP header and other headers is 7 bytes. The 12.2 kbps voice carries 244 bits per 20 ms and the overhead can be calculated as

$$10\log_{10}\left(\frac{\text{payload} + \text{IP header}}{\text{payload}}\right) = 10\log_{10}\left(\frac{244 + 7.8}{244}\right) = 0.9\,\text{dB} \qquad (12.5)$$

The overhead of 0.9 dB reduces air interface capacity $1 - 10^{\wedge}(-0.9/10) = 19\,\%$. The typical VoIP capacities are shown in Table 12.14 for three AMR modes. The small loss in air interface capacity with VoIP is compensated by the flexibility of end-to-end IP traffic in rich calls.

Table 12.14. Typical circuit switched voice and voice over IP capacity

	AMR 12.2 kbps	AMR 7.95 kbps	AMR 4.75 kbps
Circuit switched voice	66 users	90 users	134 users
Voice over IP	54 users	70 users	95 users

12.4 Capacity Trials

12.4.1 Single Cell Capacity Trials

This section presents the measurement methods and results from capacity trials. The tests are done with a single sector without interference from adjacent sectors or sites. The test environment has usually been the typical suburban case where several test mobiles are located in one or more fixed and stationary locations. The mobiles are located without line-of-sight connection but the coverage is relatively good, i.e. the thermal noise component in the total amount of noise is minimised. In terms of CPICH RSCP good coverage means better than −90 dBm conditions. Also, the base station should be configured so that the hardware and transmission resources are not becoming the limiting factor, however that is not always possible and therefore some of the pole capacity results shown in this section are extrapolated values.

12.4.1.1 AMR Voice Capacity Uplink

The AMR voice capacity test is carried out with a constant bit rate of 12.2 kbps and has additional conditions such as 100 % voice activity factor, 0.8 % BLER target and AMR unequal error protection. The received total base station power and the number of simultaneous UEs can be plotted as shown in the example in Figure 12.21. The received power level of approximately −102 dBm without any users is the thermal noise level.

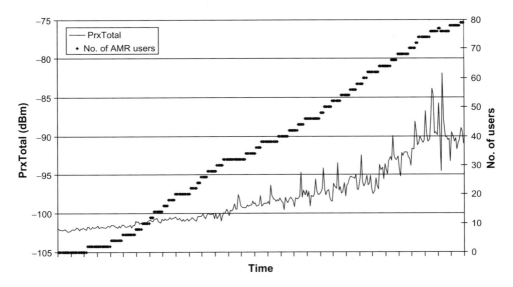

Figure 12.21. Uplink received power (Prx) and the number of simultaneous AMR users

From Figure 12.21 the average fractional load per number of connected UEs can be calculated based on the equation:

$$\text{Noise rise} = \frac{P_{\text{rxtotal}}}{P_{\text{noise}}} = \frac{1}{1 - \eta_{\text{UL}}} \tag{12.6}$$

where η_{UL} is the fractional load, P_{rxtotal} is the total received power, including noise and other cell and own cell users. The noise rise as a function of the AMR users is plotted in Figure 12.22 (a) and the fractional load in Figure 12.22 (b).

The best linear fit, fractional load as a function of number of users, can then be derived as depicted in Figure 12.22, where y is the fractional load η, x is the number of connected UEs

UL Noise Rise vs. number of AMR users

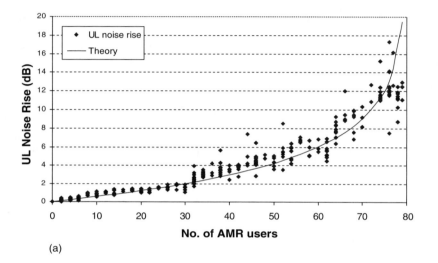

(a)

ULFractional Load vs. number of AMR users

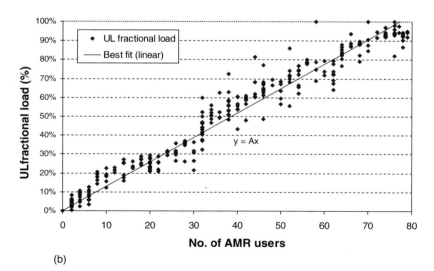

(b)

Figure 12.22. Uplink noise rise (a) and uplink fractional load (b) as a function of the number of AMR users

and A is the linear fit slope which is equivalent to the fractional load of a single user. The fractional load can also be expressed according to Equation (8.14) from Chapter 8:

$$\eta_{\text{UL}} = \frac{E_b/N_0}{W/R} \cdot N \cdot v \cdot (1+i) \qquad (12.7)$$

where N is the number of users, i is the other-to-own cell interference, W is the chip rate, E_b/N_0 is the uplink E_b/N_0 requirement for a user, R is the bit rate of a user and v is the voice activity. As all the users are using the same service 12.2 kbps AMR, the other-to-own cell interference is 0 in the single cell scenario and the voice activity is 100 %, the equation can be simplified as:

$$\eta_{\text{UL}} = \frac{E_b/N_0}{W/R} \cdot N \qquad (12.8)$$

Using the best linear fit slope A from Figure 12.22, the average E_b/N_0 can be derived as:

$$A = \frac{E_b/N_0}{W/R} \Rightarrow \frac{E_b}{N_0} = A\frac{W}{R} \qquad (12.9)$$

The pole capacity of the service can then be calculated assuming 100 % fractional load. The average achieved results for AMR speech capacity testing are presented in Section 12.4.3.

12.4.1.2 AMR Voice Capacity Downlink

The downlink capacity test has the same assumptions as the uplink test. The orthogonality is assumed to be 0.5. The total transmitted base station power and the number of UEs is presented in Figure 12.23. The transmission power of 35.5 dBm = 3.2 W is caused by the common channel powers.

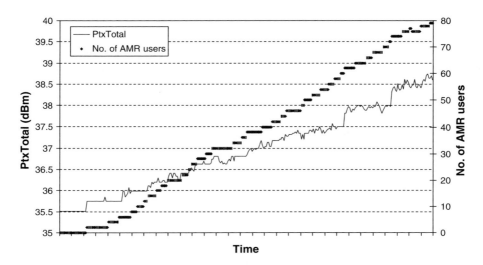

Figure 12.23. Downlink total transmitted power (Ptx) as a function of the number of AMR users

The downlink analysis starts from the common downlink equation for the connection E_b/N_0 requirement as presented in [11].

$$\left(\frac{E_b}{N_0}\right)_i = \frac{W}{R} \cdot \frac{\dfrac{P_i}{L_{m,i}}}{\dfrac{P_{tot,m}}{L_{m,j}}(1-\alpha_i) + \displaystyle\sum_{n=1,n\neq m}^{N} \frac{P_{tot,n}}{L_{n,i}} + P_N} \tag{12.10}$$

where $(E_b/N_0)_i$ is the downlink E_b/N_0 requirement, P_i is the required transmit power at base station m for the connection i, N is the number of interfering base stations, P_N is the noise power at the UE receiver, R_i is the bit rate for the UE i, W is the chip rate, $P_{tot,m}$ is the required total transmit power of the serving base station m, $P_{tot,n}$ is the required total transmit power of another neighbouring base station n, $L_{m,i}$ is the path loss from the serving base station m to UE i, $L_{n,i}$ is the path loss from another base station n to UE i and α_i is the orthogonality factor for UE i. The required transmitted power for connection i, P_i can be solved as follows:

$$P_i = \frac{\left(\dfrac{E_b}{N_0}\right)_i \cdot R}{W}\left[(1-\alpha_i) \cdot P_{tot,m} + \sum_{n=1,\,n\neq m}^{N} \frac{L_{m,i}}{L_{n,i}} \cdot P_{tot,n} + P_N \cdot L_{m,i}\right] \tag{12.11}$$

In these single cell tests, the other-to-own cell interference can be set to be zero. Also, as the testing is done in good RSCP conditions, the noise power can be assumed to be negligible (i.e. $P_{tot,n}/L_{m,i} \gg P_N$). The two assumptions can be used to simplify the equation to the following format:

$$P_i = \frac{\left(\dfrac{E_b}{N_0}\right)_i \cdot R}{W} \cdot (1-\alpha_i) \cdot P_{tot,m} \tag{12.12}$$

Then the constant common channel power is added as well as the activity factor and both sides are summed over all the connections under the cell. This gives the total base station downlink transmitted power as:

$$\sum_{i=1}^{I} P_i + P_{CCH} = P_{tot,m} \cdot \sum_{i=1}^{I} \frac{v \cdot \left(\dfrac{E_b}{N_0}\right)_i \cdot R}{W} \cdot (1-\alpha_i) + P_{CCH} \tag{12.13}$$

where I is the number of connections in the cell. The power rise over the common channels can be solved as follows

$$\frac{P_{tot,m}}{P_{CCH}} = \frac{1}{1 - \displaystyle\sum_{i=1}^{I} \frac{v \cdot \left(\dfrac{E_b}{N_0}\right)_i \cdot R}{W} \cdot (1-\alpha_i)} \tag{12.14}$$

Using the downlink load factor, η_{DL} we can write

$$\text{Power rise} = \frac{1}{1 - \eta_{DL}} \tag{12.15}$$

The power rise above the common channel power and the fractional load as a function of the number of AMR users can be plotted as in the example charts in Figure 12.24.

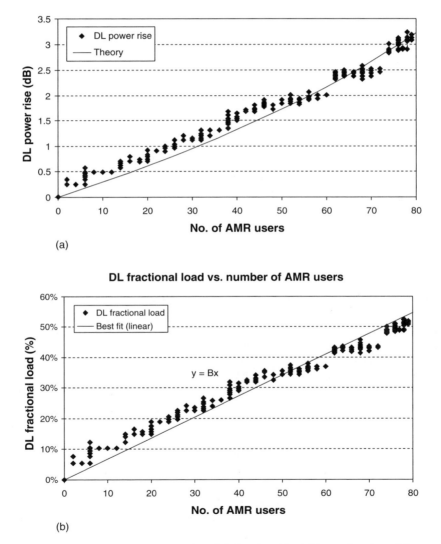

(a)

(b)

Figure 12.24. Downlink power rise (a) and downlink fractional load (b) as a function of the number of AMR users

The downlink E_b/N_0 can be defined from the downlink fractional load by using the best linear fit as:

$$\eta_{\text{DL}} = \frac{\left(\dfrac{E_b}{N_0}\right)_i \cdot R}{W} \cdot (1 - \alpha_i) \cdot I = B \cdot I$$

$$B = \frac{\left(\dfrac{E_b}{N_0}\right)_i \cdot R}{W} \cdot (1 - \alpha_i) \tag{12.16}$$

$$\left(\frac{E_b}{N_0}\right)_i = \frac{B \cdot W}{R \cdot (1 - \alpha_i)}$$

The resulting average downlink E_b/N_0 values and the calculated pole capacity are shown in Section 12.4.3. The downlink code power can be plotted as well to see the power deviation per connected UE. One example is shown in Figure 12.25.

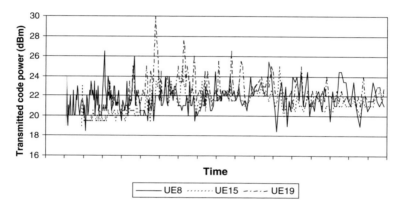

Figure 12.25. Downlink transmitted code power per AMR connection when load is increased

There is, on average, 6 dB fluctuation on the power per connection. Also, it seems that the fluctuation is increasing as the load increases. The average powers per connection during high load are shown in Figure 12.26. The average downlink transmitted code power per connection varies between 17 dBm and 24 dBm in this example, showing the variation for UEs in different locations experiencing different path losses and multipath conditions as well as different UE models. These average powers and the power fluctuations need to be considered in the network planning when setting the maximum allowed powers per connection.

In poor coverage conditions, the downlink calculation formula shown earlier in this section does not apply any more because the noise power cannot be assumed to be negligible, i.e. $P_{\text{tot},n}/L_{m,i} \gg P_N$ does not apply. In this case, the downlink E_b/N_0 can be calculated, based on the average base station transmitted code power, base station total transmitted power and received CPICH E_c/N_0, using the definition of geometry factor G:

$$G = \frac{P_i}{L_{m,i}} \Big/ (P_{\text{other}} + P_N) \tag{12.17}$$

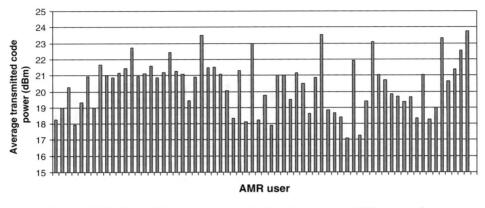

Figure 12.26. Downlink average transmitted code power per AMR connection

Equation (12.10) can now be written

$$\frac{W}{R} \cdot \frac{P_i}{P_{\text{tot},m}} \cdot \frac{1}{(1 - \alpha_i) + \dfrac{1}{G}} = \left(\frac{E_b}{N_0}\right)_i, \qquad i = 1, \ldots I \qquad (12.18)$$

On the other hand, the geometry factor can be defined using CPICH E_c/N_0 as:

$$G = \frac{1}{\dfrac{P_{tx,\text{CPICH}}}{\dfrac{P_{\text{tot},m}}{\dfrac{E_{c,\text{CPICH}}}{N_0}}} - 1} \qquad (12.19)$$

This way the noise power can be included in the calculations as well. In the uplink direction, the tests performed in poor coverage conditions tend to show better results than tests performed in good coverage conditions. This is due to the UE in poor coverage conditions not having as many transmitted power resources, therefore possible power increases due to interference are lower. In the downlink direction, the tests performed in poor coverage conditions usually lead to worse results due to the average transmission power requirement being higher, and the power deviation is higher for the UEs at the coverage border than for the UEs in good coverage conditions. This is shown as a higher E_b/N_0 requirement in poor coverage than in good coverage conditions.

As the last topic, we take a look at the CPICH E_c/N_0 values as a function of downlink load. Figure 12.27 shows CPICH E_c/N_0 as a function of the number of AMR users, and Figure 12.28 as a function of power rise. When the load of the cell increases, the received CPICH E_c/N_0 decreases and deviation becomes larger. As the power rise increases by 5 dB, the CPICH E_c/N_0 is decreased correspondingly by 5 dB on average. This scenario leads to the topic of network optimisation in high load condition, which is recommended for further study.

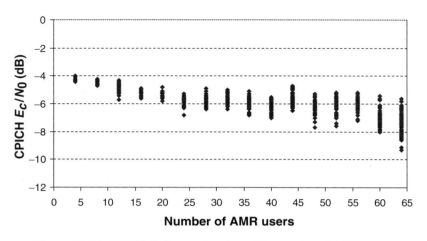

Figure 12.27. CPICH E_c/I_0 as a function of the number of AMR users

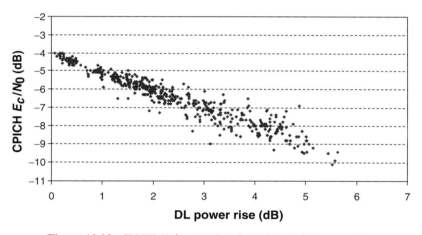

Figure 12.28. CPICH E_c/I_0 as a function of downlink power rise

12.4.1.3   Circuit Switched Video Capacity Uplink

The circuit switched video call capacity test is done with a constant bit rate of 64 kbps and has additional conditions such as: 100 % activity factor and 0.5 % BLER target. The total received power measurements per amount of connected UEs can be plotted as shown in the example in Figure 12.29.

In a similar way to the AMR speech case, the average fractional load per number of connected UEs can be calculated in good coverage conditions, i.e. assuming $P_{tot,n}/L_{m,i} \gg P_N$, and plotted as in Figure 12.30. The best linear fit, fractional load as a function of number of connections, can then be derived as depicted in Figure 12.30, where y is the fractional load η, and x is the number of connections. The average achieved results for circuit switched video call capacity testing are presented in Section 12.4.3

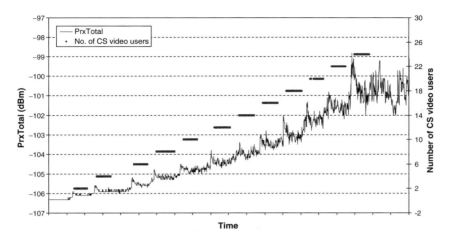

Figure 12.29. Uplink total received power (Prx) as a function of the number of circuit switched video call users

UL noise rise vs. number of CS video users

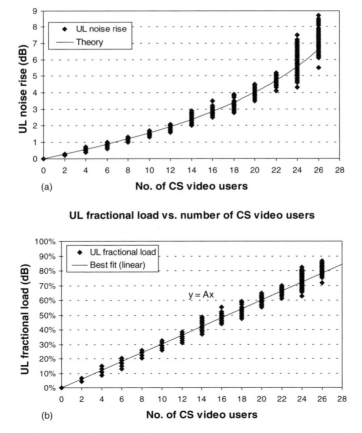

UL fractional load vs. number of CS video users

Figure 12.30. Uplink noise rise (a) and fractional load (b) as a function of the number of circuit switched video call users

12.4.1.4   Circuit Switched Video Capacity Downlink

The downlink capacity test has the same assumptions as the uplink test. It is further assumed that the orthogonality is 0.5. An example of the total transmitted power and the number of connected UEs in the tested cell is shown in Figure 12.31.

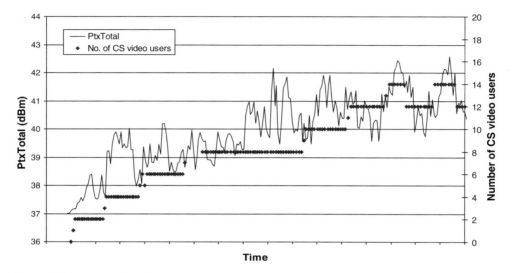

Figure 12.31. Downlink total transmitted power (Ptx) as a function of the number of circuit switched video call users

Using the same calculation method as for AMR voice, the fractional load can be plotted as depicted in Figure 12.32.

The resulting average downlink E_b/N_0 example values are shown in Section 12.4.3. The downlink E_b/N_0 for a circuit switched video call is usually lower than for an AMR speech connection in the same conditions. This is due to a higher bit rate having more efficient turbo coding. Assuming 1 dB lower E_b/N_0 for circuit switched video, and taking into account the difference in the processing gains between 64 kbps and 12.2 kbps, the expected difference in the downlink connection powers is approximately 6 dB. Example measured downlink code powers are shown in Figure 12.33 and the averaged power in Figure 12.34. The maximum downlink code power fluctuation can be seen to be, on average, 10 to 12 dB, which is higher than for an AMR speech call. This is due to the lower BLER requirement for a circuit switched video call and higher required average transmission power from the base station compared to the AMR speech call. The average downlink power per connection in Figure 12.34 varies between 21 dBm and 30 dBm in this example, showing the variation for UEs in different locations as well as different UE models. The average power difference between a circuit switched video call and an AMR speech call is approx 6–7 dB when comparing Figure 12.26 for AMR voice and Figure 12.34 for circuit switched video. That difference is an expected result based on the processing gain difference.

12.4.1.5   Packet Data Capacity Downlink

The downlink packet data capacity test is done for 384 kbps bit rate and the same analysis methodology is used as for AMR voice and circuit switched video. The BLER target is set to

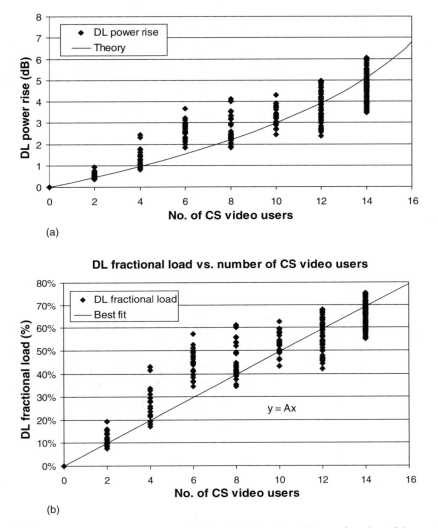

Figure 12.32. Downlink power rise (a) and downlink fractional load (b) as a function of the number of circuit switched video call users

1 %. The application used in the testing is FTP download, and for every connection the FTP throughput is recorded. The total transmitted power and the number of connected UEs in an example case are shown in Figure 12.35.

The power rise above the common channel powers and the fractional load can be plotted as in Figure 12.36.

The resulting average downlink E_b/N_0 example values are shown in Section 12.4.3. The downlink E_b/N_0 for a packet switched 384 kbps call is lower than for a AMR speech or for

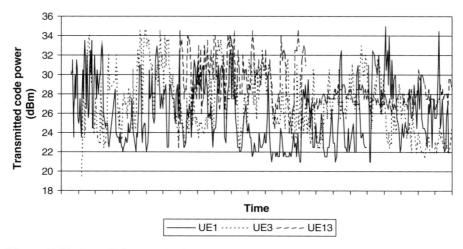

Figure 12.33. Downlink transmitted code power per circuit switched video call connection

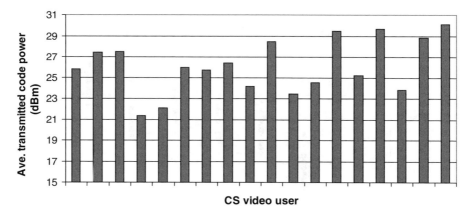

Figure 12.34. Downlink average power per circuit switched video call connection

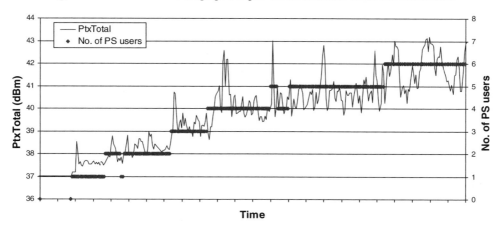

Figure 12.35. Downlink total transmitted power (Ptx) as a function of the number of packet switched 384 kbps users

DL Power Rise vs. number of PS 384kbps users

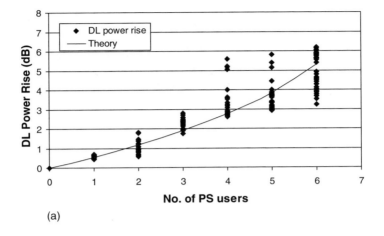

(a)

DL Fractional Load vs. number of PS 384kbps users

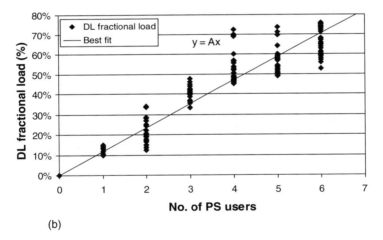

(b)

Figure 12.36. Downlink power rise (a) and downlink fractional load (b) as a function of the number of packet switched 384 kbps users

circuit switched video call. This is due to a higher bit rate service having more data to transmit, therefore the Turbo coding and interleaving become more efficient.

The downlink code power can be plotted as well to see the power deviation per connected UE. In Figure 12.37 one example is shown. The maximum downlink code power fluctuation can be seen to be in the region of 8 dB, which is slightly higher than for AMR and close that for a circuit switched video call. However, it should be noted that the transmission power in these cases is close to its maximum value, reducing the power variance. This can be seen in the case of UE2 in Figure 12.37. The maximum power per connection was set to 39 dBm.

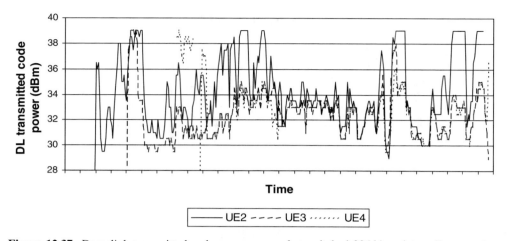

Figure 12.37. Downlink transmitted code power per packet switched 384 kbps data call connection

The average downlink code power per connection is depicted in Figure 12.38, and it can be seen that, also, the average power difference between a packet switched 384 kbps packet call and an AMR speech call is about 12–13 dB, which is an expected result taking into account the difference in the processing gain of 15 dB and the lower E_b/N_0 for packet data.

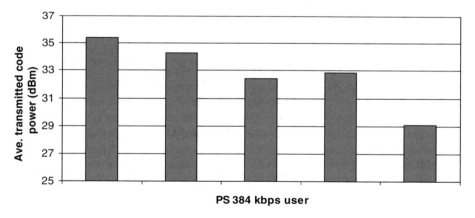

Figure 12.38. Downlink average transmitted code power per packet switched 384 kbps data call connection

The throughput of each 384 kbps connection is also recorded in the trials. Examples of FTP download throughput are shown in Figure 12.39. The throughput is sampled every second and captured by the application PC connected to each UE. Occasional fluctuation in throughput can be expected in the plots due to the very short sampling period. It can be seen that the throughput performance varies from one UE to another, depending on the location of the UE. Degradation is expected for UE requiring higher downlink transmitted code power

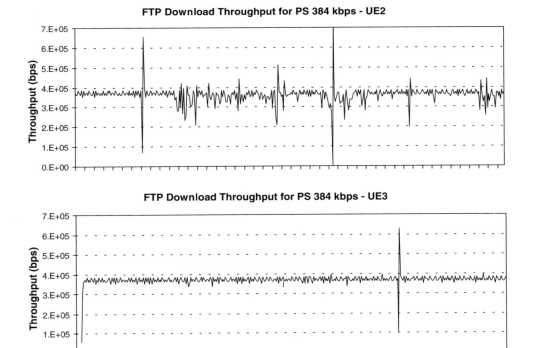

Figure 12.39. Examples of FTP download throughput during the capacity trial

under bad radio or high load conditions due to the fact that the maximum downlink transmitted code power is sometimes achieved. The upper part of Figure 12.39 shows some drops in UE2 throughput due to power limitations, as was seen in Figure 12.37. In other cases, little or no impact is seen in the throughput when the downlink transmitted code power is within the downlink power control range.

12.4.2   Multicell Capacity Trials

The multicell capacity testing is similar to the single cell capacity testing described in the previous section. The difference is that the other-to-own cell interference i has significant impact and it needs to be included in the calculations.

12.4.2.1   Uplink Methodology

The average E_b/N_0 can be estimated first by having only one UE driving inside the coverage area. In this way we can define the average E_b/N_0 in moving conditions over the whole coverage area. The effect of other-to-own cell interference can be included using Equation (12.7), where the other-to-own cell interference can be determined based on downlnk testing, as specifed in the following section.

12.4.2.2   Downlink Methodology

The downlink calculations start with Equation (12.10). The equation can be simplified by assuming that the total transmitted power of all base stations in the area is the same. This assumption gives slightly pessimistic results, as testing is done having several UEs in the measurement van and the neighbouring base station average total transmitted power is always smaller than the average total transmitted power of the serving base station. Equation (12.10) can be modified by adding the activity factor as well as the common channel power. The total transmitted power, $P_{tot,m}$, can be solved by summing over all the connections i as specified below:

$$\sum_{i=1}^{I} P_i + P_{CCH} = P_{tot,m} \cdot \sum_{i=1}^{I} \frac{\left(\frac{E_b}{N_0}\right)_i \cdot v_i}{W/R} \left[(1 - \alpha_i) + \sum_{n=1,n\neq m}^{N} \frac{L_{m,i}}{L_{n,i}}\right] + P_N \cdot \sum_{i=1}^{I} \frac{\left(\frac{E_b}{N_0}\right)_i \cdot v_i}{W/R}$$

(12.20)

The path loss $L_{x,y}$ in the above equation stands for the air interface path loss, i.e. excluding antenna gain, cable losses and soft handover gain. $P_{tot,m}$ can be solved as:

$$P_{tot,m} = \frac{P_{CCH} + P_N \cdot \sum_{i=0}^{I} \frac{\left(\frac{E_b}{N_0}\right)_i \cdot v_i}{W/R} L_{m,i}}{1 - \sum_{i=1}^{I} \frac{\left(\frac{E_b}{N_0}\right)_i \cdot v_i}{W/R} \left[(1 - \alpha_i) + \sum_{n=1,n\neq m}^{N} \frac{L_{m,i}}{L_{n,i}}\right]}$$

(12.21)

The downlink load factor, η_{DL}, can be formulated as the increase in the required base station power in order to maintain the required I connections, taking into account the non-orthogonal inter-cell interference and the non-orthogonal part of the intra-cell interference.

$$\eta_{DL} = \sum_{i=1}^{I} \frac{\left(\frac{E_b}{N_0}\right)_i \cdot v_i}{W/R} \left[(1 - \alpha_i) + \sum_{n=1,n\neq m}^{N} \frac{L_{m,i}}{L_{n,i}}\right]$$

(12.22)

The average other-to-own cell interference can be calculated in downlink as:

$$\bar{i}_{DL} = \frac{\sum_{i=1}^{I} \sum_{n=1,n\neq m}^{N} \frac{L_{m,i}}{L_{n,i}}}{I}$$

(12.23)

The path loss can be calculated from the RSCP measurements when the CPICH transmision power is known. The average other-to-own cell interference ratio can be calculated by averaging the values over all measurement points. This other-to-own cell interference can be used also to calculate the uplink load, assuming that the average other-to-own cell interference per cell is the same in uplink and in downlink. This assumption is correct as long as the traffic is uniformly distributed between the cells.

The average E_b/N_0 in downlink can be calculated in the same way as described for single cell test cases: by plotting the downlink fractional load as a function of the number of connected UEs and defining the best fit. It should be noted that the E_b/N_0 calculated this way also includes the macro diversity gain as the load measurement from the base station cell includes all the served UEs, i.e. UEs in soft handover and UEs not in soft handover.

12.4.3 Summary

The results from the capacity trials are summarised in this section. The results show limiting factors in maximum capacity, uplink and downlink pole capacity, uplink and downlink E_b/N_0 and the average downlink connection powers. The single cell test results are shown in Table 12.15. This summary includes results also from other single cell tests besides the ones introduced earlier in this chapter.

Table 12.15. Summary of single cell capacity trials

	AMR 12.2 kbps voice with 100 % voice activity	64 kbps video	384 kbps packet data
Limiting factor	Uplink noise rise	Total downlink power usually or uplink noise rise	Total downlink power
Uplink pole capacity	80–112 users	27–38 users	—
Downlink pole capacity	93–125 users	15–33 users	5.8 – 8.9 users (code limit 7 users)
Uplink E_b/N_0	4.5 dB – 5.9 dB	2.1 dB – 3.5 dB	—
Downlink E_b/N_0	6.1 dB – 7.4 dB	4.6 dB – 7.9 dB	2.6 dB – 4.4 dB

We can note that for an AMR 12.2 kbps voice service, the pole capacity is always limited by the uplink noise rise. For a 64 kbps circuit switched video service, the pole capacity is usually limited by the maximum downlink power of the cell. However, under good radio conditions, i.e. good coverage with small path loss from the cell, the uplink noise rise could be the limiting factor due to low downlink E_b/N_0, as well as the expected low downlink transmitted code power per connection. For a 384 kbps packet switched data service, the pole capacity is always limited by the maximum downlink power of the cell, as the data rate is asymmetrical. When comparing the E_b/N_0 values from the field trials to the expected values in Section 12.5, we can note that the values are quite well in line.

The expected downlink connection powers for each service are summarised in Table 12.16. The results are shown as the relative power compared to AMR voice and as absolute powers. The connection powers provide some indications for designing the downlink power control range for different services.

The single cell results are applied to estimate the maximum multicell capacities. The following assumptions are used: voice activity of 50 %, downlink orthogonality 0.5 and other-cell to own-cell interference ratio of 0.65. The voice activity of 50 % is assumed to lead to physical layer activity factor of 67 % in uplink and 58 % in downlink. The results in

Table 12.16. Comparison of the average downlink transmitted code power for different services

	AMR 12.2 kbps voice with 100 % voice activity	64 kbps video	384 kbps packet data
Average downlink code power	17 dBm – 23 dBm	21 dBm – 30 dBm	29 dBm – 35 dBm
Relative power compared to AMR voice	Reference	6 dB – 7 dB higher than AMR	12 dB – 13 dB higher than AMR

Table 12.17 show that the limiting factor in the multicell case typically is the total downlink power. The uplink may become a limiting factor for symmetric services if we want to limit the uplink noise rise to a low value to provide maximised coverage. The estimated multicell capacities are summarised in Figure 12.40.

Table 12.17. Estimated multicell pole capacities based on single cell measurements

	AMR 12.2 kbps voice with 50 % voice activity	64 kbps video	384 kbps packet data
Limiting factor	Total downlink power	Total downlink power	Total downlink power
Uplink pole capacity	72 – 101 users	16 – 23 users	—
Downlink pole capacity	70 – 94 users	7 – 14 users	2.5 – 3.9 users

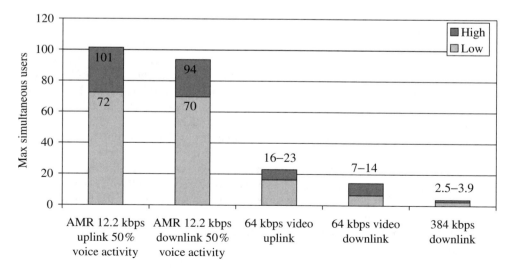

Figure 12.40. Estimated multicell pole capacities based on single cell measurements

12.5 3GPP Performance Requirements

In this section we derive typical E_b/N_0 values and receiver noise figures from 3GPP performance requirements [12, 13]. A simplified view of the receiver parts is shown in Figure 12.41. The assumed services are 12.2 kbps AMR voice and packet data 64, 128 and 384 kbps with BLER of 1 %.

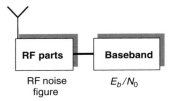

Figure 12.41. Simplified view of the receiver and the performance requirements

12.5.1 E_b/N_0 Performance

The uplink E_b/N_0 values are derived as follows:

1. Take the E_b/N_0 requirement for static and multipath channels from [12].

2. [12] includes higher layer signalling channel with 100 bits per 40 ms. We can remove that overhead since the signalling channel is active only when there is a signalling need. The overhead is $10 \cdot \log 10$(user bits per 40 ms + 100)/(user bits per 40 ms)).

3. We assume that the average base station product performs 1.5 dB better than 3GPP requirement.

4. We assume that outer loop power control causes a loss of 0.3 dB, approximately half of a typical power control step size of 0.5 dB.

5. We assume that fast power control degrades performance by 0.3 dB compared to the ideal non-power control case at 120 km/h.

6. We assume that the required E_b/N_0 is 1.0 dB lower at 3 km/h than at 120 km/h.

The calculations of the uplink E_b/N_0 values are shown in Table 12.18 for a static channel and in Table 12.19 for a multipath fading channel.

Table 12.18. Uplink static channel performance

	Voice	Data 64 kbps	Data 128 kbps	Data 384 kbps
E_b/N_0[12]	5.1 dB	1.7 dB	0.9 dB	1.0 dB
DCCH overhead	−0.8 dB	−0.2 dB	−0.1 dB	−0.03 dB
Product vs. 3GPP	−1.5 dB	−1.5 dB	−1.5 dB	−1.5 dB
Outer loop power control	+0.3 dB	+0.3 dB	+0.3 dB	+0.3 dB
E_b/N_0	3.1 dB	0.3 dB	−0.4 dB	−0.2 dB

Table 12.19. Uplink multipath fading channel (Case 3) performance

	Voice	Data 64 kbps	Data 128 kbps	Data 384 kbps
E_b/N_0[12]	7.2 dB	3.8 dB	3.2 dB	3.6 dB
DCCH overhead	−0.8 dB	−0.2 dB	−0.1 dB	−0.03 dB
Product vs. 3GPP	−1.5 dB	−1.5 dB	−1.5 dB	−1.5 dB
Outer loop power control	+0.3 dB	+0.3 dB	+0.3 dB	+0.3 dB
Fast power control	+0.3 dB	+0.3 dB	+0.3 dB	+0.3 dB
E_b/N_0 120 km/h	5.5 dB	3.0 dB	2.5 dB	3.0 dB
3 km/h vs 120 km/h	−1.0 dB	−1.0 dB	−1.0 dB	−1.0 dB
E_b/N_0 3 km/h	4.5 dB	2.0 dB	1.5 dB	2.0 dB

The downlink E_b/N_0 values are derived as follows:

1. Take the E_c/I_{or} requirement from [13]. Take also the I_{or}/I_{oc} operation point from [13].

2. Calculate E_b/N_0 with the following formula

$$\frac{E_b}{N_0} = 10 \log_{10} \left(\frac{\dfrac{\text{Chip rate}}{\text{Bit rate}} \cdot \dfrac{E_c}{I_{or}}}{1 - \text{orthogonality} + \dfrac{1}{I_{or}/I_{oc}}} \right) \qquad (12.24)$$

We assume an orthogonality of 1.0 for a static channel and 0.5 for a multipath channel.

3. We remove the effect of the signalling channel as in uplink.

4. A slot format of 11 is assumed for voice in [13]. The physical layer overhead can be decreased by 0.8 dB by using a slot format of 8.

5. We assume that the average base station product performs 1.5 dB better than the 3GPP requirement.

6. We assume that outer loop power control causes a loss of 0.3 dB, approximately half of a typical power control step size of 0.5 dB.

7. We assume that fast power control gives the same performance as the ideal non-power control case at 120 km/h.

8. We assume that the required E_b/N_0 is 0.3 dB higher at 3 km/h than at 120 km/h because of the power rise.

The calculations of the downlink E_b/N_0 values are shown in Table 12.20 for voice and in Table 12.21 for data.

The E_b/N_0 values are summarised in Figure 12.42 for uplink and in Figure 12.43 for downlink. We can note that the uplink E_b/N_0 is typically 2–3 dB lower than that for the downlink. The reason is antenna diversity, that is assumed in uplink but not in downlink. The downlink transmit diversity would improve the downlink E_b/N_0 values.

The calculated values can be compared to the values obtained from field measurements. Since stationary mobiles were used in the measurements, we compare the results to the

Table 12.20. Downlink static channel performance

	Voice	Data 64 kbps	Data 128 kbps	Data 384 kbps
E_c/I_{or} from [13]	−16.6 dB	−12.8 dB	−9.8 dB	−5.5 dB
I_{or}/I_{oc}	−1.0 dB	−1.0 dB	−1.0 dB	−1.0 dB
E_b/N_0 from Eq. (12.24)	7.4 dB	3.7 dB	3.4 dB	3.4 dB
DCCH overhead	−0.8 dB	−0.2 dB	−0.1 dB	−0.03 dB
Slot format 8	−0.8 dB	—	—	—
Product vs. 3GPP	−1.5 dB	−1.5 dB	−1.5 dB	−1.5 dB
Outer loop power control	+0.3 dB	+0.3 dB	+0.3 dB	+0.3 dB
E_b/N_0	4.6 dB	2.6 dB	2.7 dB	2.3 dB

Table 12.21. Downlink multipath fading channel (Case 3) performance

	Voice	Data 64 kbps	Data 128 kbps	Data 384 kbps
E_c/I_{or} from [13]	−11.8 dB	−7.4 dB	−8.5 dB	−5.1 dB
I_{or}/I_{oc}	−3.0 dB	−3.0 dB	3.0 dB	6.0 dB
E_b/N_0 from Eq. (12.24)	9.2 dB	5.7 dB	5.3 dB	5.3 dB
DCCH overhead	−0.8 dB	−0.2 dB	−0.1 dB	−0.03 dB
Slot format 8	−0.8 dB	—	—	—
Product vs. 3GPP	−1.5 dB	−1.5 dB	−1.5 dB	−1.5 dB
Outer loop power control	+0.3 dB	+0.3 dB	+0.3 dB	+0.3 dB
Fast power control	+0.0 dB	+0.0 dB	+0.0 dB	+0.0 dB
E_b/N_0 120 km/h	6.4 dB	5.0 dB	5.0 dB	4.9 dB
3 km/h vs 120 km/h	+0.3 dB	+0.3 dB	+0.3 dB	+0.3 dB
E_b/N_0 3 km/h	6.7 dB	5.3 dB	5.3 dB	5.2 dB

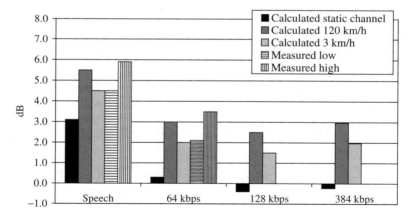

Figure 12.42. Summary of uplink E_b/N_0

calculated values at 3 km/h. The measured lowest values are similar to those calculated in uplink for voice and 64 kbps, but the measured highest values are 1.5 dB higher than the calculated ones. It is to be expected that the measured worst case E_b/N_0 values are higher than the calculated ones. In the case of 64 kbps, part of the difference is caused by the different BLER requirements: calculations are based on packet data BLER 1 %, while the

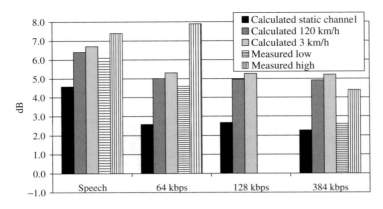

Figure 12.43. Summary of downlink E_b/N_0

measurements are based on BLER 0.5 %. The measured lowest voice E_b/N_0 values in downlink are 0.6 dB lower, and the measured highest values 0.6 dB higher than the calculated ones. The lowest measured 64 kbps E_b/N_0 values are slightly below the calculated ones, but the highest values are clearly above the calculated ones. The difference in BLER target also affects the benchmarking here. The measured 384 kbps values are clearly below the calculated ones. The low measured values are partly explained by the 384 kbps link transmission power hitting its maximum value in the measurements, see Figure 12.37.

This book has typically used the following E_b/N_0 values for low mobile speed cases: uplink voice 4.0–5.0 dB, uplink data 1.5–2.0 dB, downlink voice 7.0 dB and downlink data 5.0 dB.

12.5.2   RF Noise Figure

The E_b/N_0 values represent mainly the baseband performance of the receiver. The receiver RF performance is described with the RF noise figure, which represents the loss of the signal power in the receiver RF parts. The required reference sensitivity level for 12.2 kbps voice for the base station is −121 dBm and for the mobile station −117 dBm [13]. The RF noise figure can be calculated when the baseband E_b/N_0 performance is known:

$$\text{RF noise figure} = \text{Reference sensitivity} - \left(-174\text{dBm} + 10\log_{10}(12.2e3) + E_b/N_0\right)$$

$$(12.25)$$

The RF noise figure calculations are shown in Table 12.22. The required base station noise figure is estimated to be 4.5 dB and the mobile station 8.7 dB. The RF noise figure

Table 12.22. RF noise figures

	Base station	Mobile station
3GPP sensitivity requirement for voice	−121 dBm [13]	−117 dBm [1]
3GPP E_b/N_0 requirement in static channel	5.1 dB + 2.5 dB[1] from Table 12.18	7.4 dB from [1]
RF noise figure from Equation (12.25)	4.5 dB	8.7 dB

[1]The sensitivity of −121 dBm is without antenna diversity. The 2.5 dB factor is the estimated antenna diversity gain.

requirement can be more relaxed if the baseband E_b/N_0 performance is better than the assumption here. On the other hand, the average product is better than the requirement due to the production variations. In this book we have typically assumed an average base station noise figure of 4.0 dB and an average mobile station noise figure of 7.0 dB.

12.6 Performance Enhancements

WCDMA performance enhancements with advanced antenna structures and with baseband multiuser detection are described in this section. Both solutions are considered here mainly for the base station for improving uplink performance.

12.6.1 Smart Antenna Solutions

The use of antenna arrays at the Node B can help increase the capacity of terrestrial cellular systems significantly, or alternatively improve the coverage [14, 15, 16]. The basic principle is to multiply the signals at the different antenna branches with complex weight factors before the signals are transmitted, or before the received signals are summed, as illustrated in Figure 12.44. This set-up can be regarded as a spatial filter, where the signals at the different antenna branches represent spatial samples of the radio channel, and the complex weight factors are filter coefficients. The set of complex weight factors (also known as weight vectors) is typically different for each user.

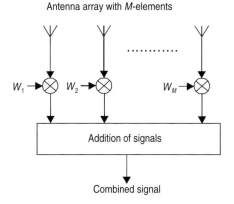

Figure 12.44. Basic principle for antenna array processing. The example is for the case where the antenna array is used for reception

Antenna array systems can basically be designed to operate in one of two distinct modes: (1) diversity mode or (2) beamforming mode. Diversity techniques help increase the signal-to-interference ratio and reduce the likelihood of deep fades, provided that there is statistical independence between the signals at the antenna elements. Beamforming techniques, on the other hand, are applied when there is coherence between the signals at the antenna elements, so a narrow beam can be created towards the desired user. Such schemes are known to provide an average spatial interference suppression gain, which basically depends on the effective beamwidth and the sidelobe level, assuming a scenario with a large number of users compared to the number of antenna elements. Coherence between the antenna signals in a compact array with half a wavelength element spacing can be expected in environments such as typical urban macro cells, where the azimuth dispersion of the radio channel seen at the antenna array is typically small. As an example, Figure 12.45 illustrates the physical dimensions of a four-element linear antenna array for WCDMA, as well as an example of the use of such arrays in a three-sector site configuration.

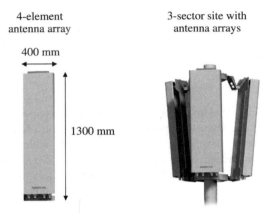

Figure 12.45. A four-element linear antenna array

During the last two decades, many beamforming algorithms have been proposed and analysed for selection of complex weight vectors; see results for uplink vector Rake receivers in [17] and downlink beamforming techniques in [18, 19, 20], among others. More degrees of freedom are typically available for uplink beamforming, where the radio channel can be more accurately estimated and different beamformer weights can be applied for the different multipath delays. On the contrary, downlink beamforming from the Node B is subject to more constraints, since the downlink radio channel is unknown, only the average direction towards the UEs can be estimated. In the following sub-sections, we will further address the use of beamforming techniques in UMTS, and shortly discuss how implementation of such techniques impacts on the radio resource management algorithms. Performance results for the gain of beamforming schemes are also presented.

12.6.1.1 Beamforming Options

Two different beamforming modes are possible for the downlink of UMTS within one logical cell: (1) user-specific beamforming and (2) fixed beamforming. User-specific beamforming allows for the generation of individual beams to each UE without any restrictions on the selection of weight vectors. However, this prevents the UE from using the primary common pilot channel (P-CPICH) for phase reference, since the P-CPICH must be transmitted on a sector beam which provides coverage in the entire cell area. The phase rotation of the P-CPICH received at the UE is therefore likely to be different from the phase rotation of the signal transmitted under the user-specific beam. UEs receiving signals subject to user-specific beamforming are therefore informed via higher layer protocols to use dedicated pilot symbols for phase reference instead of the P-CPICH. Fixed beamforming refers to the case where a finite set of beams are synthesised at the Node B, so multiple UEs may receive signals transmitted under the same beam. Even though this beamforming mode is referred to as fixed beamforming, the complex weight factors used to synthesise the beams may be varied over time to facilitate a slow adaptation of the directional beams. Each of the beams is associated with a unique secondary common pilot channel (S-CPICH), which the UEs are informed to use for phase reference via higher layer protocols. Table 6.5 in Chapter 6 shows an overview of the beamforming options for different downlink channels. Notice that beamforming is not allowed on all channels, and some channels support only fixed beamforming, while user-specific beamforming is optional. In this context, optional means that this beamforming mode is not mandatory for the UE, and therefore the network cannot assume that all UEs in the system support, for instance, user-specific beamforming on HS-DSCH. As the HS-DSCH and HS-SCCH belong to the high-speed downlink packet access (HSDPA) concept, the scope of applicability for the combination of user-specific beamforming and HSDPA depends on the UE implementation. Figure 12.46 shows an example with the mapping of physical channels onto a grid of six directional fixed beams plus one sector beam. Notice that, despite the Node B using beamforming, it still needs to offer a sector beam for channels which are not allowed to use beamforming. It is assumed that transmission of transport channels towards one UE from a Node B is conducted via one directional beam per cell only. Reception from two directional beams would involve the use of two S-CPICHs for phase reference at the UE (one for each directional beam), which is not allowed according to the current UMTS specifications. Link level results for both user-specific beamforming and fixed beamforming are available in [18 and 19].

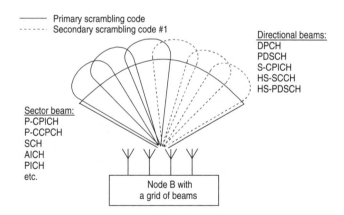

Figure 12.46. Mapping of physical channels onto a grid of fixed beams for the case where four elements are used to form six directional beams, plus one wide sector beam. For the sake of simplicity, the sidelobes on the directional beams are not illustrated

12.6.1.2 Higher Order Sectorisation via Beamforming

Instead of synthesising beams within one logical cell, the antenna array can also be exploited to create individual cells where the coverage area of each beam represents a logical cell. This implies that each beam is transmitted under a unique primary scrambling code. In addition, each beam will have its own P-CPICH, BCH, PCH, etc. An example is shown in Figure 12.47, where three uniform linear antenna arrays with four elements each are used to create

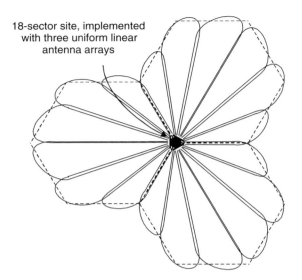

18-sector site, implemented
with three uniform linear
antenna arrays

Figure 12.47. An example of an 18-sector site implemented with beamforming on three uniform linear antenna arrays. The figure shows the approximate coverage areas of each beam (sector)

18 logical sectors. Provided that the beamforming network is implemented digitally, this solution allows for adaptive cell sectorisation to maximise the coverage and capacity in coherence with the traffic distribution in the network. Using this configuration, the pilot overhead from having an S-CPICH per beam is avoided compared to synthesising a grid of fixed beams in a logical cell. However, removal of the S-CPICH overhead is obtained at the expense of reduced downlink orthogonality, since each beam is transmitted under a unique primary scrambling code. Finally, higher order sectorisation via beamforming allows for flexible allocation of transmit power in the different sectors, compared to the scheme where each sector has its own panel antenna and power amplifier.

12.6.1.3 Node B Measurements for Beamforming Support

As discussed in Chapters 9 and 10, the family of radio resource management algorithms such as admission control, packet scheduling, handover control, and congestion control are primarily implemented in the RNC according to the UMTS architecture. As a consequence of this functional split, these algorithms must rely on standardised measurements collected at the Node B and reported via the open Iub interface to the RNC. For this reason the Node B informs the RNC via a configuration message whether it uses fixed beamforming, user-specific beamforming, or no beamforming.

In order to provide measurements of the spatial load distribution in cells with beamforming capabilities, new measurements have been proposed for the Node B per *cell portion*, as well as the corresponding reporting over the Iub interface to the RNC [21]. A cell portion is defined as a geographical part of a cell for which a Node B measurement can be reported to the RNC, and for which the RNC can allocate a phase reference for a UE that is within the cell portion. A cell portion is semi-static, and identical for both the uplink and the downlink. Within a cell, a cell portion is uniquely identified by a cell portion ID [21]. Hence, assuming that fixed beamforming as discussed in Section 12.6.1.1. is applied, the cell portions are typically configured to be identical to the directional beams.

A pseudo uplink direction-of-arrival (DoA) measurement is included in 3GPP Release 5 for Node B to facilitate beam switching in the downlink. The Node B measures the average uplink signal-to-interference ratio (SIR) on the dedicated physical control channel (DPCCH) received from each UE in all the cell portions [21]. The four highest SIR values and the corresponding cell portion IDs are sent to the RNC. These measurements are used to implement the beam switching functionality that is required when fixed beamforming is used in the downlink. Based on the cell portion specific SIR measurements, the RNC will typically inform the Node B to transmit the data to a UE under the beam (cell portion) corresponding to the highest uplink measured SIR, as well as informing the UE which S-CPICH it should use for phase reference via a radio resource control (RRC) message, as illustrated in Figure 12.48. Notice that beam switching only includes measurements from the Node B, i.e. no measurements from the UE are used to trigger beam switching. The beam switching operation is discussed in more detail in [20].

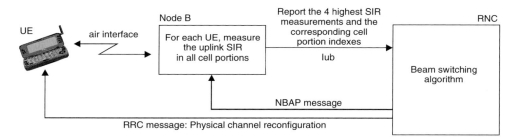

Figure 12.48. Simplified sketch of the Node B measurments, and the signalling flow from the RNC to the Node B and the UE during beam switching

For Release 6, measurements of the carrier transmit power and total received wideband power per cell portion are also defined for the Node B. These measurements can be regarded as downlink/uplink measurements of the load per cell portion, i.e. equivalent to directional load measurements. These beamforming-specific measurements make it possible to take full system performance benefits from the beamforming capabilities. As an example, directional power-based radio resource management algorithms are proposed in [22 and 23], to make sure that the spatial interference suppression gain offered by beamforming techniques is mapped into a capacity gain in the network.

12.6.1.4   Capacity Results From Dynamic Network Simulations
The results presented in this section are obtained from downlink dynamic network simulations with multiple UEs and 33 cells, with a site-to-site distance of 2.0 km. The simulations include accurate modelling of the radio link performance towards each UE, mobility of the UEs, radio resource management algorithms, traffic models with 64 kbps circuit switched services on dedicated channels, etc. The detailed simulation methodology is described in [20]. The reference configuration is a standard three-sector network topology, where a panel antenna is used for each sector, with a 3 dB beamwidth of 65 degrees, as illustrated in Figure 12.49. The considered beamforming configuration assumes a four-element linear antenna array per sector, which is exploited to synthesise six directional

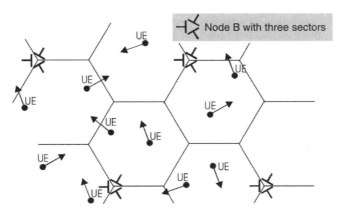

Figure 12.49. Network topology for the reference configuration

beams with a separate S-CPICH assigned to each beam. Finally, a scenario with higher order sectorisation, as illustrated in Figure 12.47, is also considered, where three uniform linear antenna arrays with four elements are used to synthesise a total of 18 sectors per Node B.

Assuming that only one scrambling code is allocated per sector, the relative capacity gain is found be a factor of 1.9 for the beamforming case. For this configuration the capacity gain is severely limited by the shortage of channelisation codes, since it happens with a probability of 28 % that an incoming call is rejected because there are no available channelisation codes. Hence, under these conditions the capacity is hard limited rather than interference/power limited, so the interference suppression gain offered by the use of beamforming techniques cannot be fully exploited. By introducing a secondary scrambling code in the cell, the capacity gain increases to a factor of 2.4 and the likelihood of channelisation code shortage is practically reduced to zero. This indicates that the spatial interference suppression gain from using beamforming techniques is effectively mapped into a capacity gain by using two scrambling codes per cell for the considered scenario. Notice that the capacity gain of 2.4 includes the overhead from the S-CPICH per beam. The use of antenna arrays with more than four elements may require more than two scrambling codes per cell, due to the larger spatial interference suppression gain.

Additional simulation results for the three considered network configurations are sum-marised in Figure 12.50. The presented results for the net capacity gain and soft handover are normalised with the results for the reference scenario. Two scrambling codes are enabled for the scenario where six directional beams are formed within each logical cell. The largest capacity gain is achieved for the three-sector configuration with an antenna array in each sector, without increasing the soft handover overhead. The capacity gain for the configura-tion with 18 sectors is approximately 20 % less than this, in spite of the fact that both configurations use three antenna arrays with four elements to create the same number of beams. The configuration with 18 sectors suffers from a slightly worse equivalent downlink orthogonality factor, since different primary scrambling codes are used for each sector (i.e. 18 scrambling codes per site), while the three-sector configuration with beamforming uses only two scrambling codes per sector, or equivalently six scrambling codes per site. In addition, the 18-sector configuration suffers from a large soft handover overhead, which is more than twice as large as the overhead observed for the reference configuration

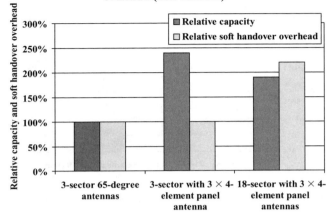

Figure 12.50. Comparison of the downlink capacity gain for dedicated channel and SHOO for different site configurations

with three-sector sites. The larger overhead not only indicates the need for additional baseband resources and transmission capacity over the Iub, but it also represents potential problems in managing fast updates of the UE's active set for fast moving UEs.

The results above are obtained for the downlink assuming that the traffic is carried on dedicated channels. The equivalent beamforming capacity gain is typically found to be 10–15 % higher for the uplink compared to the downlink assuming single antenna uplink reception, as there is no additional pilot overhead for the uplink and different beams can be formed for each multipath delay in the radio channel. However, when compared against Node B configuration two uplink receive antennas and maximal ratio combining, the uplink beamforming capacity gain with a four-element antenna array is found to be of the order of a factor 1.7.

The later results were all obtained for capacity limited scenarios. However, beamforming techniques can also be applied to improve the coverage. Ideally, the additional beamforming gain equals $10 \log(M)$, where M is the number of antenna elements in the array. However, due to azimuthal dispersion in the radio channel and other imperfections, the actual beamforming gain is typically slightly smaller, i.e. for $M = 4$ antennas, a more realistic estimate of the beamforming antenna gain is of the order of 5 dB, compared to 6 dB in the ideal case. Assuming a path loss exponent of -3.5, a beamforming antenna gain of 5 dB is equivalent to approximately a 38 % increase in range extension.

12.6.1.5 Beamforming for HSDPA

The combination of fixed beamforming and HSDPA is also feasible. This is possible by transmitting to one UE under each beam during each TTI, so the number of served UEs in each TTI equals the number of beams. It is the packet scheduler in the Node B (in the MAC-hs) that determines which UEs should be scheduled under the different beams. As discussed in [24], independent packet schedulers for each of the beams can be used, such as, proportional fair packet scheduling or simple round robin packet scheduling. Results for

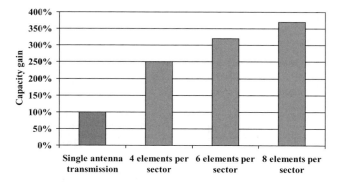

Figure 12.51. HSDPA cell capacity gain versus the number of antenna elements

the HSDPA beamforming cell capacity gain are shown in Figure 12.51 as a function of the number of antenna elements in the array. These results are obtained from dynamic macro cellular network simulations, assuming an ITU Vehicular-A power delay profile and proportional fair packet scheduling [24].

12.6.2 Multiuser Detection

The uplink performance improvements with base station multiuser detection (MUD) are discussed in this section. The aim is to give an overview of different multiuser detection algorithms, and references to more detailed information. Another target is to assess capacity and coverage gains of an MUD receiver. Section 12.6.2.1 gives an overview of MUD receivers. Section 12.6.2.2 describes a practical parallel interference canceller (PIC). Section 12.6.2.3 contemplates PIC efficiency and derives equations for capacity and coverage gains that PIC can offer. Section 12.6.2.4 presents numerical results of soft quantisation PIC (SQ-PIC) for different propagation channels and bit rates for one, two and four diversity antennas.

12.6.2.1 Overview

CDMA systems are inherently interference-limited from both the receiver performance and system capacity points of view [32–34]. From the receiver perspective this means that, if the number of users is large enough, an increase in signal-to-noise ratio yields no improvement in bit or frame error rate. From the system capacity view, it means that the larger the signal-to-interference-plus-noise ratio required for the desired quality of service, the fewer users can be accommodated in the communication channel.

The interference-limited nature of CDMA systems results from the receiver design. In CDMA systems, the core of the receiver is a spreading code matched filter (MF) or correlator [34]. Since the received spreading codes are usually not completely orthogonal, multiple access interference (MAI) is generated in the receiver. If the spreading factor is moderate, the received powers of users are equal (no near–far problem), and the number of interfering users is large (>10), by the central limit theorem the multiple access interference can be modelled as increased background noise with a Gaussian distribution. This approximation has led to the conclusion that the matched filter followed by decoding is the optimal receiver for CDMA systems in additive white Gaussian noise (AWGN) channels. In frequency-selective channels, the Rake receiver [35] can be considered optimal with corresponding reasoning.

Although multiple access interference can be approximated as AWGN, it inherently consists of received signals of CDMA users. Thus, multiple access interference is very structured, and can be taken into consideration in the receiver. This observation led Verdú [36] to analyse the optimal multiuser detectors (MUDs) for multiple access communications. Verdú was able to show that CDMA is not inherently interference-limited, but that is a limitation of the conventional matched filter receiver.

The optimal multiuser detectors [36] can use either maximum *a posteriori* (MAP) detection or maximum likelihood sequence detection (MLSD). In other words, techniques similar to those applied in channels with inter-symbol interference [35] can be used to combat multiple access interference. The drawback of both the MLSD- and MAP-based multiuser detectors is that their implementation complexity is an exponential function of the number of users. Thus, they are not feasible for most practical CDMA receivers. This fact, together with Verdú's observation that CDMA with an MLSD receiver is not interference-limited, has triggered an avalanche of papers on sub-optimal multiuser receivers. A brief summary of the sub-optimal multiuser detection techniques is given below. For a more complete treatment, the reader is referred to the overview paper by Juntti and Glisic [37] and to the book by Verdú [38].

The existing sub-optimal multiuser detection techniques can be categorised in several ways. One way is to classify the detection algorithms as centralised multiuser detection or decentralised single user detection algorithms. The centralised algorithms perform real multiuser joint detection, i.e. they detect jointly each user's data symbols; they can be considered practical in base station receivers. The decentralised algorithms detect the data symbols of a single user based on the received signal observed in a multiuser environment containing multiple access interference; the single user detection algorithms are applicable to both base station and terminal receivers.

In addition to one kind of implementation-based categorisation on multiuser and single user detectors, the multiuser detectors can be classified based on the method applied. Two main classes in this category can be identified: linear equalisers and subtractive interference cancellation (IC) receivers. Linear equalisers are linear filters suppressing multiple access interference. The most widely studied equalisers include the zero-forcing (ZF) or decorrelating detector [39–41] and the minimum mean square error (MMSE) detector [42, 43]. The IC receivers attempt to explicitly estimate the multiple access interference component, after which it is subtracted from the received signal. Thus, the decisions become more reliable. Multiple access interference cancellation can be performed in parallel to all users, resulting in parallel interference cancellation (PIC) [44, 45]. Interference cancellation can also be performed in a serial fashion, resulting in serial interference cancellation (SIC) [46, 47].

Both linear equalisers and interference cancellation receivers can be applied in centralised receivers. The linear equalisers can also be implemented adaptively as single-user type decentralised detectors. This is possible if the spreading sequences of the users are periodic over a symbol interval so that multiple access interference becomes cyclostationary. Various adaptive implementations based on training sequences of the MMSE detectors have been studied [48–52]. So-called blind adaptive detectors not requiring training sequences have also been considered [51, 52, 53].

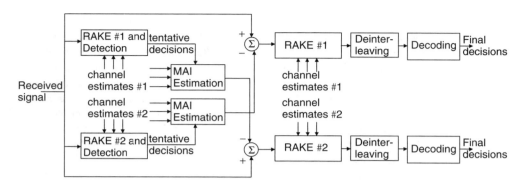

Figure 12.52. Parallel interference cancellation receiver for two users

The choice of multiuser detection techniques for WCDMA base station receivers has been studied [55–58]. Both the receiver performance and implementation complexity have been considered. The conclusion of the studies is that a multiuser receiver based on multistage parallel interference cancellation (PIC) is currently the most suitable method to be applied in CDMA systems with a single spreading factor. The PIC receiver principle with one cancellation stage for a two-user CDMA system is illustrated in Figure 12.52. The parallel interference cancellation means that interference is cancelled from all users simultaneously, i.e. in parallel. The cancellation performance can be improved by reusing the decisions made after interference cancellation in a new IC stage. This results in a multistage interference cancellation receiver, which is illustrated in Figure 12.53.

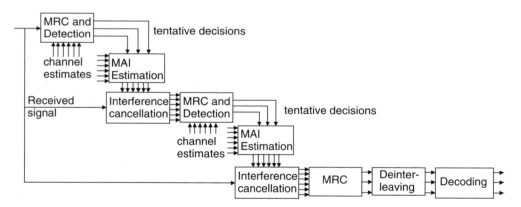

Figure 12.53. Multistage interference cancellation receiver

The choice of multiuser detection for multiservice CDMA systems with a variable spreading factor access has been considered by Ojanperä [59]. For such a system, a groupwise serial interference cancellation (GSIC) [60–63] receiver seems attractive. In the GSIC receiver, the users with a certain spreading factor are detected in parallel, after which multiple access interference caused by them is subtracted from the users with other spreading factors. A main reason for the GSIC being efficient is that the power of users depends on the spreading factor. By starting the cancellation from the users with the lowest spreading factor, the highest power users (the most severe interferers) are cancelled first.

The adaptive linear equalisers can be applied only if the spreading sequences of users are periodic over a relatively short time, such as over the symbol interval. Therefore, by using the short scrambling code option in a WCDMA uplink, the adaptive receivers could be utilised therein. In the WCDMA downlink, the spreading codes are periodic over one radio frame, whose duration is 10 ms. The period is so long that conventional adaptive receivers are practically useless. The problem can be partially overcome by introducing chip equalisers [64–69]. The idea here is to equalise the impact of the frequency-selective multipath channel on a chip-interval level. This suppresses inter-path interference (IPI) of the signals and also retains the orthogonality of spreading codes of users within one cell. The latter impact is possible, since synchronous transmission with orthogonal signature waveforms is applied in the downlink. In other words, multiple access interference in the downlink is caused by the multipath propagation, which can now be compensated for by the equaliser. The effect of multipath propagation on downlink performance without any interference suppression receivers is presented in Section 12.3.1.1. For a complete overview of the techniques available and the relevant literature, we refer the reader to [68]. The advanced receiver algorithms can be applied to WCDMA/HSDPA terminals to improve the end user data rates and system capacity. The performance of advanced HSDPA receivers is discussed in Chapter 11.

12.6.2.2   Structure of PIC

The PIC algorithm presented here is the so-called residual PIC in Figure 12.54. The first stage is a Rake bank, which makes tentative bit decisions for each user. Also the channel

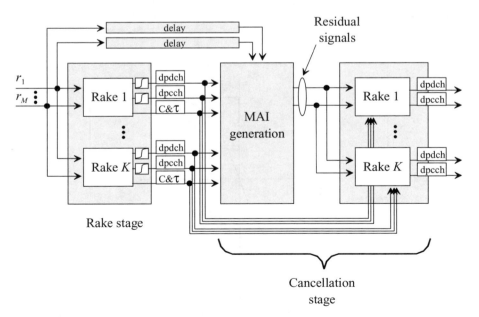

Figure 12.54. PIC receiver with M diversity antennas

estimation is done at the Rake stage. After the Rake stage, there are one or several cancellation stages. A cancellation stage comprises of a multiple access interference estimator and a Rake receiver with an interference canceller. The tentative decisions and channel estimates are fed to the interference estimator, which computes an estimate of the received wideband signal. This estimate is subtracted from the delayed received signal, and the result is called a residual. The residual is then despread and after that, bit decisions from the previous Rake stage, multiplied by channel estimates, are added. After the Rake cancellation stage we get final bit decisions from which interference has been removed.

The reason for using residual PIC lies in the implementation architecture, i.e. only one wideband signal needs to be distributed from the Rake stage to the PIC stage. In conventional PIC there are separate interference estimates for every user. Note that the performance of residual PIC is the same as that of conventional PIC.

A problem of the conventional hard decision PIC is that, when a tentative decision is wrong, the interference from that case is doubled in the cancellation stage. One way to overcome this problem is to use a null zone hard decision device [70] or to use an adaptive decision threshold [71], where the threshold is based on the statistic of matched filter output. The cancellation can also be made only to those signals which are reliable, as in [72]. The reliability of decisions is also covered in [73], where Divsalar *et al.* propose the use of a hyperbolic tangent function as a decision device instead of a sign function. These two functions are depicted in Figure 12.55. A sign function makes hard decisions (either +1 or −1); a hyperbolic tangent makes soft quantised decisions (anything between +1 and −1). Ideally, the horizontal axis in Figure 12.55 should be instantaneous signal-to-noise-and-interference ratio (SINR), but in practice, we have to estimate average SINR or assume interference + noise constant and use the power of the symbol. Hence, the reliability of the decision is taken into account. Note that a soft quantised decision is different from a soft decision, where there is no reliability weighting and the values are not limited to +1 and −1. In the following, a modified version of this receiver is called SQ-PIC.

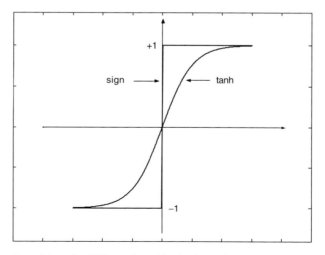

Figure 12.55. Non-linearities of a PIC receiver. The horizontal axis is the input signal level and the vertical axis, the output signal level

The 3GPP standard and implementation costs of the receiver set basic requirements for a PIC receiver. The strong channel coding specified in the standard and usually short spreading factor mean that SINR can be quite low at a receiver, resulting in unreliable tentative decisions and high BER before decoding, up to 15 %. This means that the hard decision-based PIC cannot work very well, since for every wrong decision, the corresponding interference is doubled. SQ-PIC tries to overcome this by using a reliability measure to weight the tentative decisions as mentioned. In order to minimise costs, only one PIC stage is suggested as the performance improvement from the second or third stage is not that large.

12.6.2.3   PIC Efficiency and Derivation of Network Level Gains

We define PIC efficiency β as the amount of own-cell interference it can remove. (It is assumed that PIC cannot remove other inter-cell interference.) We can write I_{total} for Rake and PIC as

$$I_{total,rake} = I_{own} + iI_{own} + N_0 = (1+i)K_{rake}P_j + P_N \tag{12.26a}$$

$$I_{total,PIC} = (1-\beta)I_{own} + iI_{own} + N_0 = (1+i-\beta)K_{pic}P_j + P_N \tag{12.26b}$$

where i is the ratio of other-cell interference to own-cell interference, P_N is the thermal noise power and we have assumed that users are homogenous, each having the same received power P_j. K_{rake} and K_{pic} are the number of users for Rake and PIC, respectively. K_{pic} is selected so that $I_{total,rake}$ and $I_{total,pic}$ are equal i.e. noise rises of Rake and PIC are the same. We can solve capacity gain $G_{cap} = K_{pic}/K_{rake}$ from the equations above:

$$G_{cap} = \frac{K_{pic}}{K_{rake}} = \frac{1+i}{1+i-\beta} \tag{12.27}$$

Coverage gain is defined as the ratio of required E_b/N_0s for Rake and PIC when the number of users, K, is kept constant:

$$G_{cov} = \frac{\{E_b/N_0\}_{rake}}{\{E_b/N_0\}_{pic}} \tag{12.28}$$

Using the definition of E_b/N_0 and using Equations (12.26) we get:

$$\{E_b/N_0\}_K = \frac{W}{R}\frac{P_j}{P_N} = \frac{W}{R}\frac{1}{\frac{1}{L_j} - (1+i-\beta)K} \tag{12.29}$$

Note that β is zero for a Rake receiver. From Equations (12.28) and (12.29) we obtain:

$$G_{cov} = \frac{\{E_b/N_0\}_{rake}}{\{E_b/N_0\}_{pic}} = \frac{\dfrac{W}{R}\dfrac{1}{\frac{1}{L_j} - (1+i)K}}{\dfrac{W}{R}\dfrac{1}{\frac{1}{L_j} - (1+i-\beta)K}} = \frac{\dfrac{1}{L_j} - (1+i-\beta)K}{\dfrac{1}{L_j} - (1+i)K} \tag{12.30}$$

We can solve K from the noise rise equation, Equation (8.9):

$$K = \frac{\dfrac{1}{L_j} - \dfrac{1}{L_j\Delta I_{rake}}}{1+i} \tag{12.31}$$

where ΔI_{rake} is the noise rise with a Rake receiver. We can now solve G_{cov} as a function of i, β and ΔI_{rake}:

$$G_{cov} = 1 + \frac{\beta \cdot (\Delta I_{rake} - 1)}{1 + i} \qquad (12.32)$$

Examples of the capacity gain in Equation (12.27) and the coverage gain in Equation (12.28) as functions of i are depicted in Figure 12.56 and Figure 12.57. In coverage gain examples, the noise rise of rake ΔI_{rake} is also a parameter. The gains increase as i decreases, which is

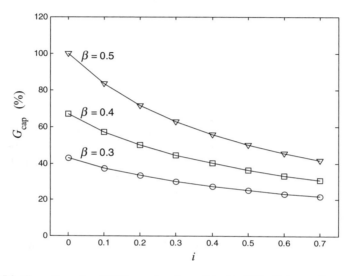

Figure 12.56. Capacity gain of PIC as a function of i with PIC efficiency β as a parameter

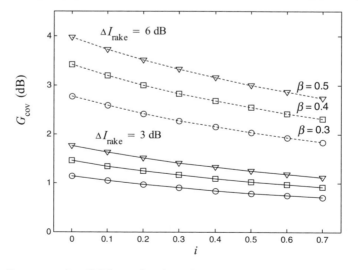

Figure 12.57. Coverage gain of PIC as a function of inter-cell interference i, multiuser efficiency β and noise rise ΔI_{Rake}

expected since PIC cannot cancel other-cell interference. The gains naturally also depend on β. The dependence is particularly strong in single cell without any other-cell interference, $i = 0$. Coverage gain also depends on the target noise rise with a Rake receiver: the higher the interference level without PIC, the higher the gain from PIC.

12.6.2.4   Performance of SQ-PIC

The performance of SQ-PIC was evaluated by Monte Carlo simulations in the link level without inter-cell interference [74]. Two propagation channels were considered, namely a pedestrian type of environment (Case 1 in 3GPP TS 25.141) and a vehicular type of environment (Case 3 in 3GPP TS 25.141). In the pedestrian channel, the UE velocity was 3 km/h, and in the vehicular channel, 120 km/h. The 12.2 kbps speech and 384 kbps data services were studied.

We estimate the PIC efficiency β from the simulation results by finding the value of β that gives the best fit to the simulation results. The results are shown in Table 12.23. The PIC

Table 12.23. PIC efficiency

Propagation channel	Number of diversity antennas (M)	PIC efficiency β	
		384 kbps data	12.2 kbps speech
Case 1 $v = 3$ km/h	1	32 %	24 %
	2	36 %	26 %
	4	37 %	33 %
Case 3 $v = 120$ km/h	1	40 %	35 %
	2	41 %	32 %
	4	36 %	28 %

efficiency is between 24 % and 41 %. The highest efficiency is obtained at high mobile speed with high data rate 384 kbps. The fast power control cannot keep the received power level exactly constant at high speed and there are larger power differences that can be cancelled by PIC. A high data rate provides better efficiency, since the number of simultaneous users is lower and the number of estimated parameters by PIC is lower, resulting in more accurate estimates. With a data service, the small spreading factor results also in high cross correlation between users, making the performance of Rake poor and hence allowing higher potential gain for SQ-PIC. The lowest gain is obtained at low mobile speed with voice users.

Table 12.24 shows capacity and coverage gains with the estimated efficiencies. The capacity gains are 26–35 % in a typical macro cell with $i = 0.55$. The coverage gain is 1.6–2.5 dB. This coverage gain assumes an initial noise rise of 6 dB. If the initial noise rise was 3 dB, the corresponding coverage gain would be 0.6–1.0 dB.

Numerical results for a data service are depicted in Figure 12.58 for channel case 3. The number of diversity antennas was one, two or four for both Rake and SQ-PIC. Rake with diversity antennas can be seen as an alternative to PIC, since increasing the order of diversity also provides substantial gains. The reason for this is Rake's ability to average MAI over diversity antennas, as the interference components from different diversity channels are independent. The results show that increasing the number of antennas with a Rake receiver

Table 12.24. Capacity and coverage gains with simulated PIC efficiencies (with outer-to-own cell interference ratio $i = 0.55$)

Propagation channel	M	Capacity gain (%)		Coverage gain (dB) with 6 dB noise rise	
		384 kbps data	12.2 kbps speech	384 kbps data	12.2 kbps speech
Case 1 $v = 3$ km/h	1	26 %	26 %	2.1 dB	1.6 dB
	2	30 %	30 %	2.3 dB	1.8 dB
	4	31 %	31 %	2.3 dB	2.1 dB
Case 3 $v = 120$ km/h	1	35 %	35 %	2.5 dB	2.2 dB
	2	36 %	36 %	2.5 dB	2.1 dB
	4	30 %	30 %	2.3 dB	1.9 dB

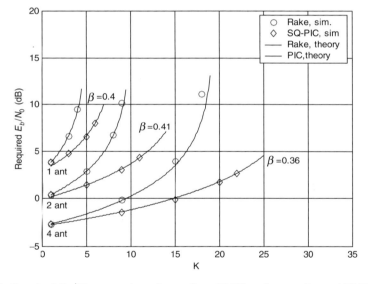

Figure 12.58. Required E_b/N_0 vs. number of users for a 384 kbps data service and BLER target 10 % in case 3 channel (120 km/h), the number of diversity antennas 1, 2 and 4

provides higher capacity and better coverage than introducing PIC. In the case of a low number of users, the gain from any interference cancellation is low, while more antennas provide clear coverage benefits. On the other hand, adding more antennas and antenna cables may not be possible from the site solution point of view, while the introduction of interference cancellation as the baseband processing enhancement is easier.

Interference cancellation, namely SQ-PIC, is the most promising method for improving base station receiver performance, as well as system capacity and coverage. Uplink interference cancellation may provide further gains in end user throughput when the uplink peak data rates exceed 1 Mbps in High-Speed Uplink Packet Access, HSUPA, which is a 3GPP study item in Release 6. For more details see Chapter 11.

References

[1] Holma, H., Soldani, D. and Sipilä, K. 'Simulated and Measured WCDMA Uplink Performance', *Proceedings VTC 2001* Fall, Atlantic City, NJ, USA, pp. 1148–1152.

[2] UMTS, Selection Procedures for the Choice of Radio Transmission Technologies of the UMTS, ETSI, v.3.1.0, 1997.

[3] Laiho-Steffens, J. and Lempiäinen, J., 'Impact of the Mobile Antenna Inclinations on the Polarisation Diversity Gain in DCS1800 Network', *Proceedings of PIMRC'97*, Helsinki, Finland, September 1997, pp. 580–583.

[4] Lempiäinen, J. and Laiho-Steffens, J., 'The Performance of Polarisation Diversity Schemes at a Base-Station in Small/Micro Cells at 1800MHz', *IEEE Trans. Vehic. Tech.*, Vol. 47, No. 3, August 1998, pp. 1087–1092.

[5] Sorensen, T.B., Nielsen, A.O., Mogensen, P.E., Tolstrup, M. and Steffensen, K., 'Performance of Two-Branch Polarisation Antenna Diversity in an Operational GSM Network', *Proceedings of VTC'98*, Ottawa, Canada, 18–21 May 1998, pp. 741–746.

[6] Grandell, J. and Salonaho, O. 'Macro Cell Measurements with the Nokia WCDMA Experimental System', *IEE Conference on Antennas and Propagation*, UK, 2001, pp. 516–520.

[7] Westman, T. and Holma, H., 'CDMA System for UMTS High Bit Rate Services', *Proceedings of VTC'97*, Phoenix, AZ, May 1997, pp. 825–829.

[8] Pehkonen, K., Holma, H., Keskitalo, I., Nikula, E. and Westman, T., 'A Performance Analysis of TDMA and CDMA Based Air Interface Solutions for UMTS High Bit Rate Services', *Proceedings of PIMRC'97*, Helsinki, Finland, September 1997, pp. 22–26.

[9] Ojanperä, T. and Prasad, R., *Wideband CDMA for Third Generation Mobile Communications*, Artech House, 1998, 439.

[10] Holma, H., 'A Study of UMTS Terrestrial Radio Access Performance', Doctoral thesis, Communications Laboratory, Helsinki University of Technology, Espoo, Finland, 2003.

[11] Sipila, K., Honkasalo, Z.C., Laiho-Steffens, J., Wacker, A. 'Estimation of Capacity and Required Transmission Power of WCDMA Downlink Based on a Downlink Pole Equation', *Proceedings of VTC 2000 Spring.*, Tokyo, Japan, May 2000, pp. 1002–1005.

[12] 3GPP Technical Specification 25.104 'UTRA (BS) FDD; Radio transmission and Reception'.

[13] 3GPP Technical Specification 25.101 'UE Radio Transmission and Reception (FDD)'.

[14] Winters, J.H., 'Smart Antennas for Wireless Systems', *IEEE Personal Communications*, Vol. 5, Issue 1, February 1998, pp. 23–27.

[15] Paulraj, A. and Chong-Ng, B. 'Space–Time Modems for Wireless Personal Communications', *IEEE Personal Communications*, February 1998, pp. 36–49.

[16] Anderson, S., Hagerman, B., Dam, H., Forssen, U., Karlsson, J., Kronested, F., Mazur, S. and Molnar, K. 'Adaptive Antennas for GSM and TDMA Systems', *IEEE Personal Communications*, June 1999, pp. 74–86.

[17] Pedersen, K.I. and Mogensen, P.E. 'Evaluation of Vector-Rake Receivers Using Different Antenna Array Configurations and Combining Schemes', *Internal Journal on Wireless Information Networks*, 1999, vol. 6, pp. 181–194.

[18] Osseiran, A., Ericson, M., Barta, J., Goransson, B. and Hagerman, B. 'Downlink Capacity Comparison Between Different Smart Antenna Concepts in a Mixed Service WCDMA System', *IEEE Proc. Vehicular Technology Conference*, September 2001, pp. 1528–1532.

[19] Tiirola, E. and Ylitalo, J. 'Performance Evaluation of Fixed-Beam Beamforming in WCDMA Downlink', *Proc. Vehicular Technology Conference*, Tokyo, Japan, May 2000, pp. 700–704.

[20] Pedersen, K.I., Mogensen, P.E. and Ramiro-Moreno, J. 'Application and Performance of Downlink Beamforming Techniques in UMTS', *IEEE Communications Magazine*, October 2003, pp. 134–143.

[21] 3GPP Technical Specification 25.215 'Physical Layer Measurements', v.5.4.

[22] Pedersen, K.I. and Mogensen, P.E. 'Directional Power Based Admission Control for WCDMA Systems Using Beamforming Antenna Array Systems', *IEEE Trans. on Vehicular Technology*, November 2002, Vol. 51, No. 6, pp. 1294–1303.

[23] Osseiran, A. and Ericson, M. 'On downlink admission control with fixed multi-beam antennas for WCDMA systems', *Proc. IEEE Vehicular Technology Conference, VTC-2003-Spring*, April 2003, pp. 1203–1207.

[24] Pedersen, K.I. and Mogensen, P.E. 'Performance of WCDMA HSDPA in a Beamforming Environment Under Code Constraints', *Proc. IEEE Vehicular Technology Conference VTC-2003-fall*, October 2003.

[25] Klingenbrunn, T., '*Downlink Capacity Enhancements of UTRA FDD Networks*', Ph.D. Thesis, Aalborg University, 2001.

[26] Godara, L.C., 'Application of Antenna Arrays to Mobile Communications, Part I: Performance Improvement, Feasibility, and System Considerations', *Proc. IEEE*, Vol. 85, No. 7, 1997, pp. 1031–1060.

[27] Godara, L.C., 'Application of Antenna Arrays to Mobile Communications, Part II: Beam-forming and Direction-of-Arrival Considerations', *Proc. IEEE*, Vol. 85, No. 8, 1997, pp. 1195–1245.

[28] Jakes, W.J. (ed.), *Microwave Mobile Communications*, IEEE Press, New Jersey, 1974.

[29] Muszynski, P., 'Interference Rejection Rake-Combining for WCDMA', *Proceedings of WPMC'98*, Yokosuka, Japan, November 1998, pp. 93–97.

[30] Winters, J.H., 'Optimum Combining in Digital Mobile Radio with Co-channel Interference', *IEEE Trans. Vehic. Tech.*, Vol. 33, No. 3, 1984, pp. 144–155.

[31] Monzingo, R.A. and Miller, T.W., *Introduction to Adaptive Arrays*, John Wiley & Sons, New York, 1980.

[32] Pursley, M.B., 'Performance Evaluation for Phase-Coded Spread-Spectrum Multiple-Access Communication – Part I: System Analysis', *IEEE Trans. Commun.*, Vol. 25, No. 8, 1977, pp. 795–799.

[33] Gilhousen, K.S., Jacobs, I.M., Padovani, R., Viterbi, A.J., Weaver, L.A. and Wheatley III, C. E., 'On the Capacity of a Cellular CDMA System', *IEEE Trans. Vehic. Tech.*, Vol. 40, No. 2, 1991, pp. 303–312.

[34] Viterbi, A.J., *CDMA: Principles of Spread Spectrum Communication*, Addison-Wesley Wireless Communications Series, Addison-Wesley, Reading, MA, 1995.

[35] Proakis, J.G., *Digital Communications*, 3rd edn, McGraw-Hill, New York, 1995.

[36] Verdú, S., 'Minimum Probability of Error for Asynchronous Gaussian Multiple-Access Channels', *IEEE Trans. Inform. Th.*, Vol. 32, No. 1, 1986, pp. 85–96.

[37] Juntti, M. and Glisic, S., 'Advanced CDMA for Wireless Communications', in *Wireless Communications: TDMA Versus CDMA*, ed. S. Glisic and P. Leppänen, Chapter 4, pp. 447–490, Kluwer, 1997.

[38] Verdú, S., *Multiuser Detection*, Cambridge University Press, Cambridge, UK, 1998.

[39] Lupas, R. and Verdú, S., 'Near–Far Resistance of Multiuser Detectors in Asynchronous Channels', *IEEE Trans. Commun.*, Vol. 38, No. 4, 1990, pp. 496–508.

[40] Klein, A. and Baier, P.W., 'Linear Unbiased Data Estimation in Mobile Radio Systems Applying CDMA', *IEEE J. Select. Areas Commun.*, Vol. 12, No. 7, 1999, pp. 1058–1066.

[41] Zvonar, Z., 'Multiuser Detection in Asynchronous CDMA Frequency-Selective Fading Channels', *Wireless Personal Communications*, Kluwer, Vol. 3, No. 3–4, 1996, pp. 373–392.

[42] Xie, Z., Short, R.T. and Rushforth, C.K., 'A Family of Suboptimum Detectors for Coherent Multiuser Communications', *IEEE J. Select. Areas Commun.*, Vol. 8, No. 4, 1990, pp. 683–690.

[43] Klein, A., Kaleh, G.K. and Baier, P.W., 'Zero Forcing and Minimum Mean-Square-Error Equalization for Multiuser Detection in Code-Division Multiple Access Channels', *IEEE Trans. Vehic. Tech.*, Vol. 45, No. 2, 1996, pp. 276–287.

[44] Varanasi, M.K. and Aazhang, B., 'Multistage Detection in Asynchronous Code-Division Multiple-Access Communications', *IEEE Trans. Commun.*, Vol. 38, No. 4, 1990, pp. 509–519.

[45] Kohno, R., Imai, H., Hatori, M. and Pasupathy, S., 'Combination of an Adaptive Array Antenna and a Canceller of Interference for Direct-Sequence Spread-Spectrum Multiple-Access System', *IEEE J. Select. Areas Commun.*, Vol. 8, No. 4, 1990, pp. 675–682.

[46] Viterbi, A. J., 'Very Low Rate Convolutional Codes for Maximum Theoretical Performance of Spread-Spectrum Multiple-Access Channels', *IEEE J. Select. Areas Commun.*, Vol. 8, No. 4, 1990, pp. 641–649.

[47] Patel, P. and Holtzman, J., 'Analysis of a Simple Successive Interference Cancellation Scheme in a DS/CDMA System', *IEEE J. Select. Areas Commun.*, Vol. 12, No. 10, 1994, pp. 796–807.

[48] Madhow, U. and Honig, M.L., 'MMSE Interference Suppression for Direct-Sequence Spread-Spectrum CDMA', *IEEE Trans. Commun.*, Vol. 42, No. 12, 1994, pp. 3178–3188.

[49] Rapajic, P. B. and Vucetic, B. S., 'Linear Adaptive Transmitter–Receiver Structures for Asynchronous CDMA Systems', *European Trans. Telecommun.*, Vol. 6, No. 1, 1995, pp. 21–27.

[50] Miller, S. L., 'An Adaptive Direct-Sequence Code-Division Multiple-Access Receiver for Multiuser Interference Rejection', *IEEE Trans. Commun.*, Vol. 43, No. 2/3/4, 1995, pp. 1746–1755.

[51] Latva-aho, M., 'Advanced Receivers for Wideband CDMA Systems', Vol. C125 of *Acta Universitatis Ouluensis*, Doctoral thesis, University of Oulu Press, Oulu, Finland, 1998.

[52] Latva-aho, M. and Juntti, M., 'Modified LMMSE Receiver for DS-CDMA – Part I: Performance Analysis and Adaptive Implementations', *Proceedings of ISSSTA'98*, Sun City, South Africa, September 1998, pp. 652–657.

[53] Honig, M., Madhow, U. and Verdú, S., 'Blind Adaptive Multiuser Detection', *IEEE Trans. Inform. Th.*, Vol. 41, No. 3, 1995, pp. 944–960.

[54] Latva-aho, M., 'LMMSE Receivers for DS-CDMA Systems in Frequency-Selective Fading Channels', in *CDMA Techniques for 3rd Generation Mobile Systems*, ed. F. Swarts, P. van Rooyen, I. Oppermann and M. Lötter, Chapter 13, Kluwer, 1998.

[55] Juntti, M. and Latva-aho, M., 'Multiuser Receivers for CDMA Systems in Rayleigh Fading Channels', *IEEE Trans. Vehic. Tech.*, Vol. 49, No. 3, 2000, pp. 885–889.

[56] Correal, N.S., Swanchara, S.F. and Woerner, B.D., 'Implementation Issues for Multiuser DS-CDMA Receivers', *Int. J. Wireless Inform. Networks*, Vol. 5, No. 3, 1998, pp. 257–279.

[57] Juntti, M., 'Multiuser Demodulation for DS-CDMA Systems in Fading Channels', Vol. C106 of *Acta Universitatis Ouluensis*, Doctoral thesis, University of Oulu Press, Oulu, Finland, 1997.

[58] Ojanperä, T., Prasad, R. and Harada, H., 'Qualitative Comparison of Some Multiuser Detector Algorithms for Wideband CDMA', *Proceedings of VTC'98*, Ottawa, Canada, May 1998, pp. 46–50.

[59] Ojanperä, T., *Multirate Multiuser Detectors for Wideband CDMA*, Ph.D. thesis, Technical University of Delft, Delft, The Netherlands, 1999.

[60] Johansson, A.-L., *Successive Interference Cancellation in DS-CDMA Systems*, Doctoral thesis, Chalmers University of Technology, Göteborg, Sweden, 1998.

[61] Juntti, M., 'Multiuser Detector Performance Comparisons in Multirate CDMA Systems', *Proceedings of VTC'98*, Ottawa, Canada, May 1998, pp. 36–40.

[62] Wijting, C.S., Ojanperä, T., Juntti, M. J., Kansanen, K. and Prasad, R., 'Groupwise Serial Multiuser Detectors for Multirate DS-CDMA', *Proceedings of VTC'99*, Houston, TX, May 1999, pp. 836–840.

[63] Juntti, M., 'Performance of Multiuser Detection in Multirate CDMA Systems', *Wireless Pers. Commun.*, Kluwer, Vol. 11, No. 3, 1999, pp. 293–311.

[64] Werner, S. and Lillberg, J., 'Downlink Channel Decorrelation in CDMA Systems with Long Codes', *Proceedings of VTC'99*, Houston, TX, May 1999, pp. 836–840.

[65] Hooli, K., Latva-aho, M. and Juntti, M., 'Multiple Access Interference Suppression with Linear Chip Equalizers in WCDMA Downlink Receivers', *Proceedings of Globecom'99*, Rio de Janeiro, Brazil, December 1999, pp. 467–471.

[66] Hooli, K., Juntti, M. and Latva-aho, M., 'Performance Evaluation of Adaptive Chip-Level Channel Equalizers in {WCDMA} Downlink', *Proceedings of ICC'01*, Helsinki, Finland, June 2001, pp. 1974–1979.

[67] Komulainen, P. and Heikkilä, M., 'Adaptive Channel Equalization Based on Chip Separation for CDMA Downlink', *Proceedings of PIMRC'99*, Osaka, Japan, September 1999, pp. 1114–1118.

[68] Hooli, K. *Equalization in WCDMA Terminals*. Doctoral thesis. Acta Universitatis Ouluensis C 192, University of Oulu Press, Oulu, Finland, 2003.

[69] Grant, P.M., Spangenberg, S.M., Cruickshank, G.M., McLaughlin, S. and Mulgrew, B., 'New Adaptive Multiuser Detection Technique for CDMA Mobile Receivers', *Proceedings of PIMRC'99*, Osaka, Japan, September 1999, pp. 52–54.

[70] Divsalar, D. and Simon, M. K., 'Improved CDMA performance using parallel interference cancellation', *Proc. IEEE MILCOM'94*, Fort Monmouth, N. J. USA, Oct. 2–5, 1994, pp. 911–917.

[71] Cho, Bong Youl and Lee, Jae Hong, 'Nonlinear parallel interference cancellation with partial cancellation for a DS-CDMA system', *IEICE Trans. On Communications*, Vol. E83-B, September 2000.

[72] Bae, J., Song, I. and Won, D. H., 'A selective and adaptive interference cancellation scheme for code division multiple access systems', *Signal Processing*, Vol. 83, No. 2, February 2003, Elsevier Science B.V.

[73] Divsalar, D., Simon, M.K., and Raphaeli, D., 'Improved Parallel Interference Cancellation for CDMA', *IEEE Trans. Commun.*, Vol. 46, No. 2, February 1998, pp. 258–268.

[74] Vihriälä, J. and Horneman, K., 'Impacts of SQ-PIC to Capacity and Coverage in WCDMA Uplink', to appear in ISSSTA'04, September 2004.

13

UTRA TDD Modes

Antti Toskala, Harri Holma, Otto Lehtinen and Heli Väätäjä

13.1 Introduction

The UTRA TDD modes are intended to operate in the unpaired spectrum, as shown in Figure 1.2 in Chapter 1, illustrating the spectrum allocations in various regions. As can be seen from Figure 1.2, there is no TDD spectrum available in all regions. The background of UTRA TDD was described in Chapter 4. During the standardisation process in ETSI and 3GPP, the major parameters were harmonised between UTRA FDD and TDD modes, including chip rate of 3.84 Mcps and modulation, for the Release '99 specifications. During the Release 4 work, the low chip rate TDD with 1.28 Mcps (TD-SCDMA) was introduced, following the same principles as the 3.84 Mcps TDD but with a few additional features, such as uplink synchronisation, as well as the mandatory differences arising from the different chip rate. Both TDD modes are covered in the physical layer specifications for the 3rd Generation Partnership Project (3GPP), the documents TS 25.221–TS 25.224 and TS 25.102 [1–5] are especially valuable references for obtaining information on the exact details. For Release 5, the High-Speed Downlink Packet Access (HSDPA) presented for FDD in Chapter 11 has been included for TDD as well. The TDD operation of HSDPA includes similar ARQ operation, use of 16 QAM modulation and fast Node B based scheduling, as described in Chapter 11.

This chapter first introduces TDD as a duplex method on a general level. The physical layer and related procedures of the UTRA TDD modes are introduced in Section 13.2. UTRA TDD interference issues are evaluated in Section 13.3. The HSDPA operation principles with TDD modes are covered in Section 13.4.

13.1.1 Time Division Duplex (TDD)

Three different duplex transmission methods are used in telecommunications: frequency division duplex (FDD), time division duplex (TDD) and space division duplex (SDD). The FDD method is the most common duplex method in the cellular systems. It is used, for example, in GSM, as well as with the WCDMA terminals currently commerciallly deployed in the UMTS frequency bands. The FDD method requires separate frequency bands for both

WCDMA for UMTS, third edition.   Edited by Harri Holma and Antti Toskala
© 2004 John Wiley & Sons, Ltd   ISBN: 0-470-87096-6

uplink and downlink. The TDD method uses the same frequency band but alternates the transmission direction in time. TDD is used, for example, for the digital enhanced cordless telephone (DECT). The SDD method is used in fixed-point transmission where directive antennas can be used. It is not used in cellular terminals, however, the use of beamforming techniques with FDD or TDD can be considered an SDD application as well.

Figure 13.1 illustrates the operating principles of the FDD and TDD methods. The term downlink or forward link refers to transmission from the base station (fixed network side) to the mobile terminal (user equipment), and the term uplink or reverse link refers to transmission from the mobile terminal to the base station.

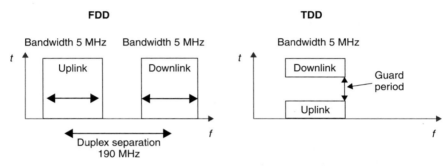

Figure 13.1. Principles of FDD and TDD operation

There are some characteristics peculiar to the TDD system and these are listed below.

- **Utilisation of unpaired band**. The TDD system can be implemented on an unpaired band while the FDD system always requires a pair of bands.

- **Discontinuous transmission**. Switching between transmission directions requires time, and the switching transients must be controlled. To avoid corrupted transmission, the uplink and downlink transmissions require a common means of agreeing on transmission direction and allowed time to transmit. Corruption of transmission is avoided by allocating a guard period which allows uncorrupted propagation to counter the propagation delay. Discontinuous transmission may also cause audible interference to audio equipment that does not comply with electromagnetic susceptibility requirements.

- **Interference between uplink and downlink**. Since uplink and downlink share the same frequency band, the signals in these two transmission directions can interfere with each other. In FDD, this interference is completely avoided by the duplex separation of 190 MHz. In UTRA TDD, individual base stations need to be synchronised to each other at frame level to avoid this interference. This interference is further analysed in Section 13.3.

- **Asymmetric uplink/downlink capacity allocation**. In TDD operation, uplink and downlink are divided in the time domain. It is possible to change the duplex switching point and move capacity from uplink to downlink, or vice versa, depending on the capacity requirement between uplink and downlink.

- **Reciprocal channel**. The fast fading depends on the frequency, and therefore, in FDD systems, the fast fading is uncorrelated between uplink and downlink. As the same frequency is used for both uplink and downlink in TDD, the fast fading is the same in uplink and in downlink. Based on the received signal, the TDD transceiver can estimate the fast fading, which will affect its transmission. Knowledge of the fast fading can be utilised in power control and in adaptive antenna techniques in TDD.

13.1.2   Differences in the Network Level Architecture

The UTRA TDD differs from the FDD mode operation for those issues that are related to the different Layer 1. The protocols have been devised with the principle that there are typically common messages, but the information elements are specific for the mode (FDD or TDD) being used. The use of the Iur interface is not needed for soft handover purposes with TDD, as only one Node B is transmitting for one user. If the TDD system uses relocation, then there is no need to transfer user data over the Iur interface at all in practice. Figure 13.2

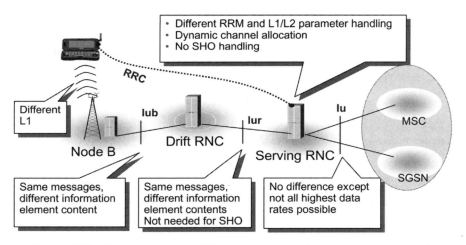

Figure 13.2. TDD differences to FDD from the network architecture point of view

illustrates the key differences for TDD compared to the architecture in Chapter 5. The core network does not typically see any difference, except that some of the highest data rate capabilities are not likely to be available, depending on the sum of the uplink and downlink data rates being used. From the RNC point of view, the biggest issue to cope with is the radio resource management (RRM), which in TDD is based on different measurements and resource allocation principles, along with the use of dynamic channel allocation (DCA).

13.2   UTRA TDD Physical Layer

The UTRA TDD mode uses a combined time division and code division multiple access (TD/CDMA) scheme that adds a CDMA component to a TDMA system. The different user signals are separated in both time and code domains. Table 13.1 presents a summary of the UTRA physical layer parameters. All the major RF parameters are harmonised within

Table 13.1. Comparison of UTRA FDD and TDD physical layer key parameters (Release 4)

	UTRA TDD	UTRA FDD
Multiple access method	TDMA, CDMA (inherent FDMA)	CDMA (inherent FDMA)
Duplex method	TDD	FDD
Channel spacing (nominal)	5 MHz/1.66 MHz	5 MHz
Carrier chip rate	3.84 Mcps/1.28 Mcps	3.84 Mcps
Time slot structure	15/14 slots/frame	15 slots/frame
Frame length	10 ms	
Multirate concept	Multicode, multislot and orthogonal variable spreading factor (OVSF)	Multicode and OVSF
Forward error correction (FEC) codes	Convolutional coding $R = \frac{1}{2}$ or 1/3, constraint length $K = 9$, turbo coding (8-state PCCC $R = 1/3$) or service-specific coding	
Interleaving	Inter-frame interleaving (10, 20, 40 and 80 ms)	
Modulation	QPSK/8PSK	QPSK
Burst types	Three types: traffic bursts, random access and synchronisation burst	Not applicable
Detection	Coherent, based on midamble	Coherent, based on pilot symbols
Dedicated channel power control	Uplink: open loop; 100 Hz or 200 Hz Downlink: closed loop; rate $\leq$ 800 Hz	Fast closed loop; rate = 1500 Hz
Intra-frequency handover	Hard handover	Soft handover
Inter-frequency handover	Hard handover	
Channel allocation	DCA supported	No DCA required
Intra-cell interference cancellation	Support for joint detection	Support for advanced receivers at base station
Spreading factors	1 ... 16	4 ... 512

UTRA for FDD and 3.84 Mcps TDD mode, with 1.28 Mcps TDD, the resulting RF parameters are obviously different due to the different bandwidth.

13.2.1 Transport and Physical Channels

UTRA TDD mode transport channels can be divided into dedicated and common channels. Dedicated channels (DCH) are characterised in basically the same way as in the FDD mode. Common channels can be further divided into common control channels (CCCH), the random access channel (RACH), the downlink shared channel (DSCH) in the downlink, and the uplink shared channel (USCH) in the uplink. Each of these transport channels is then mapped to the corresponding physical channel.

The physical channels of UTRA TDD are the dedicated physical channel (DPCH), common control physical channel (CCPCH), physical random access channel (PRACH), paging indicator channel (PICH) and synchronisation channel (SCH). For the SCH and PICH there do not exist corresponding transport channels. The mapping of the different transport channels to the physical channels and all the way to the bursts is shown in

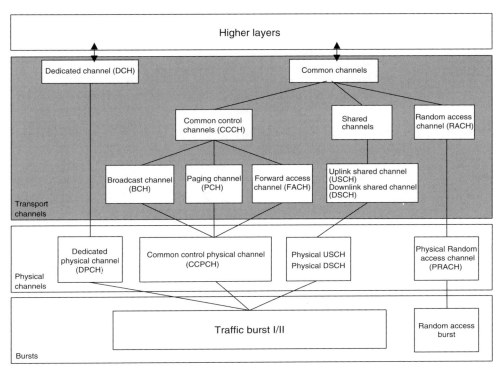

Figure 13.3. Mapping of the Release 4 UTRA TDD transport channels to physical channels

Figure 13.3. The physical channel structure is discussed in the following section in more detail.

13.2.2 Modulation and Spreading

The data modulation scheme in UTRA TDD is QPSK, additionally, 8PSK was added in the 1.28 Mcps TDD to enable the theoretical 2 Mbps peak rate to be reached. The modulated data symbols are spread with a specific channelisation code of length 1–16. The modulated and spread data is finally scrambled by a pseudorandom sequence of length 16. The same type of orthogonal channelisation codes are used in the UTRA FDD system (see Section 6.3). Data spreading is followed by scrambling with a cell- or source-specific scrambling sequence; the scrambling process is chip-by-chip multiplication. The combination of multiplying with channelisation code and the cell-specific scrambling code is a user- or cell-specific spreading procedure. Finally, pulse shape filtering is applied to each chip at the transmitter: each chip is filtered with a root raised cosine filter with roll-off factor $\alpha = 0.22$, identical to UTRA FDD.

13.2.3 Physical Channel Structures, Slot and Frame Format

The physical frame structure is similar to that of the UTRA FDD mode. The frame length is 10 ms and it has two different forms, depending on the chip rate. The 3.84 Mcps TDD

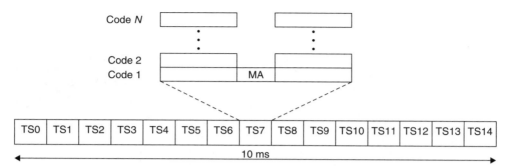

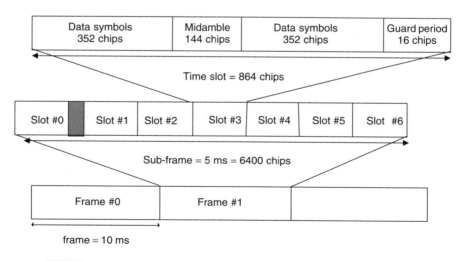

Figure 13.4. Frame structure of UTRA TDD. The number of code channels that may be used within a single time slot varies depending on the propagation conditions (MA = midamble)

frame is divided into 15 time slots, each of 2560 chips, i.e., the time slot duration is 666 µs. Figure 13.4 shows the frame structure for 3.84 Mcps TDD.

Each of the 15 time slots within a 10 ms frame is allocated to either uplink or downlink. Multiple switching points for different transmission directions per frame allow closed loop power control and a physical synchronisation channel (PSCH) in dedicated downlink slots to speed cell search. On the other hand, to be able to cover dynamic asymmetric services, the flexibility in slot allocation in the downlink/uplink direction guarantees efficient use of the spectrum. To maintain maximum flexibility while allowing closed loop power control whenever useful, the SCH has two time slots per frame for downlink transmission. The PSCH is mapped to two downlink slots.

For the 1.28 Mcps TDD the 15 slots principle was not chosen, as the resulting number of chips is not evenly divisible by 15. Instead, 14 slots was decided upon for the frame content. These 14 slots are divided into 5 ms sub-frames as indicated in Figure 13.5.

Figure 13.5. 1.28 Mcps TDD frame, sub-frame and downlink slot structure

Since the TDMA transmission in UTRA TDD is discontinuous, the average transmission power is reduced by a factor of $10 \times \log_{10}(n/15)$, where n is the number of active time slots per frame. For example, to provide the same coverage with UTRA TDD using a single time slot for 144 kbps requires at least four times more base station sites than with UTRA FDD. This 12 dB reduction in the average power would result in a typical macro cell environment to reduce the cell range more than into half, and thus, the cell area to a quarter. When utilising the same hardware in the UE, the TDMA discontinuous transmission with low duty cycle leads to reduced uplink range. With higher data rates the coverage difference to FDD reduces. Due to these properties, TDD should be used in small cell environments where power is not a limiting factor and data rates used for the coverage planning are higher.

13.2.3.1 Burst Types

There are three bursts defined for UTRA TDD. All of them are used for dedicated channels while common channels typically use only a subset of them.

Burst types I and II are usable for both uplink and downlink directions with the difference between the type I and II being the midamble length. Figure 13.6 illustrates the general downlink, and Figure 13.7 the uplink burst structure with transmission power control (TPC) and transmission format combination indicator (TFCI).

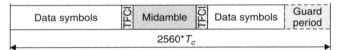

Figure 13.6. Generalised UTRA TDD downlink burst structure

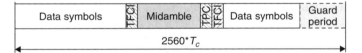

Figure 13.7. Generalised uplink burst structure

The burst types with two variants of midamble length can be used for all services up to 2 Mbps. The logical traffic channel (TCH), which contains user data, is mapped to a burst. Additionally, for the 1.28 Mcps TDD there is physical layer control information added in the downlink to support the uplink synchronisation procedure.

The data fields are separated by a midamble which is used for channel estimation. The transport format combination indicator (TFCI) is used to indicate the combination of used transport channels in the dedicated physical channel (DPCH) and is sent only once per frame. The TFCI uses in-band signalling and has its own coding. The number of TFCI bits is variable and is set at the beginning of the call.

For the uplink burst in Figure 13.7, both transmission power control (TPC) and TFCI are present. Both TFCI and TPC are transmitted in the same physical channel and use in-band signalling. The length of the TPC command is one symbol.

The burst contains two data fields separated by a midamble and followed by a guard period. The duration of a burst is one time slot. The midamble is used for both channel

Table 13.2. Burst type field structures for 3.84 Mcps TDD

Burst name	Data field 1 length	Training sequence length	Data field 2 length	Guard period length
Burst type I	976 chips	512 chips	976 chips	96 chips
Burst type II	1104 chips	256 chips	1104 chips	96 chips
Burst type III	976 chips	512 chips	880 chips	192 chips

equalisation and coherent detection at the receiver. The midamble reduces the user data payload. Table 13.2 shows the burst type I and II structures in detail.

Due to the longer midamble, burst type I is applicable for estimating 16 different uplink channel impulse responses. Burst type II can be used for the downlink independently of the number of active users. If there are fewer than four users within a time slot, burst type II can also be used for the uplink.

For the 1.28 Mcps the different burst types offer typically from 32 to 44 bits per data field, depending on the use of TPC or synchronisation. The smaller number of chips and resulting payload comes directly from the relationship of the chip rates between the different modes, though some of the overheads take a relatively larger amount in 1.28 Mcps TDD. Also, the use of uplink synchronisation requires additional bits for physical layer signalling in 1.28 Mcps TDD.

The midambles, i.e. the training sequences of different users, are time-shifted versions of one periodic basic code. Different cells use different periodic basic codes, i.e. different midamble sets. Due to the generation of midambles from the same periodic basic code, channel estimation of all active users within one time slot can be performed jointly, for example by one single cyclic correlator. Channel impulse response estimates of different users are obtained sequentially in time at the output of the correlator [6].

The downlink uses either a spreading factor of 16 with the possibility of multicode transmission, or a spreading factor of 1 for high bit rate applications in case such a capability is supported by the terminals. In the uplink, orthogonal variable spreading factor (OVSF) codes with spreading factors from 1 to 16 are used. The total number of the burst formats is 20 in the downlink and 90 in the uplink.

The burst type III is used in the uplink direction only. This has developed for the needs for the PRACH, as well as to facilitate handover in cases when timing advance is needed. The guard time of 192 chips (50 μs) equals a cell radius of 7.5 km.

13.2.3.2 Physical Random Access Channel (PRACH)

The logical random access channel (RACH) is mapped to a physical random access (PRACH) channel. Table 13.2 shows the burst type III used with PRACH, and Figure 13.8 illustrates the burst type III structure. Spreading factor values of 16 and 8 are used for PRACH. With PRACH there are typically no TPC or TFCI bits used, as shown in Figure 13.8.

Figure 13.8. UTRA TDD burst type III when used with PRACH

13.2.3.3 Synchronisation Channel (SCH)

The time division duplex creates some special needs for the synchronisation channel. A capturing problem arises due to the cell synchronisation, i.e. a phenomenon occurring when a stronger signal masks weaker signals. The time misalignment of the different synchronisation channels of different cells would allow for distinguishing several cells within a single time slot. For this reason a variable time offset (t_{offset}) is allocated between the SCH and the system slot timing. The offset between two consecutive shifts is $71T_c$. There exist two different SCH structures. The SCH can be mapped either to the slot number $k \in \{0 \ldots 14\}$ or to time slots k and $k + 8$, $k \in \{0 \ldots 6\}$. Figure 13.9 shows the latter SCH structure for $k = 0$. This dual-SCH-per-frame structure is intended for cellular use. The position of the SCH can vary on a long-term basis.

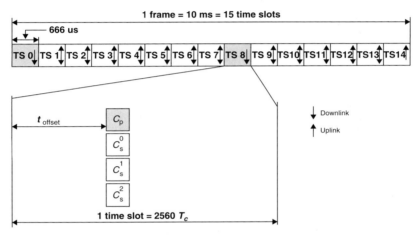

Figure 13.9. UTRA TDD SCH structure. This example has two downlink slots allocated for SCH ($k = 0$). The primary code (c_p) and three QPSK-modulated secondary codes (c_s) are transmitted simultaneously. The time offset (t_{offset}) is introduced to avoid adverse capture effects of the synchronous system. The combined transmission power of the three c_s is equal to the power of c_p

The terminal can acquire synchronisation and the coding scheme for the BCCH of the cell in one step and will be able to detect cell messaging instantly. The primary (c_p) and the three secondary (c_s) synchronisation sequences are transmitted simultaneously. Codes are 256 chips long as in the UTRA FDD mode, and the primary code is generated in the same way as in the FDD mode, as a generalised hierarchical Golay sequence. The secondary synchronisation code words (c_s) are chosen from every 16th row of the Hadamard sequence H_8, which is used also in the FDD mode. By doing this there are only 16 possible code words, in comparison to 32 of the FDD mode. The codes are QPSK modulated and the following information is indicated by the SCH:

- Base station code group out of 32 possible alternatives (5 bits);
- Position of the frame in the interleaving period (1 bit);
- Slot position in the frame (1 bit);
- Primary CCPCH locations (3 bits).

With a sequence it is possible to decode the frame synchronisation, the time offset (t_{offset}), the midamble and the spreading code set of the base station, as well as the spreading code(s) and location of the broadcast channel (BCCH).

The cell parameters within each code group are cycled over two frames to randomise interference between base stations and to enhance system performance. Also, network planning becomes easier with the averaging property of the parameter cycling.

In the 1.28 Mcps TDD, the downlink pilots, as indicated in Figure 13.10, contain the necessary synchronisation information. The Downlink Pilot Channel (DwPCH) is transmitted in each 5 ms sub-frame over the whole coverage area, in a similar way to the SCH in the 3.84 Mcps TDD. The pattern used on the 64 chips of information can have 32 different downlink synchronisation codes.

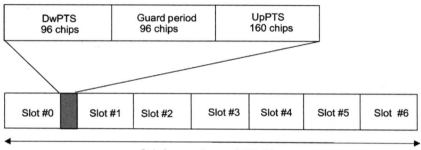

Figure 13.10. The 1.28 Mcps TDD sub-frame pilot structure

13.2.3.4   Common Control Physical Channel (CCPCH)

Once the synchronisation has been acquired, the timing and coding of the primary broadcast channel (BCH) are known. The CCPCH can be mapped to any downlink slot(s), including the PSCH slots, and this is indicated by the primary BCH.

The CCPCH is similar to the downlink dedicated physical channel (DPCH). It may be coded with more redundancy than the other channels to simplify acquisition of in formation.

13.2.3.5   UTRA TDD Shared Channels

The UTRA TDD specification also defines the Downlink Shared Channel (DSCH) and the Uplink Shared Channel (USCH). These channels use exactly the same slot structure as do the dedicated channels. The difference is that they are allocated on a temporary basis.

In the downlink, the signalling to indicate which terminals need to decode the channel can be done with TFCI, by detecting midamble in use or by higher layers. In the uplink, the USCH uses higher layer signalling and thus is not shared in practice on a frame-by-frame basis.

13.2.3.6   User Data Rates

Table 13.3 shows the UTRA TDD user bit rates with $\frac{1}{2}$-rate channel coding and spreading factor 16. The tail bits, TFCI, TPC or CRC overhead have not been taken into account. Spreading factors other than 16 (from the orthogonal variable spreading scheme) can be seen

Table 13.3. UTRA TDD 3.84 Mcps air interface user bit rates

Number of allocated codes with spreading factor 16	Number of allocated timeslots		
	1	4	13
1	13.8 kbps	55.2 kbps	179 kbps
8	110 kbps	441 kbps	1.43 Mbps
16 (or spreading factor 1)	220 kbps	883 kbps	2.87 Mbps

as subsets of spreading factor 16 (i.e. spreading factor 8 in the uplink corresponds to two parallel codes with spreading factor 16 in the downlink). When the number of needed slots exceeds seven, the corresponding data rate can be provided only for either the uplink or the downlink. The bit rates shown in Table 13.3 are time slot and code limited bit rates, the maximum interference limited bit rate can be lower. The 1.28 Mcps TDD resulting data rate is around 8 kbps with one slot per sub-frame (two slots per 10 ms), spreading factor 16 and the use of QPSK. There are 69 different uplink formats with QPSK and 24 different downlink formats that can be used to build a particular data rate.

13.2.4 UTRA TDD Physical Layer Procedures

13.2.4.1 Power Control

The purpose of power control is to minimise the interference of separate radio links. Both the uplink and downlink dedicated physical channels (DPCH) and physical random access channel (PRACH) are power controlled. The forward access channel (FACH) may be power controlled. The implementation of advanced receivers, such as the joint detector, will suppress intra-cell (own-cell) interference and reduce the need for fast power control. The optimum multiuser detector is near–far resistant [7] but in practice the limited dynamic

Table 13.4. Power control characteristics of 3.84 Mcps UTRA TDD

	Uplink	Downlink
Method	Open loop	SIR-based closed inner loop
Dynamic range	65 dB	30 dB (all the users are within 20 dB in
	Minimum power −44 dBm or less	one time slot)
	Maximum power 21 dBm	
Step size	1, 2, 3 dB	1, 2, 3 dB
Rate	Variable	From 100 Hz to approximately 750 Hz
	1–7 slots delay (2-slot PCCPCH)	
	1–14 slots delay (1-slot PCCPCH)	

range of the sub-optimum detector restricts performance. Table 13.4 shows the 3.84 Mcps UTRA TDD power control characteristics and Table 13.5 shows the 1.28 Mcps TDD power control characteristics.

In the downlink, closed loop is used after initial transmission. The reciprocity of the channel is used for open loop power control in the uplink. Based on interference level at the

Table 13.5. 1.28 Mcps TDD power control characteristics

	Uplink	Downlink
Method	Initially open loop and then SIR-based inner loop (for some control channel only open loop)	SIR-based inner loop
Rate	Closed loop: 0-200 Hz	0–200 Hz
	Open loop: variable delay depending on slot allocation	
Closed loop step sizes	1,2,3 dB	1,2,3 dB

base station and on path loss measurements of the downlink, the mobile weights the path loss measurements and sets the transmission power. The interference level and base station transmitter power are broadcast. The transmitter power of the mobile is calculated by the following equation [4]:

$$P_{UE} = \alpha L_{PCCPCH} + (1 - \alpha)L_0 + I_{BTS} + SIR_{TARGET} + C \qquad (13.1)$$

In Equation (13.1) P_{UE} is the transmitter power level in dBm, L_{PCCPCH} is the measured path loss in dB, L_0 is the long-term average of path loss in dB, I_{BTS} is the interference signal power level at the base station receiver in dBm, and α is a weighting parameter which represents the quality of path loss measurements. α is a function of the time delay between the uplink time slot and the most recent downlink PCCPCH time slot. SIR_{TARGET} is the target SNR in dB; this can be adjusted through higher layer outer loop. C is a constant value.

13.2.4.2 Data Detection

UTRA TDD requires that simultaneously active spreading codes within a time slot are separated by advanced data detection techniques. The usage of conventional detectors, i.e. matched filters or Rake, in the base station requires tight uplink power control, which is difficult to implement in a TDD system since the uplink is not continuously available. Thus, advanced data detection techniques should be used to suppress the effect of power differences between users, i.e. the near–far effect. Both inter-symbol interference (ISI) due to multipath propagation and multiple access interference (MAI) between data symbols of different users are present also in downlink. In downlink, the intra-cell interference is suppressed by the orthogonal codes, and the need for advanced detectors is lower than in uplink. In UTRA TDD the number of simultaneously active users is small and the use of relatively short scrambling codes, together with spreading, make the use of advanced receivers attractive.

The sub-optimal data detection techniques can be categorised as single user detectors and multiuser detectors (see Section 12.5.2). In UTRA TDD, single user detectors can be applied when all signals pass through the same propagation channel, i.e. they are primarily applied for the downlink [8]. Otherwise, multiuser or joint detection is applied [9, 10].

Single user detectors first equalise the received data burst to remove the distortion caused by the channel. When perfect equalisation is assumed, the orthogonality of the codes is restored after equalisation. The desired signal can now be separated by code-matched filtering. The advantages of using single user detectors are that no knowledge of the other user's active codes is required and the computational complexity is low compared to joint detection [8].

To be able to combat both MAI and ISI in UTRA TDD, equalisation based on, for example, zero-forcing (ZF) or minimum mean-square-error (MMSE) can be applied. Both equalisation methods can be applied with or without decision feedback (DF). The computational complexity of the algorithms is essentially the same, but the performance of the MMSE equalisers is better than that of the ZF equalisers [10]. The decision feedback option improves performance (about 3 dB less E_b/N_0 at practical bit error rates) and the MMSE algorithm generally performs better (less than 1 dB difference in E_b/N_0 requirements) than zero-forcing. Antenna diversity techniques can be applied with joint detection [11, 12] to further enhance the performance.

The performance of Rake, ZF equaliser, MMSE equaliser and HD-PIC (hard decision parallel interference canceller [13]) in the UTRA TDD uplink was studied using Monte Carlo computer simulations in the UTRA TDD uplink [14]. Eight users with spreading factor of 16 occupy one time slot within a 10 ms frame. A two-path channel with tap gains of 0 dB and -9.7 dB, and with a mobile speed of 3 km/h is considered. Channel estimation and power control are assumed to be ideal and channel coding is omitted. The performance of Rake, ZF, MMSE, and one- and two-stage HD-PIC are shown in Figure 13.11. The results

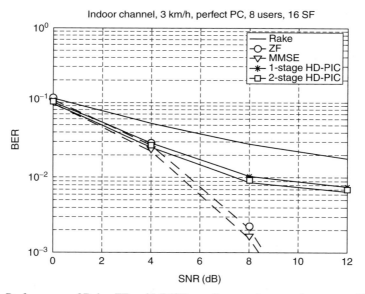

Figure 13.11. Performance of Rake, ZF and MMSE equalisers and one- and two-stage HD-PIC in the 3.84 Mcps UTRA TDD uplink

show that the advanced base station receivers give a clear gain compared to the Rake receiver in UTRA TDD, even with ideal power control. As the signal-to-noise ratio (SNR) increases, the performance of ZF and MMSE is better than the performance of HD-PIC. Channel coding typically increases the differences between the performance of different detectors. For example, in the operational area of BER $= 5-10\%$ the gain from the advanced receiver structures can be up to 2 dB with perfect power control and even more with realistic power control. The difference between the presented advanced detectors is small in this operational area.

13.2.4.3   Timing Advance

To avoid interference between consecutive time slots in large cells, it is possible to use a timing advancement scheme to align the separate transmission instants in the base station receiver. The timing advance is determined by a 6-bit number with an accuracy of four chips (1.042 μs). The base station measures the required timing advance, and the terminal adjusts the transmission according to higher layer messaging. The maximum cell range is 9.2 km.

The UTRA TDD cell radius without timing advance can be calculated from the guard period of traffic burst (96 chips = 25 μs), resulting in a range of 3.75 km. This value exceeds practical TDD cell ranges (micro and pico cells) and in practice the timing advance is not likely to be needed.

13.2.4.4   Channel Allocation in TDD

In order to offer continuous coverage, a TDD system needs to use Dynamic Channel Allocation (DCA) to cope with the interference at the cell borders with reuse 1 and with the lack of soft handover. The 3GPP specifications define the RNC-controlled DCA signalling, which offers the possibility of slow DCA based on the Node B and UE measurements of the interference conditions in different time slots.

The measurement reports are passed always to the SRNC, thus, for practical operation, SRNC needs to be the same as CRNC, which is secured by means of relocation. In Release 6, additional procedures are being worked on that could enable the relaying of the measurement from the SRNC to CRNC, which removes the requirement for relocation, but on the other hand makes the DCA operation at RNC level still slower.

The fast DCA in general, referred to in earlier editions of this book as Node B terminated DCA, does not exist in practice in Release '99 or Release 4. The HSDPA in Release 5 can be considered as being limited fast DCA between HSDPA users but it does not modify, e,g, the uplink and downlink slot resource allocation or allocation to CS domain services.

13.2.4.5   Handover

UTRA TDD supports inter-system handovers and intra-system handovers (to UTRA FDD and to GSM). All these handovers are mobile-assisted hard handovers.

UTRA TDD does not use soft handover (or macro diversity). This is a clear difference from UTRA FDD, in which the protocol structure has been designed to support soft handover. The UTRA TDD protocol structure has followed the same architecture as FDD for termination points for maximum commonality above the physical layer. This means, for example, that handover protocols terminate at the same location (RNC) but consist of FDD and TDD mode-specific parameters.

13.2.4.6   UTRA TDD Transmit Diversity

UTRA TDD supports four downlink transmit diversity methods. They are comparable to those in UTRA FDD. For dedicated physical channels Switched Transmitter Diversity (STD) and Transmit Adaptive Antennas (TxAA) methods are supported. The antenna weights are calculated using the reciprocity of the radio link. In order to utilise the TxAA method, the required base station receiver and transmitter chain calibration makes the implementation more challenging.

For common channels, Time Switched Transmit Diversity (TSTD) is used for PSCH, and Block Space Time Transmit Diversity (Block STTD) is used for primary CCPCH.

For uplink at the base station, the same receiver diversity methods as in FDD are applicable to enhance the performance.

13.2.4.7   1.28 Mcps TDD-specific Physical Layer Procedures

The 1.28 Mcps TDD contains some chip rate-specific refinements to the physical layer procedures arising from differences in the physical layer structure. Procedures like power control have differences due to the use of sub-frame division, which results in different command rates. There is also a fully 1.28 Mcps-specific procedure, as, instead of timing advance, uplink synchronisation is used to try to reduce uplink interference. The principle is to have users in the uplink sharing the same scrambling code and to have uplink transmission partly orthogonal by coordinating the uplink TX timing with closed loop control and small (1/5–1/8 chip) resolution. Different resolutions are allowed in order not to force the use of any particular sampling rate in the terminal.

13.3   UTRA TDD Interference Evaluation

In this section we evaluate the effect of interference within the TDD band and between TDD and FDD. TDD–TDD interference is analysed in Section 13.3.1 and the co-existence of TDD and FDD systems in Section 13.3.2.

13.3.1   TDD–TDD Interference

Since both uplink and downlink share the same frequency in TDD, these two transmission directions can interfere with each other. By nature the TDD system is synchronous and this kind of interference occurs if the base stations are not synchronised. It is also present if different asymmetry is used between the uplink and downlink in adjacent cells even if the base stations are frame synchronised. Frame synchronisation requires an accuracy of a few symbols, not an accuracy of chips. The guard period allows more tolerance in synchronisation requirements. Figure 13.12 illustrates possible interference scenarios. The interference within the TDD band is analysed with system simulations in [15].

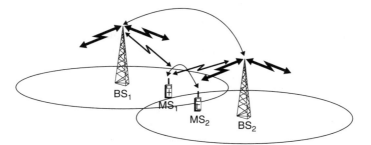

Figure 13.12. Interference between mobiles, between base stations, and between mobile and base station

Interference between uplink and downlink can also occur between adjacent carriers. Therefore, it can also take place between two operators.

In FDD operation, the duplex separation prevents interference between uplink and downlink. The interference between a mobile and a base station is the same in both TDD and FDD operation and is not considered in this chapter.

13.3.1.1 Mobile Station to Mobile Station Interference

Mobile-to-mobile interference occurs if mobile MS_2 in Figure 13.12 is transmitting and mobile MS_1 is receiving simultaneously in the same (or adjacent) frequency in adjacent cells. This type of interference is statistical because the locations of the mobiles cannot be controlled. Therefore, it cannot be avoided by network planning. Intra-operator mobile-to-mobile interference occurs especially at cell borders. Inter-operator interference between mobiles can occur anywhere where two operators' mobiles are close to each other and transmitting on fairly high power. Methods to counter mobile-to-mobile interference are:

- DCA and radio resource management;
- Power control.

13.3.1.2 Base Station to Base Station Interference

Base station to base station interference occurs if base station BS_1 in Figure 13.12 is transmitting and base station BS_2 is receiving in the same (or adjacent) frequency in adjacent cells. It depends heavily on the path loss between the two base stations and therefore can be controlled by network planning.

Intra-operator interference between base stations depends on the base station locations. Interference between base stations can be especially strong if the path loss is low between the base stations. Such cases could occur, for example, in a macro cell, if the base stations are located on masts above rooftops. The best way to avoid this interference is by careful planning to provide sufficient coupling loss between base stations.

The outage probabilities in [15] show that cooperation between TDD operators in network planning is required, or the networks need to be synchronised and the same asymmetry needs to be applied. Sharing base station sites between operators will be very problematic, if not impossible. The situation would change if operators had inter-network synchronisation and identical uplink/downlink splits in their systems.

From the synchronisation and coordination point of view, the higher the transmission power levels and the larger the intended coverage area, the more difficult will be the coordination for interference management. In particular, the locations of antennas of the macro cell type tend to result in line-of-sight connections between base stations, causing strong interference. Operating TDD in indoor and micro/pico cell environments will mean lower power levels and will reduce the problems illustrated.

13.3.2 TDD and FDD Co-existence

The UTRA FDD and TDD have spectrum allocations that meet at the border at 1920 MHz, and therefore TDD and FDD deployment cannot be considered independently: see Figure 13.13. The regional allocations were shown in Figure 1.2 in Chapter 1. Dynamic channel allocation (DCA) can be used to avoid TDD–TDD interference, but DCA is not effective between TDD and FDD, since FDD has continuous transmission and reception. The possible interference scenarios between TDD and FDD are summarised in Figure 13.14.

13.3.2.1 Co-siting of UTRA FDD and TDD Base Stations

From the network deployment perspective, the co-siting of FDD and TDD base stations looks an interesting alternative. There are, however, problems due to the close proximity of the frequency bands. The lower TDD band, 1900–1920 MHz, is located adjacent to the FDD

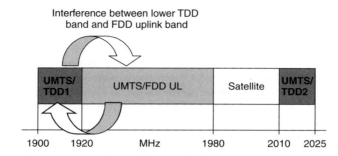

Figure 13.13. Interference between lower TDD band and FDD uplink band

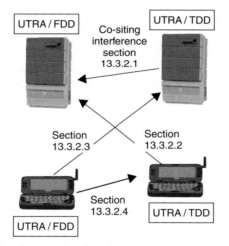

Figure 13.14. Possible interference situations between lower TDD band and FDD uplink band

uplink band, 1920–1980 MHz. The resulting filtering requirements in TDD base stations are expected to be such that co-siting a TDD base station in the 1900–1920 MHz band with an FDD base station is not considered technically and commercially a viable solution. Table 13.6 illustrates the situation. The output power of 24 dBm corresponds to a small pico base station and 43 dBm to a macro cell base station.

Table 13.6. Coupling loss analysis between TDD and FDD base stations in adjacent frequencies at 1920 MHz

TDD base station output power (pico/macro)	24/43 dBm
Adjacent channel power ratio	−45 dBc
Isolation between antennas (separate antennas for FDD and TDD base stations)	−30 dB
Leakage power into FDD base station receiver	−51/−32 dBm
Allowed leakage power	−110 dBm
Required attenuation	59/78 dB

The required attenuation between TDD macro cell base stations is 78 dB. If we introduce the 5 MHz guard band, with centre frequencies 10 MHz apart, the additional frequency separation of 5 MHz would increase the channel protection by 5 dB. The co-siting (co-located RF parts) is not an attractive alternative with today's technology.

The micro and pico cell environments change the situation, since the TDD base station power level will be reduced to as low as 24 dBm in small pico cells. On the other hand, the assumption of 30 dB antenna-to-antenna separation will not hold if antennas are shared between TDD and FDD systems. Antenna sharing is important to reduce the visual impact of the base station site. Also, if the indoor coverage is provided with shared distributed antenna systems for both FDD and TDD modes, there is no isolation between the antennas. Thus, the TDD system should create a separate cell layer in UTRAN. In the pico cell TDD deployment scenario the interference between modes is easier to manage with low RF powers and separate RF parts.

13.3.2.2   Interference from UTRA TDD Mobile to UTRA FDD Base Station

UTRA TDD mobiles can interfere with a UTRA FDD base station. This interference is basically the same as that from a UTRA FDD mobile to a UTRA FDD base station on the adjacent frequency. The interference between UTRA FDD carriers is presented in Section 8.5. There is, however, a difference between these two scenarios: in pure FDD interference there is always the corresponding downlink interference, while in interference from TDD to FDD there is no downlink interference. In FDD operation the downlink interference will typically be the limiting factor, and therefore uplink interference will not occur. In the interference from a TDD mobile to an FDD base station, the downlink balancing does not exist as it does between FDD systems, since the interfering TDD mobile does not experience interference from UTRA FDD. This is illustrated in Figure 13.15.

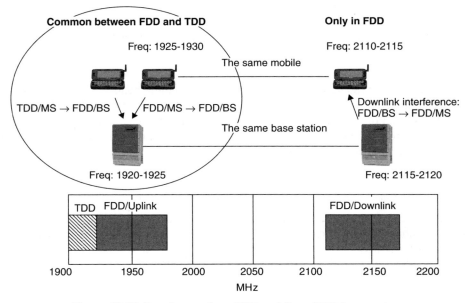

Figure 13.15. Interference from TDD mobile to FDD base station

One way to avoid uplink interference problems is to make the base station receiver less sensitive on purpose, i.e. to desensitise the receiver. For small pico cells indoors, base station sensitivity can be degraded without affecting cell size. Another solution is to place the FDD base stations so that the mobile cannot get very close to the base station antenna.

13.3.2.3   Interference from UTRA FDD Mobile to UTRA TDD Base Station

A UTRA FDD mobile operating in 1920–1980 MHz can interfere with the reception of a UTRA TDD base station operating in 1900–1920 MHz. Uplink reception may experience high interference, which is not possible in FDD-only operation. The inter-frequency and inter-system handovers alleviate the problem. The same solutions can be applied here as in Section 13.3.2.2.

13.3.2.4   Interference from UTRA FDD Mobile to UTRA TDD Mobile

A UTRA FDD mobile operating in 1920–1980 MHz can interfere with the reception of a UTRA TDD mobile operating in 1900–1920 MHz. It is not possible to use the solutions of Sections 13.3.2.2 because the locations of the mobiles cannot be controlled. One way to tackle the problem is to use downlink power control in TDD base stations to compensate for the interference from the FDD mobile. The other solution is inter-system/inter-frequency handover. This type of interference also depends on the transmission power of the FDD mobile. If the FDD mobile is not operating close to its maximum power, the interference to TDD mobiles is reduced. The relative placement of UTRA base stations has an effect on the generated interference. Inter-system handover requires multimode FDD/TDD mobiles and this cannot always be assumed.

13.3.3   Unlicensed TDD Operation

Unlicensed operation with UTRA TDD is possible if DCA techniques are applied together with TDMA components. DCA techniques cannot be applied for high bit rates since several time slots are needed. Therefore, unlicensed operation is restricted to low to medium bit rates if there are several uncoordinated base stations in one geographical area.

13.3.4   Conclusions on UTRA TDD Interference

Sections 13.3.1–13.3.3 considered those UTRA TDD interference issues that are different from UTRA FDD-only operation. The following conclusions emerge:

- Frame-level synchronisation of each operator's UTRA TDD base stations is required.
- Frame-level synchronisation of the base stations of different TDD operators is also recommended if the base stations are close to each other.
- Cell-independent asymmetric capacity allocation between uplink and downlink is not feasible for each cell in the coverage area.
- Dynamic channel allocation is needed to reduce the interference problems within the TDD band.
- Interference between the lower TDD band and the FDD uplink band can occur and cannot be avoided by dynamic channel allocation.
- Inter-system and inter-frequency handovers provide means of reducing and escaping the interference.

- Co-siting of UTRA FDD and TDD macro cell base stations is not feasible, and co-siting of pico base stations sets high requirements for UTRA TDD base station implementation.

- Co-existence of FDD and TDD can affect the FDD uplink coverage area and the TDD quality of service.

- With proper planning TDD can form a part of the UTRAN where TDD complements FDD.

According to [16], TDD operation should not be prohibited in the FDD uplink band. Based on the interference results in this chapter, there is very little practical sense in such an arrangement, nor is it foreseen to be supported by the equipment offered for the market.

13.4   HSDPA Operation with TDD

The HSDPA operation covered in Chapter 11 is supported with TDD as well. The same principles of Node B-based scheduling and HARQ with physical layer-based feedback and link adaptation are present with TDD modes as well. The resulting physical layer is slightly different and has a different channel arrangement, especially for signalling purposes. The data transmission itself is more or less similar to Release '99 principles, the difference being the possibility to use 16 QAM for the data part of the burst for the HS-DSCH.

The signalling in the downlink is also similar to the FDD, the High-speed Shared Control Channel (HS-SCCH) is transmitted by the Node B, containing information to which UE, and with what transmission parameters (HARQ process, modulation, etc.), the data is coming. In the uplink direction, there is a difference as the uplink signalling is also on a shared resource there, the Shared Information Channel for HS-DSCH (HS-SICH) is used, as shown in Figure 13.16. This was chosen as the TDD system needs to pay more attention to the code

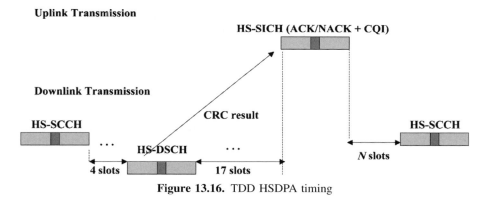

Figure 13.16. TDD HSDPA timing

resources in the uplink, thus, having a large number of users with dedicated resources would have eaten up too much of the code resources, and a similar solution to the FDD was not considered viable. The timings shown in Figure 13.16 are valid for 3.84 Mcps TDD, with 1.28 Mcps TDD, the values are three slots and nine slots respectively. The latter time for HS-DSCH decoding is larger as there is more data to decode compared to HS-SCCH decoding. The values are the minimum ones allowed, the actual slot resource allocation may result in longer values in reality.

The resulting peak data rate with 3.84 Mcps TDD is of the order of 10 Mbps, and with 1.28 Mcps TDD, the peak rate is of the order of 2.8 Mbps. The resulting increase in the peak rate is smaller for 1.28 Mcps TDD, as with Release 4, higher order modulation (8PSK) was already used to reach 2 Mbps without channel coding, and adding 16 QAM will not contribute such a big difference to peak rate as with FDD or 3.84 Mcps TDD. The resulting TDD capabilities can be found from [17].

13.5 Concluding Remarks and Future Outlook on UTRA TDD

This chapter covered UTRA TDD. The focus was on the physical layer issues, since the higher layer specifications are common, to a large extent, with UTRA FDD. In an actual implementation the algorithms for both the receiver and radio resource management differ between UTRA FDD and TDD, as the physical layers have different parameters to control. Especially in the TDD base station, advanced receivers are needed, while for mobile stations, the required receiver solution will depend on the details of performance requirements.

From the service point of view, both UTRA TDD and FDD can provide both low and high data rate services with similar QoS. The only exception for UTRA TDD is that after a certain point the highest data rates are asymmetric. The coverage of UTRA TDD will be smaller for low and medium data rate services than the comparable UTRA FDD service, due to the TDMA duty cycle. Also, to avoid interference, smaller cells provide a better starting point. Therefore, UTRA TDD is most suited for small cells and high data rate services.

Interference aspects for UTRA TDD were analysed and will need careful consideration for deployment. With proper planning, UTRA TDD can complement the UTRA FDD network, the biggest benefit being the separate frequency band that can be utilised only with TDD operation.

At the time of writing this chapter, there is no indication of the commercial use of either 1.28 Mcps TDD or 3.84 Mcps TDD technology in IMT-2000 bands. Thus it remains to be seen when we will see commercial operation in the 1900–1920 MHz band, or whether this band will be used by other technologies, as is being currently evaluated. In China, the 1.28 Mcps TDD or named as TD-SCDMA has been under discussion to be used, together with the FDD technologies, but no decisions have been announced by the Chinese operators on deployment plans or schedules.

Furthermore, consideration is currently being given in 3GPP as to whether there should be a third chip rate specified for UTRA TDD, with the proposed value of 7.68 Mcps and roughly 10 MHz channel spacing. The conclusions of the study, i.e. whether to start the specification work or not, are expected to be reached during the second half of 2004. This larger chip rate is not expected to be of any practical use for the UMTS band in Europe as the operators at most hold a single 5 MHz frequency block for TDD. For the forthcoming 2.6 GHz allocation there is likely to be also TDD allocation.

References

[1] 3GPP Technical Specification 25.221 V3.1.0, Physical Channels and Mapping of Transport Channels onto Physical Channels (TDD).
[2] 3GPP Technical Specification 25.222 V3.1.0, Multiplexing and Channel Coding (TDD).
[3] 3GPP Technical Specification 25.223 V3.1.0, Spreading and Modulation (TDD).

[4] 3GPP Technical Specification 25.224 V3.1.0, Physical Layer Procedures (TDD).

[5] 3GPP Technical Specification 25.102 V3.1.0, UTRA (UE) TDD; Radio Transmission and Reception.

[6] Steiner, B. and Jung, P., 'Optimum and suboptimum channel estimation for the uplink of CDMA mobile radio systems with joint detection', *European Transactions on Telecommunications and Related Techniques*, Vol. 5, 1994, pp. 39–50.

[7] Lupas, R. and Verdu, S., 'Near–far resistance of multiuser detectors in asynchronous channels', *IEEE Transactions on Communications,* Vol. 38, no. 4, 1990, pp. 496–508.

[8] Klein, A., 'Data detection algorithms specially designed for the downlink of CDMA mobile radio systems', in *Proceedings of IEEE Vehicular Technology Conference*, Phoenix, AZ, 1997, pp. 203–207.

[9] Klein, A. and Baier, P.W., 'Linear unbiased data estimation in mobile radio systems applying CDMA', *IEEE Journal on Selected Areas in Communications*, Vol. 11, no. 7, 1993, pp. 1058–1066.

[10] Klein, A., Kaleh, G.K. and Baier P.W., 'Zero forcing and minimum mean square-error equalisation for multiuser detection in code-division multiple-access channels', *IEEE Transactions on Vehicular Technology*, Vol. 45, no. 2, 1996, pp. 276–287.

[11] Jung, P. and Blanz, J.J., 'Joint detection with coherent receiver antenna diversity in CDMA mobile radio systems', *IEEE Transactions on Vehicular Technology*, Vol. 44, 1995, pp. 76–88.

[12] Papathanassiou, A., Haardt, M., Furio, I. and Blanz J.J., 'Multi-user direction of arrival and channel estimation for time-slotted CDMA with joint detection', in *Proceedings of the 1997 13th International Conference on Digital Signal Processing*, Santorini, Greece, 1997, pp. 375–378.

[13] Varanasi, M.K. and Aazhang, B., 'Multistage detection in asynchronous code-division multiple-access communications', *IEEE Transactions on Communications*, Vol. 38, no. 4, 1990, pp. 509–519.

[14] Väätäjä, H., Juntti, M. and Kuosmanen, P., 'Performance of multiuser detection in TD-CDMA uplink', EUSIPCO-2000, 5–8 September 2000, Tampere, Finland.

[15] Holma, H., Povey, G. and Toskala, A., 'Evaluation of interference between uplink and downlink in UTRA TDD', *VTC'99/Fall*, Amsterdam, 1999, pp. 2616–2620.

[16] ERC TG1 decision (98)183, February 1999.

[17] 3GPP Technical Specification, 25.306 V5.6.0, UE Radio Access Capabilities.

14

cdma2000

Antti Toskala

14.1 Introduction

As explained in Chapter 4, in addition to the described UTRA FDD and TDD modes in the global ITU-R IMT-2000 CDMA framework, the third mode is the Multicarrier (MC) CDMA mode, based on the cdma2000 multicarrier option being standardised by the 3rd Generation Partnership Project 2 (3GPP2)3GPP2. The key MC mode standards [1–4] scheduled were completed at the end of 1999 and facilitate connection to the IS-41 based core network. Later, the necessary extensions are planned to be specified to support connecting the MC mode to GSM-MAP based core networks as well. An overview of the cdma2000 physical layer can be found in [5]. The latest development in technology alignment between 3GPP and 3GPP2 is the decision to adopt, in the 3GPP2 core network side, the IP Multimedia Sub-system (IMS) developed for 3GPP Release 5 specifications, as described in Chapter 5.

The MC mode, part of IS-2000 specification series, was considered as the way for 3G evolution for operators with an existing IS-95 (or 1X) network, especially if a third generation network is to be deployed on the same frequency spectrum as an existing IS-95 network. This kind of spectrum *refarming* approach is foreseen in countries where there is no separate IMT-2000 spectrum, following the North American PCS spectrum allocation.

The name for the MC mode comes from the downlink transmission direction, where, instead of a single wideband carrier, multiple (up to 12) parallel narrowband CDMA carriers are transmitted from each base station. Each carrier's chip rate is 1.2288 Mcps, equal to the IS-95 chip rate. The uplink direction is direct spread, very similar to UTRA FDD, with multiple chip rate of 1.2288 Mcps. The first ITU release of cdma2000 will adopt up to three carriers (known as 3X mode) at a chip rate up to 3.6864 Mcps. The term 'MC mode' hereafter will refer to the MC mode (3X) as defined in the cdma2000 standard.

The MC mode has been considered to provide an evolutionary path for existing IS-95 systems. As illustrated in Figure 14.1, three narrowband IS-95/1X carriers, each with 1.25 MHz, are bundled to form a multicarrier transmission in the downlink with approximately 3.75 MHz (3X) bandwidth in a 5 MHz deployment. Currently, there do not seem to be commercial commitments for actually adopting the MC mode, but instead the focus has

WCDMA for UMTS, third edition.   Edited by Harri Holma and Antti Toskala
© 2004 John Wiley & Sons, Ltd   ISBN: 0-470-87096-6

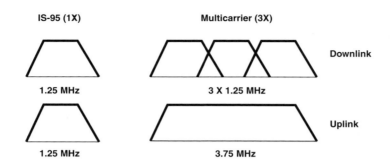

Figure 14.1. Relationship between the MC mode and IS-95 in spectrum usage

been more on the further development of narrowband operation. The technical principles presented are, in most cases, also valid for the narrowband track of IS-95 evolution, 1X, which in most cases in the downlink direction corresponds to what is defined as being on a single carrier with 3X. The 1X development work has produced different versions of the specifications, with the latest work being done on the 1xEV-DO as a data only system and 1xEV-DV. 1xEV-DO (DO = Data Only) is a data only system that uses a separate carrier from voice or other circuit switched services, while 1xEV-DV (DV = Data and Voice) uses similar technology as High-speed Downlink Packet Access (HSDPA) in WCDMA for mixed voice and packet data operation on the same carrier. The 1xEV-DO and 1xEV-DV have peak rate capabilities of the order of 2.4 Mbps and 3 Mbps. If there is a need to extend the capabilities beyond that, then wider bandwidth is necessary, like evolving to WCDMA with High-speed Downlink Packet Access (HSDPA). The HSDPA, as covered in Chapter 11, could provide 10 Mbps, or theoretically up to 14 Mbps, in the downlink direction. Evolution as a function of the downlink data rates achievable in the network is illustrated in Figure 14.2.

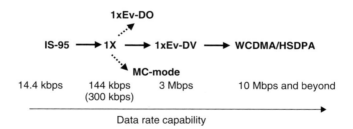

Figure 14.2. Air interface evolution as a function of downlink peak data rate

Following the Release C version improvements, like 1xEV-DV, the Release D standard (with target date the end of 2003) is keeping the downlink data rates as they are, but improving the uplink peak data rates and performance. The methods are similar to those described in Chapter 11 for WCDMA with HSUPA: HARQ, with fast (10 ms frame basis) scheduling, as well as some of the measurement related to terminal buffer status and power resources.

In terms of signal bandwidth, there is not much difference between the MC mode's multicarrier (uplink) chip rate of 3.6864 Mcps and UTRA FDD's 3.84 Mcps. The following sections describe the main characteristics of the physical layer of the MC mode and illustrate the most important differences from UTRA FDD.

With respect to the higher layers, it is worth noting that, although the protocol structures are largely similar, there are differences, such as certain protocols being implemented with less clear solutions through the protocol layers. This means that in practice, modifications to the MC mode protocol structure are required when considering interfacing the MC mode with GSM-based networks.

In this chapter the focus is on the key principles of the IS-2000 MC mode physical layer, but we also cover the key principles used in the IS-95 physical layer.

14.2 Logical Channels

Corresponding to the UTRA term 'transport channels', which carry data over the air and are mapped directly to the physical channels, the term 'logical channels' is used for cdma2000. The following logical channels are defined in the cdma2000 (Release A) specification, which was completed at the end of 1999:

- Dedicated Traffic Channel (f/r-dtch). A point-to-point logical channel that carries data or voice traffic over a dedicated physical channel; it corresponds to the dedicated transport channel in UTRA. As in UTRA, dtch is intended for use by a single terminal.

- Common Control Channels (f/r-cmch control). These are used to carry MAC messages with shared access for several terminals.

- Dedicated Signalling Channel (f/r-dsch). A point-to-point logical channel that carries upper layer signalling traffic over a dedicated physical channel, for a single terminal.

- Common Signalling Channel (f/r-csch). A point-to-multipoint logical channel that carries upper layer signalling traffic over a common physical channel, with shared access for several terminals.

One difference in terminology worth noting is the use of the term 'reverse link' instead of 'uplink', and of 'forward link' instead of 'downlink', in cdma2000 documentation. For convenience and for consistency between the different chapters in this book, this chapter adopts the terms used in UTRA. For example, in physical channel terminology the terms Forward (F) and Reverse (R) link are not used, but are replaced with downlink and uplink respectively.

14.2.1 Physical Channels

The MC mode provides basically the same functionality as does UTRA FDD. Functions such as the Broadcast Channel, Random Access Channel, and so on, are essential to the basic operation of all cellular systems. Also, the paging channel is needed to page the mobiles in the system. The physical layer contains slightly more differences, due to different design philosophies in some areas.

The common channel types more specific to UTRA, such as shared channels, uplink Common Packet Channel, and so on, have no direct counterparts in the MC mode, but the same functionalities are implemented by means of different arrangements.

The corresponding channel for data use is the Supplemental Channel, which, in the downlink, has similarities with the Downlink Shared Channel (DSCH) in UTRA FDD. The Supplemental Channel also exists in the uplink in the MC mode, but not as an enhancement for the Random Access Channel like the Common Packet Channel in UTRA FDD. The characteristics of the Supplemental Channel are covered in more detail in connection with user data transmission.

The Access Channel differs due to the differences in the higher layer protocols. The typical duration of the MC mode random access message is longer than in UTRA, since in the latter case the change to a dedicated channel takes place earlier. Thus, the Random Access Channel transmission may last over several frames. To avoid CDMA near–far problems, a common power control channel may be used to transmit power control information for the MC mode uplink random access procedure.

The MC mode employs a quick paging channel that is used to indicate to the terminal when to listen to the actual paging channel itself. This bears some similarities to the Paging Indicator Channel (PICH) in both UTRA FDD and TDD modes.

14.3 Multicarrier Mode Spreading and Modulation

14.3.1 Uplink Spreading and Modulation

Uplink modulation is very similar to that of UTRA FDD in the sense that different channels are provided in either the I or Q branch and then experience a complex-valued scrambling operation after spreading, to balance the I and Q branch powers. This results in rather similar requirements for amplifier linearity as in UTRA FDD. The multicode transmission is employed earlier than in UTRA FDD when the data rate increases. When higher data rates are desired with the MC mode, the Supplemental Channel is used in parallel with the Fundamental Channel, which provides only a limited set of possible lower data rates. Release D adds QPSK modulation as an alternative with 1xEV-DV uplink.

The uplink spreading is done with Walsh functions, while in UTRA FDD, OVSF codes are used. Variable rate spreading is not used in the MC mode during the connection on a frame-by-frame basis, as no rate information is provided in the physical layer signalling.

The uplink long code used for scrambling has a period of $2^{42} - 1$ chips. This is significantly longer than in UTRA FDD, where the code period is 38 400 chips for the dedicated channels, and the code length is only 256 chips for short scrambling codes. With a period of 38 400 chips, no degradation is expected, while the code length of 256 chips without advanced receivers usually results in some degradation due to the reduced cross-correlation averaging effect. The Access Channels have a specific scrambling code with a period of 2^{15} chips.

For the adjacent channel attenuation, 40 dBm attenuation in the signal level should be reached outside 4.44 MHz bandwidth. The 3.75 MHz signal bandwidth is not a practical value to be used for frequency planning.

14.3.2 Downlink Spreading and Modulation

The downlink modulation is obviously characterised by its multicarrier nature. The downlink carriers can be operated independently, or the terminal can demodulate them all. The benefit of receiving on all carriers is the frequency diversity that is improved over a single

1.2288 Mcps carrier. As each carrier contains a pilot channel for channel estimation, they can also be sent from different antennas if desired, to allow additional diversity. This is similar to the transmission diversity methods in UTRA.

The channel on each carrier is spread with Walsh functions using a constant spreading factor during the connection, in a similar way to UTRA with a few exceptions. Like the OVSF codes in UTRA, the Walsh functions separate channels from the same source and have similar orthogonality for transmission from a single source. The spreading factors for data transmission range from 256 down to 4. Downlink modulation consisting of three carriers is illustrated in Figure 14.3. Note that PN sequences and Walsh functions on parallel carriers are the same.

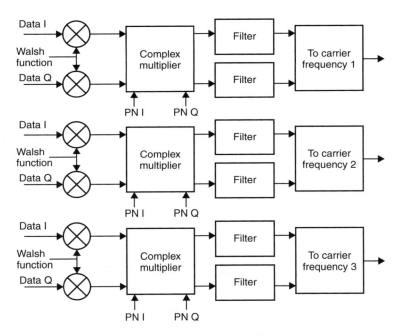

Figure 14.3. Downlink multicarrier spreading and scrambling

Downlink scrambling is characterised by the use of a single code throughout the system. Since the MC mode is operated with synchronised base stations, a single code is used, the different base stations using a different phase of the same code. The number of available phases is 512, corresponding to the number of UTRA FDD primary scrambling codes. In practical networks, the phases with minimum separation are often avoided in order to relax the requirements associated with timing issues in the network planning process.

The MC mode pulse shaping has been specified with exact filter coefficients. Based on these, mean-squared error criteria have been defined that should be met for the filter implementation. Although the single carrier bandwidth discussed has often been 1.25 MHz, the bandwidth that has been defined for a single carrier spectrum mask with 40 dB attenuation for the power level is 1.48 MHz for the base station transmission.

14.4 User Data Transmission

This section describes the key principles of user data transmission in the MC mode, highlighting the main differences from UTRA FDD operation. One general difference from UTRA specifications is that for the MC mode the different data rates have been defined exactly in terms of puncturing or repetition factors, while in UTRA, repetition and puncturing rules are given that can generate rate matching for any arbitrary rate. This does not cause practical differences, unless higher layers need to do a lot of padding or other operations to provide the necessary data rate if the physical layer data rate available does not suit the needs of the application. The data rates expected to be added to the MC mode are those needed to support the AMR voice codec used in UTRA and the GSM side, since at low rates the overhead may become significant for implementing the AMR voice codec data rates from the predefined MC mode data rate set.

14.4.1 Uplink Data Transmission

The Fundamental Channel in the MC mode is specified in detail for a given set of data rates, with a maximum data rate of 14.4 kbits/s. It can change the momentary data rate with changes in the repetition, but the symbol rate is not changed. This allows the use of blind rate detection in the base station. The Fundamental Channel and pilot channel structure is illustrated in Figure 14.4, where the pilot channel also contains the power control symbols with a 1.25 ms interval. This allows downlink fast power control at a rate of 800 Hz.

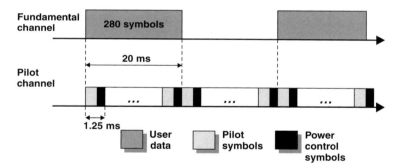

Figure 14.4. Uplink Fundamental Channel structure

As the data rate increases, larger data rates are not to be introduced on the same channel as in UTRA FDD, but rather with the Supplemental Channel. This is a parallel code channel separated with a different Walsh function from the Fundamental Channel. Uplink transmission may contain one or two Supplemental Channels with data rates ranging from a few kbps up to 1 Mbps, depending on the radio configuration. A typical radio configuration defines ten or fewer different data rates for a Supplemental Channel. Rates above 14.4 kbps use turbo coding.

Thus, while in UTRA FDD the physical layer control channel maintains constant parameters for the spreading factor, and so on, in the MC mode the parameters for the Fundamental Channel are fixed and then one or more Supplemental Channels can be added (with a fixed spreading factor for a given data rate). There is a natural background to this difference. The MC mode does not contain physical layer control information to inform the receiver of the change of data rate, or more generally of the change of Transport Format Combination (TFI) as termed in UTRA. Thus, the in-band signalling on the Fundamental Channel must carry all such information, so the parameters of the Fundamental Channel itself cannot be altered on a frame-by-frame basis.

For user data the radio frame length is 20 ms, while in UTRA it is 10 ms. In any case, speech services in both the MC mode and UTRA use at least 20 ms interleaving, since the AMR speech codec, as well as existing GSM speech codecs, provide data in 20 ms intervals, so that using 10 ms interleaving does not result in a shorter delay. In Release D with 1xEV-DV uplink improvements, a 10 ms frame structure is taken into use.

In the MC mode, there is no concept corresponding to UTRA's uplink Common Packet Channel (CPCH). However, the enhanced access channel that does the RACH functionality can be used for sending small packets, like the RACH in UTRA. The payload sizes defined for the enhanced access channel range from 172 to 744 information bits. There are three options for the radio frame length: 5, 10 or 20 ms. This contrasts with data transmission on the Fundamental or Supplemental Channel, where a 20 ms frame is always used.

14.4.2   Downlink Data Transmission

In the downlink direction the MC mode shows a major difference from UTRA FDD. The user data is divided between the three parallel CDMA sub-carriers, each with a chip rate of 1.2288 Mcps. As in the uplink, the lower data rates are implemented with the Fundamental Channel and higher data rates with the Supplemental Channel. A terminal could receive only one of the carriers. However, this would limit the data rate and would not be beneficial from a system capacity point of view, since frequency diversity would be minimised and transmitter antenna diversity unavailable.

The symbol rate for the traffic channels after channel coding and interleaving is multiplied by a factor of three. The power control symbol is then inserted with puncturing and the data demultiplexed to the three different sub-carriers, as shown in Figure 14.5.

The Walsh functions allocated for the Fundamental Channel carry user data with a fixed spreading factor, typically 256 or 128 for the lower data rates. For higher data rates, smaller spreading factors are used.

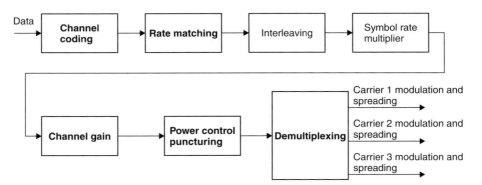

Figure 14.5. User data multiplexing to sub-carriers

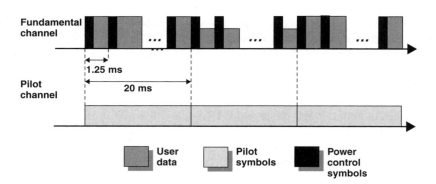

Figure 14.6. Downlink physical layer control multiplexing with user data for one sub-carrier

While the uplink direction has power control information multiplexed with the pilot channel, the downlink direction has only the common pilot, and the power control information is multiplexed with the data stream by puncturing with the rate of 800 Hz. The power control symbols are transmitted at a constant power level, as indicated in Figure 14.6, and serve as the only power reference for power control operation. The data symbols have a varying power level, since the rate matching is done with repetition or puncturing and the channel symbol rate is kept constant. The common pilot channel used as a phase reference is similar to UTRA CPICH. The power control symbols are not parallel in the time domain for different users in the MC mode downlink. This is to limit the resulting envelope variations by randomising the times when users have their power control symbols, as with DTX, only power control symbols are active on the Fundamental Channel.

There are some fundamental differences from UTRA FDD, in addition to the multicarrier structure. The Fundamental Channel does not carry any pilot symbols or rate information data. This means that blind rate detection with a variable rate connection is necessary. As the different data rates are formed with repetition or puncturing, channel decoding has to be carried out for data rate combinations in order to find out what the transmitted data rate was. This can be done for relatively small data rates, as in the case of the MC mode Fundamental Channel. The higher data rates are implemented with the Supplemental Channel and higher layer signalling is used to indicate the changes in the data rate of the Supplemental Channel.

14.4.3   Channel Coding for User Data

From the channel coding perspective, the Fundamental Channel always uses convolutional coding, while on the Supplemental Channel, turbo coding is applied. The 8-state turbo encoder and decoder are identical to the solutions in UTRA, but the turbo interleaving is different, as well as the channel interleaver. The latter would be different in any case, due to the differences in the number of symbols per frame with different spreading factors, and other differences in the frame structure. In the channel encoding there are differences also in the turbo coding rates: rates of $\frac{1}{4}$ and $\frac{1}{2}$ are used in the MC mode, while in UTRA, $\frac{1}{4}$ is not applied. Also, in UTRA the coding rate corresponding to $\frac{1}{2}$-rate turbo coding is generated by rate matching from the 1/3 rate code.

Differences in convolutional coding also exist: in addition to UTRA's $\frac{1}{2}$ and 1/3 rates, $\frac{1}{4}$ and 1/6 rate convolutional codes have been specified in the MC mode. The constraint length of 9 is the same as in UTRA.

The differences in the turbo code internal interleaver result from the different kinds of optimisation used in the selection process by the standardisation bodies. The MC mode turbo interleaver was optimised for a fixed set of data rates, while the UTRA interleaver was selected against more generic criteria for a large variety of data rates. The practical differences in performance resulting from this are rather marginal.

14.5 Signalling

The way signalling channels have been defined shows some fundamental differences between UTRA FDD and the MC mode. In the MC mode, all the signalling channels are defined as channels of their own, while UTRA FDD uses the concept of Secondary Common Control Physical Channel (CCPCH), which carries in the physical layer channels such as the paging channel or forward access channel. Another difference is that in UTRA the common channels may use the TFCI for varying the data rate, while the transmission rates in the MC mode are fixed.

14.5.1 Pilot Channel

The MC mode has a separate common pilot channel for each carrier. This pilot channel is used in a similar way as in UTRA FDD; functions such as channel estimation and measurements for handover or cell selection and reselection are also similar. As each of the three carriers has its own pilot channel, they can be sent from separate transmission antennas if desired.

For beamforming, the MC mode always uses beam-specific pilot channels, called Auxiliary Pilot Channels, as there are no other known symbols sent on the dedicated channels to provide the phase reference. To save code space, auxiliary pilots may be generated from the Walsh functions with extended length. The maximum Walsh function length used in the MC mode is 512. The most noticeable difference is when operating with user-specific antenna beams, where UTRA uses pilot symbols on the dedicated channels, while in the MC mode, a separate pilot channel is provided for each beam.

The MC mode downlink does not use the same kind of transmit diversity as UTRA, but the carriers may be transmitted from different antennas, since each has its own common pilot channel active. For each carrier also Orthogonal Transmit Diversity (OTD) can be used, where the data is 'copied' to two antennas and transmitted from both of them.

14.5.2 Synch Channel

This channel is special to the MC mode. It helps the terminals to acquire initial timing synchronisation. The Synch Channel is a low rate channel with three frames per 80 ms period. The symbol rate is 1.2 kbits/s.

14.5.3 Broadcast Channel

The Broadcast Channel in the MC mode is similar to the UTRA Primary Common Control Channel (PCCCH), which carries the broadcast information in UTRA. Typical information sent on the Broadcast Channel is the availability of Access Channels or Enhanced Access Channels for random access purposes. The MC mode Broadcast Channel is likewise a fixed rate channel with 19.2 kbps. The coding method on the Broadcast Channel is 1/3 rate convolutional coding.

14.5.4   Quick Paging Channel

The Quick Paging Channel in the MC mode is similar to the Paging Indicator Channel in UTRA. It indicates to mobile stations whether they are expected to receive the paging information or information in the Forward Common Control Channel. It is divided into slots and subdivided into paging indicators, as well as indicators that show change in the configuration.

14.5.5   Common Power Control Channel

The Common Power Control Channel provides power control information for a number of uplink channels that do not have as a pair a Fundamental Channel providing power control information. Channels that are power controlled in this way are the Reverse Common Control Channel and Enhanced Access Channels.

The Common Power Control Channel provides three different types of common power control group with 200, 400 or 800 Hz command rates. I and Q branches both provide a command stream of 9.6 kbits/s for power control purposes.

14.5.6   Common and Dedicated Control Channels

The MC mode also contains the concept of common and dedicated control channels in the uplink and downlink directions. These channels are designed to carry higher layer control information for one or more terminals. As in the uplink direction, they are intended to transmit control information for the base station from a single terminal or from multiple terminals when the Reverse Traffic Channel is not used.

The additional higher layer common control channel is the Common Assignment Channel. This carries in the downlink the resource allocation messages for a number of terminals.

14.5.7   Random Access Channel (RACH) for Signalling Transmission

The RACH channel, or Access Channel in the MC mode, performs similar functions as in UTRA, though the detailed RACH procedure differs, with different options depending on whether or not preamble ramping is used. Several access channel frames can be transmitted in the MC mode, while in UTRA only 10 or 20 ms message lengths are used in the RACH. The UTRA Common Packet Channel (CPCH) with a longer message duration corresponds more closely to the MC mode access procedure with closed loop power control.

For longer messages there is a power control channel that may be used to provide fast power control during the access procedure. The detailed access procedure for the MC mode is covered in the next section. Like the access channel, the enhanced access channel can convey signalling information for random access purposes.

14.6   Physical Layer Procedures

14.6.1   Power Control Procedure

The basic power control procedure is rather similar in the MC mode and UTRA FDD. Fast closed loop power control is available in both uplink and downlink. Many of the details are different, however. First of all, the power control command rates are different: 1500 Hz with a normal step size of 1 dB in UTRA FDD, and 800 Hz in the MC mode. In the MC mode the fast closed loop power control does not operate on its own in the uplink, but open loop power control is also active.

The open loop power control monitors the received signal strength in the downlink. If threshold values are exceeded it can alter the terminal transmission power. The open loop power control has a large uncertainty, since terminals operate on a different frequency band for reception and cannot measure the absolute power level very accurately (the open loop power control needs to compare the terminal transmission power to the received power level in the downlink). The pairing of open loop power control with closed loop power control can be turned off by the network; this option is often used in the existing IS-95 networks as well.

For the algorithm in the terminal, one difference is caused by the pilot solution. In the MC mode, the pilot symbols do not exist on the dedicated channel, thus the only symbols that can be used to aid the SIR estimation are the power control symbols as they preserve the power level unchanged with respect to changes in the data rate. The resulting power offset as such does not have a major impact on the downlink peak-to-average due to the large number of parallel transmissions in the downlink. This is also further reduced, as the position for the power control sub-channel is not the same for all users in the slot, so that the envelope variation effects are more averaged over the slot.

14.6.2   Cell Search Procedure

There are essential differences in the cell search procedure between the MC mode and UTRA FDD. As stated earlier, the MC mode uses time-shifted versions of the single scrambling code for all base stations in the network. Upon powering on, the terminal starts searching for the single sequence with a proper receiver, which could be, for example, correlator or matched filter based. The search will continue until one or more code phases have been detected.

As all the cells are synchronised, only a single sequence is needed in the system, and the terminal can search for the different phases of that single sequence. While in UTRA there are 512 different cell-specific scrambling codes, a similar procedure would be too complex or too time-consuming and therefore the search in UTRA starts from the synchronisation code word common to all cells.

As in UTRA, the cell search differs depending on whether an initial search is considered or a search is done for target cells for handover purposes. In the connected mode the terminal will get a list of the neighbouring cells and will perform the search based on the information on the PN-offset of the target cells. The list of neighbouring cells needs to contain the PN-offsets of the cells to be searched, otherwise the terminals in active mode cannot recognise the PN-offset they pick up.

14.6.3   Random Access Procedure

The random access procedure on the Enhanced Access Channel has, like UTRA, a power ramping feature, though there are several different options for the random access operation. The preamble preceding message is of varying length, as is the message itself, with the possibility of trading off preamble length and available base station resources for carrying

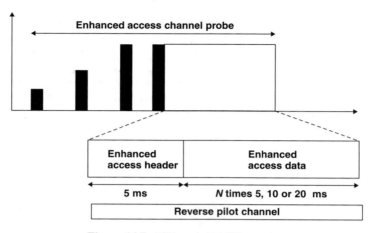

Figure 14.7. MC mode RACH ramping

out the search for the message. Ramping with Enhanced Access Channel Preamble, with the additional preamble active, is illustrated in Figure 14.7. The access probes following the initial access probe are sent with increased power level until the maximum number of access probes is reached, or higher layers allow sending of the actual message. If the option of additional preamble is selected, the ramping procedure is ended with an additional preamble to aid channel estimation at the base station.

Once the terminal is allowed to send the actual message, it can have a 5 ms Enhanced Access Header before the Enhanced Access Data, as illustrated in Figure 14.7. The Enhanced Access Data can be operated with 5, 10 or 20 ms frame length.

There also exists a power controlled access mode, where the common power control channel is used together with a random access procedure for controlling the transmission when sending the actual message. Terminals listen to the power control channel; once that starts to send down commands, terminals no longer increase the power but follow the power control command stream. This holds the power ramping for the particular channel.

14.6.4   Handover Measurements Procedure

The UTRA FDD handover included many measurements for the purposes of handover to various systems. The MC mode also provides for multimode terminals to hand over to other systems, such as IS-95. Releases later than Release '99 are expected to include the necessary enhancements for measuring and handing over to GSM or UTRA FDD from the physical layer perspective. On the network side, the degree of difficulty will depend on whether one is dealing with a simple voice call only or whether advanced data services are included. Possible commercial interest for such equipment remains to be seen as well.

Unlike in UTRA, the MC mode does not offer methods like compressed mode for inter-frequency measurements. The terminal either has to be dual receiver one, or the terminal will do measurements by simply ignoring the data sent on the downlink direction. This aspect is also addressed in [5].

References

[1] 3GPP2 IS-2002.2, Physical Layer Standard for cdma2000 Spread Spectrum Systems.
[2] 3GPP2 IS-2002.3, Medium Access Control (MAC) Standard for cdma2000 Spread Spectrum Systems.
[3] 3GPP2 IS-2002.4, Signaling Layer 2 Standard for cdma2000 Spread Spectrum Systems.
[4] 3GPP2 IS-2002.5, Upper Layer (Layer 3) Signaling Standard for cdma2000 Spread Spectrum Systems.
[5] Willenegger, S., 'cdma2000 Physical Layer: An Overview', *Journal of Communications and Networks*, Vol. 2, No. 1, March 2000, pp. 5–17.

Index

3GPP, 67
3GPP2, 70
16-QAM 312, 313

access service class, ASC, 153
acquisition indication channel, AICH,
    104, 119, 131
active set update, 177, 245
ACTS, 61
adaptive antennas, 143, 391
adjacent channel interference, ACLR, 113, 289
admission control, 264
AMR speech codec, 12
    ACELP, 12
    capacity, 367, 385
    coverage, 352
    wideband, 14
antenna diversity 356
ARIB, 65

beamforming, 143, 391
broadcast channel, 102, 128
broadcast control channel, BCCH, 153
broadcast/multicast control protocol, BMC, 161

Call session control function, CSCF, 97
capacity
    AMR speech codec, 367, 385
    downlink, 360
cdma2000, 433
    1X, 433
    1xEV-DO, 434
    1xEV-DV, 434
    3X, 433
    standardisation, 70
    multicarrier, 433
cell radio network temporary identity,
    C-RNTI, 152

cell update, 166
channel coding
    AMR, 126
    common channels, 127
    convolutional, 126
    dedicated channel, 126
    TFCI, 127
    turbo, 126
channelisation codes, 105
channel quality indicator
ciphering
    MAC layer, 153
    RLC layer, 158
co-existence
    TDD-FDD, 304
common control channel, CCCH, 153
common control physical channel
    primary, 128
    secondary, 130
common packet channel, CPCH,
    103, 120
common pilot channel, 127
    diversity, 128
    primary, 127
    secondary, 127
common traffic channel, CTCH, 153
common transport channels, 101
congestion control, 267
control plane, 81
core network 75
coverage
    AMR speech codec, 352
    downlink, 359
    effect of antenna diversity, 356
    effect of bit rate, 350
    link budget, 187
CS networks, 77
Cu Interface, 76

CWTS, 67
cyclic redundancy check, CRC, 117

data link layer, 149
data rates
    downlink, 122
    uplink, 116
    HSDPA 320
    TDD
dedicated channel
    downlink, 120
    uplink, 114
dedicated control channel, DCCH, 153
dedicated physical control channel, DPCCH, 114,
    120
dedicated physical data channel, DPDCH, 114,
    120
dedicated traffic channel, DTCH, 153
dedicated transport channel, DCH, 101
despreading, 49
dimensioning, 186
downlink modulation, 110
    synchronisation channel SCH, 112
downlink scrambling, 112
    primary scrambling codes, 112
    secondary scrambling codes, 112
downlink spreading, 111
    downlink shared channel, DSCH, 111
    synchronisation channel, SCH, 112
dual channel QPSK, 107
dynamic channel allocation, 300

Enhanced uplink DCH, E-DCH, 339
EDGE, 2, 68
EGSM 2
Erlang, 204
ETSI, 61

fast fading margin, 187
fast power control, 232
forward access channel, FACH 102, 130,
    153, 197
    user data transmission, 125
frame protocol, FP, 89, 92
FRAMES, 61
frequency accuracy, 114

GGSN, 78
GMSC, 77
GSM core network, CN, 77

handover, 245
    active set update rate 254
    algorithm, 245
    compressed mode, 140
    GSM, 140

inter-frequency, 258
inter-mode, 140
inter-system, 140, 254
intra-mode, 139
measurement procedure, 139
soft handover, 139
soft handover gains, 249, 354
soft handover probabilities, 252
hard handover, 176
home location register, HLR, 77
home subscriber server, HSS, 96
high speed dedicated physical control channel,
    HS-DPCCH, 312
high speed downlink packet access, HSDPA, 307
high speed downlink shared channel, HS-DSCH,
    308, 312
high speed shared control channel HS-SCCH, 312
high speed uplink packet access, HSUPA, 339

IMT-2000, 70
IMT-RSCP, 71
    ITU-R TG8/1, 70
integrity protection of signalling, 169
interference cancellation, 398
interference margin, 187
interleaving
    first, 118
    second, 119
inter-system cell reselection, 176
inter-system handover, 176
IP multimedia subsystem, IMS, 31, 96
IP RAN, 94
I-Q multiplexing, 107
IS-2000. *See* cdma2000
IS-95, cdmaOne, 5
Iu Interface, 78, 82
    Iu CS, 82
    Iu PS, 82
    protocol stacks, 82
Iub Interface, 78, 91
Iupc interface 94
Iur Interface, 78, 88
Iur-g interface 94

joint detection, *See* multiuser detection.

link budget, 187
load
    estimation of downlink load, 263
    estimation of uplink load, 261
load control, 267
load factor 190
logical channels
    broadcast control channel, BCCH, 153
    common control channel, CCCH, 153
    common traffic channel, CTCH, 153

dedicated control channel, DCCH, 153
dedicated traffic channel, DTCH, 153
mapping between logical channels and transport channels, 153
paging control channel, PCCH, 153
logical O&M, 92

measurement control, 169
measurement reporting, 169
measurements, 142
Media resource function, MRF, 97
Media gateway control function, MGCF, 97
medium access control protocol, MAC, 150, 151
mobile equipment, ME 76
mobile services switching centre/visitor location register , MSC/VLR, 77
maximal ratio combining, MRC, 54
multi-carrier CDMA. See cdma2000
multimedia broadcast multicast service, MBMS, 26, 162
multipath diversity 353
multiplexing
    downlink, 122
    uplink, 117
multiuser detection, 398

NBAP, 91
network planning
    dimensioning, 186
    GSM co-planning, 287
    inter-operator interference, 289
    optimisation, 208
node B, 76
noise rise, 187, 192

ODMA, 64
OFDMA, 64
Okumura-Hata, 189
open loop power control, 134, 182
orthogonal codes, 360
outer loop power control, 182, 239
OVSF codes, 105

packet data convergence protocol, PDCP, 150, 160
paging
    channel, 102, 130
    control channel, 153
    indicator channel, 131
    procedure, 134
payload unit, PU, 159
PLMN, 76
Positioning 40
    network assisted GPS, 42
    A-GPS 44

power control, 36
    compressed mode, 133
    CPCH, 136
    fast, 232
    fast closed loop, 133
    headroom, 142
    open loop, 134
    outer loop, 239
    RACH, 135
power rise, 187
processing gain, 51
PS networks, 77
Push-to-talk over cellular, PoC, 19, 296

Quality of service, QoS, 31

radio link control protocol, RLC, 155
    acknowledged mode, AM, 156
    transparent mode, Tr, 156
    unacknowledged mode, UM, 156
radio network controller, RNC, 78
    Drift RNC, 79
    Serving RNC, 80
radio network sub-system, RNS, 78
radio resource control protocol, RRC, 164
Rake receiver, 52
RANAP, 85
random access channel, RACH, 102, 119, 131
rate matching
    downlink 123
    uplink, 117
Rayleigh fading, 32
receiver
    downlink, 116
    uplink, 115
reciprocal channel, 290
refarming, 4
round trip time, 276
RNSAP, 88

scrambling, 104
SDD, 411
SGSN, 77
shared channel,
    downlink 124
    high speed, 307
soft capacity, 204
soft(er) handover, 58, 245
spectrum allocation, 3
spreading, 49
synchronisation channel, 112, 128

TDD, 411
    Handovers, 424
    user data rates, 420
TD-SCDMA, 411
terminal power class, 113

time division duplex. *See* TDD
TR46.1, 67
traffic termination point, 91
transmission control protocol, TCP, 269
transmission time interval, TTI, 116, 118
transmit diversity, 266
    closed loop, 138
    open loop, 121
    TDD, 302
    time-switched, 113
transport channels, 100
transport format and resource combination,
    TFRC, 321
transport format combination indicator, TFCI,
    100, 116
transport format indicator,TFI, 100
transport network control plane, 81
transport network user plane, 82
TTA, 65

User equipment, UE, 75
UMTS subscriber identity module,USIM, 76

uplink modulation, 105
uplink scrambling, 109
uplink spreading, 109
URA update, 166
user plane, 81
UTRA, 67
UTRA TDD. *See* TDD
UTRAN, 75
UTRAN architecture, 76
UTRAN radio network temporary identity,
    U-RNTI, 155
Uu Interface, 78
UWC-136, 66

Virtual private network, VPN, 28
Voice over IP, VoIP, 21, 299

WARC, world administrative radio conference, 3
WCDMA N/A, 66
wideband TDMA, 63
wideband TDMA/CDMA, 63
WP-CDMA, 67